Parasites in Ecological Communities

From Interactions to Ecosystems

Interactions between competitors, predators and their prey have traditionally been viewed as the foundation of community structure. Parasites – long ignored in community ecology – are now recognised as playing an important part in influencing species interactions and consequently affecting ecosystem function.

Parasitism can interact with other ecological drivers, resulting in both detrimental and beneficial effects on biodiversity and ecosystem health. Species interactions involving parasites are also key to understanding many biological invasions and emerging infectious diseases. This book bridges the gap between community ecology and epidemiology to create a wide-ranging examination of how parasites and pathogens affect all aspects of ecological communities, enabling the new generation of ecologists to include parasites as a key consideration in their studies.

This comprehensive guide to a newly emerging field is of relevance to academics, practitioners and graduates in biodiversity, conservation and population management, and animal and human health.

MELANIE J. HATCHER is Visiting Research Fellow, School of Biology, University of Bristol, UK, and Senior Research Fellow, Institute of Integrative and Comparative Biology, University of Leeds, UK.

ALISON M. DUNN is Reader in Evolutionary Ecology, Institute of Integrative and Comparative Biology, Faculty of Biological Sciences, University of Leeds, UK.

ECOLOGY, BIODIVERSITY AND CONSERVATION

The world's biological diversity faces unprecedented threats. The urgent challenge facing the concerned biologist is to understand ecological processes well enough to maintain their functioning in the face of the pressures resulting from human population growth. Those concerned with the conservation of biodiversity and with restoration also need to be acquainted with the political, social, historical, economic and legal frameworks within which ecological and conservation practice must be developed. The new *Ecology, Biodiversity and Conservation* series will present balanced, comprehensive, up-to-date and critical reviews of selected topics within the sciences of ecology and conservation biology, both botanical and zoological, and both 'pure' and 'applied'. It is aimed at advanced final-year undergraduates, graduate students, researchers and university teachers, as well as ecologists and conservationists in industry, government and the voluntary sectors. The series encompasses a wide range of approaches and scales (spatial, temporal and taxonomic), including quantitative, theoretical, population, community, ecosystem, landscape, historical, experimental, behavioural and evolutionary studies. The emphasis is on science related to the real world of plants and animals rather than on purely theoretical abstractions and mathematical models. Books in this series will, wherever possible, consider issues from a broad perspective. Some books will challenge existing paradigms and present new ecological concepts, empirical or theoretical models and testable hypotheses. Other books will explore new approaches and present syntheses on topics of ecological importance.

Ecology and Control of Introduced Plants
Judith H. Myers and Dawn Bazely

Invertebrate Conservation and Agricultural Ecosystems
T. R. New

Risks and Decisions for Conservation and Environmental Management
Mark Burgman

Ecology of Populations
Esa Ranta, Per Lundberg and Veijo Kaitala

Parasites in Ecological Communities

From Interactions to Ecosystems

MELANIE J. HATCHER
University of Bristol, UK

ALISON M. DUNN
University of Leeds, UK

CAMBRIDGE
UNIVERSITY PRESS

CAMBRIDGE UNIVERSITY PRESS
Cambridge, New York, Melbourne, Madrid, Cape Town,
Singapore, São Paulo, Delhi, Tokyo, Mexico City

Cambridge University Press
The Edinburgh Building, Cambridge CB2 8RU, UK

Published in the United States of America by Cambridge University Press, New York

www.cambridge.org
Information on this title: www.cambridge.org/9780521889704

First published 2011

Printed in the United Kingdom at the University Press, Cambridge

A catalogue record for this publication is available from the British Library

Library of Congress Cataloguing-in-Publication data
Hatcher, Melanie J.
    Parasites in ecological communities : from interactions to ecosystems / Melanie
J. Hatcher, Alison M. Dunn.
        p.   cm. – (Ecology, biodiversity and conservation)
    ISBN 978-0-521-88970-4 (Hardback) – ISBN 978-0-521-71822-6 (Paperback)
    1. Parasites–Ecology.   2. Parasites–Behavior.   3. Host-parasite relationships.
4. Parasitology.   5. Biotic communities.   I. Dunn, Alison M.   II. Title.   III. Series.
    QL757.H34 2011
    577.8′57–dc22

                                                                                    2010050212

ISBN 978-0-521-88970-4 Hardback
ISBN 978-0-521-71822-6 Paperback

Contents

x · **Contents**

Acknowledgements

The collaboration leading to this book was fostered in the smoky bars and backstreets of Oxford in the 1980s, where the authors first met whilst studying Zoology as undergraduates. In that respect, we would like to thank Sir John Krebs for (just about) tolerating us.

The book has been fermenting for many years and we thank the following for funding various research projects and studentships which have contributed to the ideas and examples in the book: the Natural Environment Research Council (particularly current grants NE/D011000/1 and NE/G015201/1); the Royal Society; the Biotechnology and Biological Sciences Research Council; the Leverhulme Trust; the European Science Foundation; Tarmac; the Yorkshire Dales National Parks Authority; the Natural History Museum; the European Union; and the Environment Agency.

For discussion on many of the ideas in this book, often over many years, we thank the late Anne Keymer (who first sparked AMD's interest in parasites), Judith Smith, Jonathan Adams, John Lawton, Thierry Rigaud, Pete Hudson, Andrew Read, Andy Dobson, Greg Hurst, Jack Werren, Charles Godfray, Jaimie Dick and Chris Tofts; also, all the fellow travellers at ecology, evolution and parasitology conferences with whom we have had many stimulating and drunken conversations.

Various of AMD's lab group members receive thanks for stimulating conversations and dedicated research, with several of them also providing helpful comments on chapter drafts: Andy Kelly, Aurore Dubuffet, Emily Imhoff, Neal Haddaway, Katie Arundell, Stephanie Peay, Paula Rosewarne, Mandy Bunke, Calum MacNeil and Jolene Slothouber Galbreath. We also thank Emily Imhoff for her lovely cartoon in Chapter 6.

A number of colleagues made time in their busy schedules for critical comment on individual chapters, which improved the work immensely. We thank Paul Hatcher, Steve Sait, Bryan Shorrocks, Jaimie Dick, Bill Kunin, Dan Tompkins, Robert Poulin, Simon Goodman, Rupert

Quinnell and Chris Tofts. For advice and critical comment on the entire work over its period of gestation, we thank Michael Usher, without whose support this work would never have been produced. Thanks are also due to Emma Walker, Megan Waddington and Dominic Lewis, who have all shown patience and understanding over our plight as novice book writers.

We thank the following publishers for kindly granting us permission to re-use or modify figures for this book: the American Association for the Advancement of Science; Cambridge University Press; Center for Disease Control and Prevention; Chicago Press; Ecological Society of America; Elsevier; National Academy of Sciences (USA); Nature Publishing Group; the Public Library of Science; the Royal Society; and Wiley & Sons.

For untiring support, intellectual, emotional, practical and technical, MJH especially wishes to thank Chris Tofts. For grandparenting services beyond the call of duty she thanks Eileen and Maurice Tofts; for a room with a view, Nicci Tofts; for advice and practical help in sustainable resource management, Paul Hatcher; and for engendering an enquiring mind, Erica and Michael Hatcher.

For their support and humour at times of crisis, AMD would like to thank various friends and family, especially Moira and Robin Dunn for support and help with fieldwork over the years (and no, Dad, we *still* don't know everything about those little shrimps) and to Claire Dunn, David Watkin, Linda Bracey Aitchison and Claire Caddell for their slightly bemused support of yet another late-night writing session.

Finally, we need to thank our own maddening and wonderful little parasites, James and Jennifer Tofts, Ethan and Aidan Dunn. Without you the book may have been sooner but less inspired. Without the book you would have done less ballet and sports camp. You will thank us one day.

Abbreviations

ABA	abscisic acid
AMF	arbuscular mycorrhizal fungi
Bd	*Batrachochytrium dendrobatidis*
BYDVs	barley yellow dwarf viruses
CDC	Center for Disease Control
CDV	canine distemper virus
CPV	canine parvovirus
EICA	evolution of increased competitive ability
EID	emerging infectious disease
EM	ectomycorrhizal
ERH	enemy release hypothesis
ET	ethylene
FIV	feline immunodeficiency virus
FLV	feline leukaemia virus
FPLV	feline panleukopenia virus
FPV	feline parvovirus
HA	haemaglutinin
HIV	human immunodeficiency virus
IBMs	individual-based models
IG	intraguild
IGP	intraguild predation
ISR	induced systemic resistance
JA	jasmonic acid
LIV	louping ill virus
MSY	maximum sustainable yield
MVP	minimum viable population
NA	neuraminidase
PDV	phocine distemper virus
PIVP	parasite-induced vulnerability to predation
PSF	plant–soil feed back
RHD	rabbit haemorrhagic disease
SA	salicylic acid
SIV	simian immunodeficiency virus

SQPV	squirrel pox virus
TMIEs	trait-mediated indirect effects
VOCs	volatile organic compounds
WAIFW	who acquires infection from whom
WNV	West Nile virus

1 · *Introduction*

Ring a ring of roses,
A pocket full of posies,
Atishoo, atishoo,
We all fall down.

Sporadic reports appear of a mysterious disease afflicting people in a far-off country. Within weeks the disease has spread to towns and cities and is reaching epidemic proportions in that country, and within months it has circulated around the world. The origins of this new disease are initially unclear but, mysteriously, large-scale die-offs of wildlife and domestic animals presage the outbreaks in several countries. Many people and animals die; furthermore, we start to see changes throughout natural communities, involving the resources and consumers of afflicted species. Eventually the disease dies out in humans and domestic stock, and the infection, if it persists, goes largely unnoticed in a handful of wildlife species. What was going on? Could we prevent it happening again, and will there be long-term consequences for natural communities? This is the plot behind many B-movies, but also something that happens in reality all the time. For example, the recent sporadic outbreaks of highly pathogenic avian flu involve transmission though a suite of wildfowl and domestic bird species, with occasional spillover into man. At the turn of this new century, as West Nile virus (WNV) spread throughout the United States, its arrival in a new county was heralded by reports of dead and dying birds, also host to the virus. The spread of chestnut blight through the deciduous forests of northern America at the turn of the previous century changed the landscape forever and affected many species associated directly or indirectly with these magnificent trees; similar effects were observed with the emergence of Dutch elm disease in northern Europe in the middle of the last century. Currently, the extinction of many hundreds of amphibian species seems a real possibility with the ongoing spread of a new fungal disease, chytridiomycosis. Similar events have happened throughout history; the English nursery rhyme 'Ring a ring of roses' is thought by many to refer to outbreaks of plague (perhaps bubonic plague in Europe in the 1340s, or the Great Plague of London in 1665), the onset of which was signified by symptoms of a rash, followed by sneezing and rapid death. As the story goes, people tried in vain to protect themselves from infection by carrying various nostrums or posies of scented flowers.

Parasites are involved in many other processes within host populations and communities that can ultimately feed through to influence species coexistence and ecosystem function. For instance, parasites play a key role in honeybee colony loss, which is an emerging threat to biodiversity and agricultural production in Europe and America. Studies of parasites in Californian salt marshes reveal that parasites, which account for a substantial proportion of the biomass in these ecosystems (equivalent to a small herd of elephants per hectare), alter food web structure, dramatically enhancing the density of trophic links in the web. This can have implications for ecosystem health, as densely linked food webs are more robust to perturbation. Biological invasions are a major driver of biodiversity loss and, through their effects on the interactions of their hosts with other species, parasites can influence the outcome of biological invasions in a diversity of species ranging from plants to crustaceans to mammals. For instance, by reducing growth and survival of perennial bunchgrasses, barley yellow dwarf virus and its variants facilitate the replacement of native bunchgrasses by annual grasses in the prairies of the United States. By reducing the predation rates and survival of infected hosts, fungal and microsporidian parasites modify the interaction between native and invasive crayfish, facilitating the extirpation of the native white-clawed crayfish (*Austropotamobius pallipes*) in the freshwaters of England. By causing high mortality in red squirrels, squirrel pox virus alters competition between red and grey squirrels, facilitating the invasion of grey squirrels and the replacement of red squirrels in the United Kingdom.

Work in our laboratories on the amphipod *Gammarus* and its parasites over the last two decades reveals the range of effects parasites can have, from the level of parasite effects on individual host fitness; through parasite mediation of host–host interactions, including competition and predation which may determine which species can coexist; to altering the functional role of species within ecosystems and influencing the success of biological invasions (Box 1.1).

Box 1.1 Parasitism in freshwater amphipod ecosystems

Amphipod crustacea are keystone species in freshwater ecosystems. Through processing nutrients and providing prey for larger inverte-brates and vertebrates, they provide important ecosystem services. They process the primary basal energy resource (leafy detritus) through shredding, with strong impacts on community structure. They also predate smaller species in the food web, influencing macroinvertebrate diversity and species richness. Furthermore, they are key prey for commercial and recreational fish stocks and for wildfowl. Therefore, the impact of parasitism on amphipod popula-tion dynamics, and on their competitive and predatory interactions, could have profound ramifications for the diversity and structure of aquatic communities, as well as having economic costs.

In rivers and lakes in Northern Ireland, a suite of gammarid amphipods occur. The native *Gammarus duebeni celticus* is subject to invasions by at least three species of invader; *G. pulex*, *G. tigrinus* and *Crangonyx pseudogracilis*. The native and invasive species interact through competition for prey, as well as through intraguild preda-tion (predation between species of the same guild), and these inter-actions are mediated by parasites (Fig. 1.1).

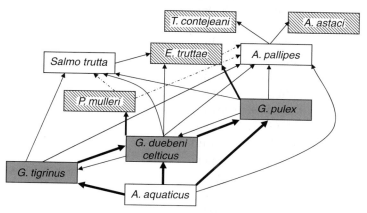

Fig. 1.1. Food web for the native and invasive *Gammarus* system studied in our laboratories. The direction of energy flow via consumption is shown by the arrows linking species (thickness indicates relative strength of interaction; dot-dashed lines depict predation of parasites by non–hosts when infected prey are eaten). Shaded boxes mark the three *Gammarus* species arranged in an intraguild predation hierarchy; stippled boxes are parasites. Brown trout (*Salmo trutta*) are definitive hosts for *Echinorhynchus truttae* and are therefore placed below this parasite in the web; other interactions as described in the text. The web is highly simplified; not all (host or parasite) species or interactions are shown.

Box 1.1 (*continued*)

These gammarids prey upon smaller macroinvertebrates, and a key prey is the isopod *Asellus aquaticus*. Infection of *G. duebeni celticus* by the microsporidian *Pleistophora mulleri* reduces the predatory impact on the isopod, thus also modifying competition with other amphipods. In contrast, the acanthocephalan parasite *Echinorhynchus truttae* increases the predatory strength of *G. pulex*, likely facilitating the exclusion of the native *G. duebeni celticus* by the invader. *G. pulex* invasions have also been found to reduce macroinvertebrate diversity and richness, hence parasite modification of predatory behaviour may have ramifications throughout the community.

Intraguild predation (IGP) is also modified by parasites. *G. duebeni celticus* is a stronger IG predator than the invader *G. tigrinus*, yet coexistence occurs in several areas and may be facilitated by the microsporidian parasite *P. mulleri* (Dunn 2009). *P. mulleri* is specific to *G. duebeni celticus* and has no discernible effect on survival. However, the infection weakens IGP by *G. duebeni celticus* on *G. tigrinus*, enhancing coexistence in field manipulations. Similarly, the acanthocephalan parasite *Echinorhynchus truttae* weakens IGP by *G. pulex* on the less predatory *G. duebeni celticus*.

Moving up through the trophic levels, gammarids are preyed upon by fish and wildfowl, and their predation risk is influenced by parasites. Whilst parasites such as the microsporidian *P. mulleri* may increase vulnerability to predation (a by-product of the infection), the trophically transmitted acanthocephalan parasites *E. truttae* and *Polymorphus minutus* enhance transmission to their definitive host (fish and wildfowl, respectively) by manipulating the antipredator behaviour of their amphipod host. *E. truttae* is likely to have a greater impact on *G. pulex* as parasite prevalence is two-fold greater than in *G. duebeni celticus*. The outcome for the predator is mixed; whilst prey might be more available (trout productivity is higher in areas of *G. pulex* invasion), the chances of infection will also increase.

Gammarids are also predated by the white-clawed crayfish (*Austropotamobius pallipes*) and the impact of this predator is mediated by parasitism. Outbreaks of crayfish plague (*Aphanomyces astaci*) can cause crayfish mortality, whilst the microsporidian *Thelohania*

Box 1.1 (*continued*)

contejeani reduces the predatory impact of the crayfish on its amphipod prey.

Hence parasites influence a variety of interspecific interactions, and may have potential effects throughout the community.

These complicated systems require a new approach that combines aspects of community ecology and parasitology, which we term *ecosystem parasitology*. Until recently, community ecology has historically ignored parasites, and parasitology has, in turn, largely ignored the community context in which infections spread. Attempts to meld these fields began in the 1980s with reviews and mathematical treaties by ecologists and epidemiologists Andrew Dobson, Pete Hudson, Peter Price, Robert Poulin, Roy Anderson and Robert May. Their papers provided an exciting route forward, but one that is only now gaining momentum. In order to integrate these disciplines, we need to combine some key concepts from community ecology and parasitology.

1.1 Concepts from epidemiology

Underlying much of modern epidemiology is the concept of R_0, the parasite's basic reproductive number, the number of secondary cases arising from each primary infection (Box 1.2). This measure predicts whether a parasite or pathogen can spread initially in a population of susceptible hosts; simply, if R_0 is greater than 1, the parasite can spread initially, if R_0 is less than 1, it cannot persist in the population. Another key concept arising from simple models of parasite spread within host populations is N_T, the threshold host population size for parasite establishment. Many (but not all) models of disease spread predict a threshold population size below which parasites and pathogens cannot become established.

However, parasites and other interactors such as competitors and predators may feed back on host population densities, and that is where the fun begins! How do parasites interact with other species, altering host population dynamics, and what are the consequences for coexistence of all the players, and for the structure and stability of communities as a whole? In order to examine these questions, we need to utilise developments from community ecology that allow epidemiological (host–parasite) models to be placed in a community context.

Box 1.2 R_0 and N_T

Basic reproductive number R_0. Whether a parasite spreads in a population depends on whether a single infection results in more than one infection in the following infection cycle. This is the concept behind the basic reproductive number R_0 (pronounced 'R nought'), defined as the average number of secondary cases produced by each primary case of infection in a completely susceptible population. For microparasites like human influenza viruses, this corresponds to the number of people infected by each infectious person. For macroparasites such as tapeworms, R_0 is the average number of tapeworms successfully reaching reproductive age produced by a single adult tapeworm.

Deriving R_0 depends on characteristics of both the parasite and host. The very simplest models of parasite–host dynamics start with the assumption that the host population is held to a constant density (which we shall call N) by factors other than disease. This is probably a reasonable assumption for many diseases in humans, for which epidemiological models were first developed. For a microparasite (such as flu or measles), we distinguish two host classes: those infected (I: for infected or infectious), and those yet to be infected (S: for susceptible). For a directly transmitted disease (i.e. one acquired directly from an infectious individual), susceptible individuals become infected at a rate dependent on contacts with infected individuals (assumed here to be proportional to population density I; this is known as density-dependent transmission; see Box 1.4 for an explanation), multiplied by the per-contact transmission efficiency of the disease (β). Once they become infected, individuals recover at a rate γ, entering the susceptible class again. Infected individuals die from the infection at rate α; infected and susceptible individuals also die from other causes ('natural' mortality) at rate b. This is one of the simplest epidemiological models we can have: all hosts are either in the S or I class and the parasite's only direct effect is to cause additional mortality to the infected class (for some of the more frequently met complications, see Box 1.4). From these definitions, we can write down the equation for the rate of change in density of the I class, in terms of the losses from and gains to this class:

Box 1.2 (*continued*)

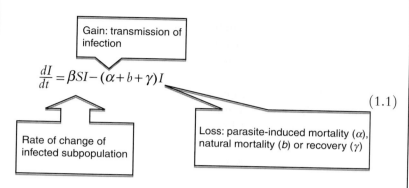

$$\frac{dI}{dt} = \beta SI - (\alpha + b + \gamma)I$$

Gain: transmission of infection

Rate of change of infected subpopulation

Loss: parasite-induced mortality (α), natural mortality (b) or recovery (γ)

$$(1.1)$$

For the infection to spread when rare, we require $dI/dt > 0$; hence

$$\beta SI > (\alpha + b + \gamma)I \Rightarrow \frac{\beta S}{\alpha + b + \gamma} > 1. \qquad (1.2)$$

This latter expression is closely related to R_0 (Anderson & May, 1981; 1991). When the disease is rare, almost all of the population are susceptible, so $S \cong N$; substituting $S = N$ into the above, we obtain:

$$R_0 = \frac{\beta N}{\alpha + b + \gamma}. \qquad (1.3)$$

This makes intuitive sense: each infected individual produces new infections in a susceptible population at a rate βN, and each infection lasts for $1/(\alpha + b + \gamma)$ on average (because duration in a class is the reciprocal of the rate of loss from that class); hence R_0 represents the average number of new infections produced by each primary infection.

Threshold population size for parasite establishment N_T. In this simple model, R_0 depends positively on host population density. Since we require $R_0 > 1$ for the parasite to spread, there is a threshold population size N_T below which the parasite cannot spread. Solving $dI/dt = 0$ (or $R_0 = 1$), we find:

$$N_T = \frac{\alpha + b + \gamma}{\beta}. \qquad (1.4)$$

This threshold means that parasites cannot invade populations smaller than N_T, but they can be maintained in larger populations.

Box 1.2 (*continued*)

It also implies that a host population cannot be driven extinct purely as a result of disease; the infection will die out once the population is reduced below N_T. Extending this concept to multi-host communities, populations of some host species may meet their species-specific N_T and others may not; the former may then act as 'reservoirs' for infection of the latter, which act as 'sinks'. Interestingly, not all epidemiological models have a threshold for parasite establishment. In one common variant (that of frequency-dependent transmission; Box 1.4), the spread of infection is independent of host population size, so parasites are predicted to spread in and potentially threaten the existence of small populations.

Force of infection is a concept related to R_0 which is sometimes easier to estimate in real populations. The force of infection (often denoted λ) is the per capita rate at which susceptibles become infected; in other words, it is a measure of the risk of becoming infected. In our simple model,

$$\lambda = \beta I. \tag{1.5}$$

Force of infection is thus dependent on the frequency of infectious individuals in the population (and hence, all other things being equal, the chances of engaging in contact that might lead to infection). In large populations with stable age structures and constant infant death rates where the parasite has reached equilibrium, the force of infection can be estimated as the reciprocal of the average age at which hosts become infected. This approach has been used for human diseases such as measles. Age at infection measures how long an individual has avoided infection (duration in the S class); its reciprocal therefore measures rate of transition into the I class.

R_0 in community ecology. R_0 is equivalent to the net reproductive rate r of organisms. Few organisms actually reproduce successfully at average rate r because other factors (density dependence in birth or death rates, for instance) intervene. The same applies to R_0 for parasites: in the initial stages of an epidemic, secondary cases are produced at a rate R_0, but as the epidemic progresses, susceptible individuals become more scarce and control measures may be taken, reducing the average number of secondary cases each infection generates. Under these changing conditions, the *effective reproductive number* for the parasite (referred to as R, or R_{int} for R

Box 1.2 (*continued*)

under intervention) also changes. If the parasite reaches equilibrium, each infection must generate exactly one new case on average ($R = 1$), otherwise the frequency of infection would change. R_0 is nevertheless a useful concept, like r. Parasites with a higher R_0 will increase more rapidly, all other things being equal. Parasites with an R_0 less than 1 will not be able to spread at all. Just as organisms are likely to have a different r in different habitats, parasites will probably have a different R_0 when infecting different host species. Hence, when we are dealing with systems involving multiple host species, parasite spread depends on a composite R_0 reflecting community composition and contacts.

1.2 Concepts from community ecology

Key to understanding how species interact within communities is a robust understanding of the types of interactions in which species engage, and how these affect the population densities of the interacting species (Box 1.3). Two species may interact directly (for example, via predation of one on the other) or indirectly via a third species (for instance, two prey species can interact indirectly via their shared natural enemy). Indirect interactions may be either density- or trait-mediated; that is, effects on the population of a focal species may result from a change in the population density of the species with which it interacts (a density-mediated effect), or from changes in the behaviour or morphology of that species (a trait-mediated effect). Parasites are prime candidates for causing trait-mediated indirect effects, because they often debilitate rather than immediately kill their hosts. However, parasites are also very good at generating density-mediated effects through their effects on host mortality, which feed through into population density.

This book is largely about the indirect effects of parasitism. Direct effects are covered in the extensive epidemiological treatments on one-host–one-parasite systems such as those by Anderson, May and Hassell. However, in order to place parasites in a community context, we need an understanding of the indirect effects of parasitism on other species. Analysis of community modules provides one approach to this. Community modules (Holt, 1997) are sets of three or more strongly interacting species. They provide a link between the artificial simplicity of

Box 1.3 Ecological interactions

Direct and indirect interactions: interactions such as interference competition and predation are considered as direct interactions because individuals of one species interact directly with the other and have direct effects on each other; these may be reciprocally negative (competition:−−), positive (mutualism ++) or beneficial to one partner and detrimental to the other (predation, parasitism: +−). Arguably of equal importance are indirect effects, which occur when the impact of one species on another is mediated by the action of a third. Pure resource (exploitation) competition (a −− interaction between the consumers mediated by the resource species) is an example; another is apparent competition (see below).

Trait- and density-mediated interactions: indirect effects (the effect of species A on species B via the actions of species C) can be density- or trait-mediated. Interactions are density-mediated when species C causes changes in A's population density, which affect its interaction with B. Trait-mediated interactions occur when C causes a change in behaviour, physiology, morphology or life history in species A, which affects its interaction with species B. This concept of trait-mediated indirect effects (TMIEs) originates in the distinction between density versus behaviourally propagated effects (Abrams, 1992), and short- versus long-term indirect effects (Holt & Kotler, 1987; see below). Most examples of TMIEs come from the behavioural ecology of predator–prey relationships; for instance, the presence of predator C can increase refuge-seeking behaviour in species A, which reduces its foraging rate, so influencing competition with species B (see Werner & Peacor, 2003 for a review). As parasites rarely kill their hosts immediately, but frequently alter host behaviour and physiology, parasitism modules are potentially a rich source of TMIEs. For instance, in our native–invasive *Gammarus* system (Box 1.1), parasites can increase or decrease the predation activity of their hosts, and increase or decrease predator avoidance by their hosts.

Apparent competition occurs when two species that do not compete for resources have reciprocal negative effects (−−) on each other via the action of a shared natural enemy. Apparent competition was first described in terms of the density-mediated effects of predators. For instance, population growth of prey A provides resource for predator C, enabling an increase in the population density of C; the consequent increase in predation has a negative impact on prey B's

Box 1.3 (*continued*)

population density (Holt, 1977). However, it can also be trait-mediated (Holt & Kotler, 1987), and it can also be mediated by parasites (Chapter 2). The knock-on effects of parasite-mediated apparent competition for ecological communities can be extensive: it underpins species coexistence/exclusion rules (Section 2.3); it can lead to unwanted side-effects in biological control (Section 2.3); and it can influence biological invasions (Section 6.5) and the dynamics of emerging infectious diseases (EIDs) (Section 8.2).

Short- and long-term effects: density- and trait-mediated indirect effects are likely to act on different timescales. Because density-mediated effects result from changes in the density of one species affecting the demographic responses of others, they should take longer to propagate to other species than the effects of direct inter-actions between species (Abrams, 1992). Behavioural and other trait-mediated effects occur on the same, or shorter, timescales as direct effects (Holt & Kotler, 1987). Because they occur rapidly, these short-term effects can, in theory, be very powerful in structur-ing communities, potentially overriding the effects of density-mediated processes (Bolker *et al.*, 2003). Short-term processes are also of interest because they produce a greater variety of interactions than those arising from density alone. For instance, in the apparent competition module (consisting of two prey and one shared enemy), all the combinations $(+-,-+,++,--)$ of indirect interactions are theoretically possible (Holt & Kotler, 1987; Abrams & Matsuda, 1996). This great variety of interactions can have knock-on effects at other trophic levels, generating (often counter-intuitive) relation-ships between species (Abrams, 1992). Thus, understanding the potential of parasites to induce short-term trait-mediated effects could change how we view their role in community ecology (Hatcher *et al.*, 2006; Tompkins *et al.*, 2010).

single-population or two-species dynamics and the intractable complex-ity of entire communities. Modules provide a basis for analysing indirect interactions mediated by a third species, and how interactions such as predation and parasitism themselves interact.

1.3 Parasites

Parasites are organisms that live on or in another host organism for all or part of their life cycle, deriving resources (usually nutrients) from the host. In contrast with predators, parasites do not immediately kill their host, although infection may increase the probability of host mortality, and some parasites eventually kill their host to facilitate their transmission to the next host. Hosts may also recover from parasites. Although ecologists have tended to ignore parasites, parasitism is in fact a major consumer strategy (Dobson *et al.*, 2008; Lafferty *et al.*, 2008a). For example, it has been estimated that 40% of known species are parasitic (Dobson *et al.*, 2008), and recent studies indicate that parasites are involved in a significant proportion of food web linkages (Lafferty *et al.*, 2006a; 2006b; Amundsen *et al.*, 2009).

Parasites can be divided into micro- and macroparasites, a distinction that not only reflects their size, but also the different modes of replication and transmission, which require different theoretical approaches (Box 1.4). Microparasites replicate directly within the host (often intracellularly), leading to increasing parasite burden within the host. Transmission occurs by direct contact between host individuals or via a vector (e.g. a mosquito). In addition to these horizontal transmission routes, some microparasites may be vertically transmitted from generation to generation of hosts. Microparasites may trigger an immune response in vertebrate hosts, leading to recovery and lifetime immunity to future infection.

In contrast, macroparasites grow within the host, but reproduction is followed by transmission to the new host, often through release of eggs and subsequent ingestion of egg or (free-living) larval parasites by the new host. Macroparasites may have direct life cycles or indirect life cycles involving two or more host species. As macroparasites have high antigenic diversity, host immunity may be delayed or may not occur.

Parasitoids are animals (usually insects) that are free-living as adults and that live on or in their host (usually an insect) during development, eventually killing the host. Hosts may recover by encapsulating and killing a parasitoid, but this does not result in future immunity from infection. Theoretical models sometimes treat parasitoids as predators; however, this fails to take into account the time delay before host death, during which time the host continues to interact with other community members as a competitor/predator/prey.

Box 1.4 Modelling parasite transmission strategies

Deciding how to represent the details of parasite transmission is not a trivial problem. Firstly, the dynamics of transmission are different for micro- and macroparasites, and parasitoids. For parasites with direct transmission, successful transmission depends on the pattern of contact between susceptible and infectious hosts. Several different models for this have been proposed, the most common of which are density- and frequency-dependent transmission. Not all parasites have direct transmission; some have free-living stages, others utilise intermediate hosts, with or without transmission via free-living stages to one or more of the hosts. Vertical (inherited) transmission requires a different modelling approach to horizontal (contagious) transmission. Finally, host characteristics need to be considered; most importantly, aspects of immunity vary widely. Hosts may (or may not) recover from infection; if so, they may (or may not) become immune to reinfection; this acquired immunity may (or may not) be life-long. Capturing these variants requires different model structures, and the population dynamics and outcomes can be strongly dependent on these; below are some of the more frequently met distinctions.

Density-dependent transmission. Here, the number of new infections per unit time is proportional to the density of infectious hosts multiplied by that of susceptible hosts. Most models of directly transmitted microparasites, like the one in Box 1.2, assume transmission is density dependent, resulting in βSI new infections per unit time (where β is the transmission efficiency, denoting the proportion of S–I contacts that lead to successful transmission). Transmission of this type is also called 'mass action', because it assumes homogeneous mixing of the S and I classes as if they were molecules contained in a closed system. Although simplistic, it appears to be a reasonable approximation for the transmission of many pathogens (Anderson & May, 1981; McCallum *et al.*, 2001).

Frequency-dependent transmission. In this model, the number of new infections per unit time is proportional to the density of susceptible hosts multiplied by the frequency (proportion) that are infected, resulting in $\beta SI/N$ new infections per unit time. This type of transmission model is more appropriate when per capita contact rates are fairly constant, irrespective of population density, in which case the per capita chance of becoming infected is proportional to

Box 1.4 (*continued*)

the prevalence of infection (I/N) in the population. This is the standard transmission model for STDs, where the number of sexual partners an individual has depends on social and mating systems rather than population density. Vector-transmitted diseases like malaria are also modelled this way, because infection risk for susceptible hosts depends on the proportion of vectors that have fed from an infected host. Basic theory for frequency-dependent transmission shows that, in contrast to density-dependent transmission (Box 1.2), there is no threshold host population size for parasite establishment (N_T). Consequently, such parasites can be maintained in small populations and could theoretically contribute to the extinction of that population. These different transmission modes have different consequences for community structure and stability (Chapter 7).

Recovery and immunity. These are not the same concept in epidemiology. Infected individuals may recover, re-entering the susceptible class, or in some cases they may become immune, in which case they are neither susceptible nor infected, and enter a new class (often called the 'removed' (R) class, because they are removed from the population available for infection). A further complication is that many diseases have a latent period after initial infection, in which individuals have contracted the infection but cannot transmit it to others. This requires a further distinction between individuals that have been exposed to infection (population class E) and those that have become infectious (I class). Compartment models are used to represent these structures; each class is represented by a compartment, and gain and loss equations like those in Box 1.2 represent the rates of transition between compartments. These models are often referred to with acronyms reflecting the ordering of compartments used (Fig. 1.2).

Macroparasites require a different approach to modelling than the SI compartment approach used for microparasites. Firstly, it is not generally appropriate to consider hosts (certainly, large vertebrate hosts) as either infected or uninfected; most hosts in a population may harbour a particular macroparasite (e.g. a helminth species), but they vary greatly in burden. Infections tend to be persistent; they are usually associated with increased morbidity rather than mortality, and the extent of deleterious effects are usually correlated with parasite burden. Macroparasites frequently

Box 1.4 (*continued*)

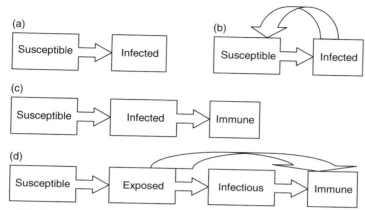

Fig. 1.2. Compartment models for microparasite infections. Each box represents a distinct stage (compartment); transitions between stages are shown by block arrows. Other loss and gain terms (birth, death, isolation or quarantine; not shown) change the fluxes into and out of compartments. Models shown are: (a) SI – there is no recovery; once infected, individuals remain so or die; (b) SIS – infected individuals can recover, re-entering the susceptible class; (c) SIR – infected individuals recover, becoming immune ('removed'); (d) SEIR – a latent (exposed) phase occurs after infection before hosts become infectious; exposed and infectious individuals can recover and acquire immunity. There are many further variations around these themes.

have an aggregated distribution among hosts: a few hosts are heavily infected (and suffer most of the fitness consequences), whereas most hosts have relatively few parasites. Macroparasite models take account of parasite distribution between hosts, and include fitness effects such as influence on fecundity, as well as mortality. Models track changes in host and parasite population densities, with parasite burden in the host described by a statistical distribution (the negative binomial distribution is often used; a clumping parameter can be set to describe the degree of parasite aggregation among hosts). Despite these differences, there are some broad similarities between the predicted effects of macro-parasites and microparasites on communities (for instance, both can cause apparent or parasite-mediated competition (Chapter 2); both can cause trophic cascades through effects on consumer species (Chapter 7)).

Box 1.4 (*continued*)

Complex life cycles. Many macroparasites have complex life cycles with obligate use of more than one host species. Transmission between host species may involve trophic transmission, where the parasite is transmitted to the definitive host (or another intermediate) via its consumption of an intermediate host; reproduction in the definitive host leads to the production of free-living stages. Models of such systems require a separate variable for each host species, and variables for each of the free-living parasite stages. Trophic transmission requires that suitable predation functions be incorporated into the model, including the possibility that parasitic infection influences the likelihood of being eaten. Aggregated parasite distribution and deleterious effects linked to burden need to be considered for most host species. Given these complications, it is perhaps not surprising that few models of parasites with complex life histories have been used to inform general questions about the roles of parasites in communities, as models tend to be tailored to specific systems.

Free-living stages. Many macro- and microparasites have free-living infective stages that survive outside the host. Some of these (baculovirus resting bodies or fungal spores, for instance) can remain viable in the environment for long periods. Free-living stages need to be represented as another class (compartment), and produce qualitatively different dynamical behaviour to the simple models of direct transmission. The time lags inherent in transmission from free-living stages can introduce cyclical dynamics. Also, for systems with multiple host species, the existence of a common pool of infective stages breaks down any community structure in transmission (the distinction between interspecific versus intraspecific transmission becomes redundant); model structure therefore bears more similarity to that for parasitoids.

Parasitoids are in some ways more like predators than microparasites, because they tend to kill their host; ovipositing females search actively for hosts and can potentially become satiated (their ova depleted), leading to similar expectations of a functional response between 'prey' density and predation/parasitism. However, they are also different from predators: they do not kill their prey instantly; they may share prey with other parasitoids or parasites; and they may be consumed during development in the host by predators, parasites or other parasitoids. Adults are effectively free-living stages (but not

Box 1.4 (*continued*)

long-lived, so not intrinsically responsible for time lags); this common pool of infective stages tends to break down transmission structures as discussed above. One consequence of this is that models of the indirect effects on hosts of shared parasitoids behave more like those of predators than parasites (Chapter 2).

Finally, a note on *variables and parameters*. The distinctions we have drawn above mostly refer to variables: that is, the different classes of individuals (S, I, E, R) which each need to be represented separately as they vary over the course of the calculations. Increasing the number of variables can make models more realistic, but impedes their analysis. There can be no general methodologies for finding solutions (identifying and characterising equilibria, for instance) for systems with more than four variables. Parameters, on the other hand, are constants, which take values depending on the differences one wishes to express about the classes under consideration. For instance, infected individuals may experience a different predation rate to uninfected hosts; this would be reflected in different parameters for predation in the equations for each variable. Well-chosen parameters reflect quantities that one might be able to measure empirically. Increasing the number of parameters can also make models more realistic, and does not make them much harder to solve analytically, but it can make them more difficult to comprehend and interpret.

1.4 Aims of this book

In this book we examine the effects of parasitism in communities, starting with the structurally simpler modules that are the basic building blocks of a community. Throughout the book, our approach is to introduce a subject area and its theoretical underpinnings, in combination with empirical examples and applications. We do not go into the detailed theory of more complicated examples, but do present basic models to show the reader how mathematics has been used to develop predictions and structure ideas. We believe that researchers and practitioners in community ecology cannot be effective without at least a basic appreciation of the theory behind their science, and it is at this level that we have aimed the mathematical presentation. This means that in the earlier chapters, where the basic principles are developed, there are a few more equations than in the latter half of the book, where these principles are applied. We hope the models discussed in the earlier chapters will allow the interested reader to interpret the assumptions and predictions of more complicated models that are beyond the scope of this book. We begin in Chapter 2 with an examination of parasitism in competition modules, first examining the interaction between parasitism and intraspecific competition, allowing a brief reprise of the one-host–one-parasite epidemiological models on which much of the subsequent theory is based. We then move onto the case of interspecific competition and parasitism, in which we see that parasites and competitors can interact in a great variety of ways. From a community perspective, indirect interactions in these modules can be regarded as horizontal effects of parasitism, as they ramify between species at the same trophic level. In Chapter 3 we examine some vertical effects, looking at the interaction between parasitism and predation. Again, there are many structurally distinct ways in which parasitism can interact with predation, and the module approach helps to clarify these differences. In Chapter 4 we consider a more complicated module – that of parasitism combined with intraguild predation (IGP, the predation of potential competitors). IGP is increasingly recognised as an important component of community structure, and doubting readers will find ample evidence in the preceding chapters as we show how many parasitism–competition and parasitism–predation interactions may also be considered under the remit of IGP.

From an analysis of these basic building blocks, we move on to examine the community-wide ramifications of parasitism, where the conservation and biodiversity implications of parasitism become clearer. In Chapter 5 we examine the community impacts of parasites and pathogens of plants. This offers the reader a reprise of how the concepts developed in the preceding chapters apply to plant systems, and how effects at the module level propagate from these basal resources up throughout the community. In Chapter 6 we examine how parasites influence biological invasions via their effects on invading or native species. These examples illustrate how the effect of parasites at the level of competition and predation propagate through the community to determine invasion outcomes. We then go on, in Chapter 7, to examine how parasites influence general ecosystem processes and properties. This rapidly developing field (which we call ecosystem parasitology) is revealing a number of keystone and structural roles for parasitism, including effects on biodiversity, the structural stability of food webs and ecosystem health. Finally, Chapter 8 investigates emerging infectious diseases from a community ecology perspective; this is a very current problem with implications for human health and wildlife conservation, and best of all exemplifies how greater collaboration and integration between parasitologists and community ecologists is needed as we enter this third millennium.

2 · *Parasites and competitors*

2.1 Introduction

As long ago as the 1940s, the potential importance of parasites in influencing competition between species had been demonstrated experimentally (Park, 1948). These laboratory experiments, which have now become a classic example in ecology, showed that infection by the shared sporozoan parasite *Adelina tribolii* reversed the outcome of competition between two species of flour beetle (*Tribolium castaneum* and *T. confusum*). When *A. tribolii* was present, *T. castaneum* was driven extinct in 66 of the 74 mixed-species cultures. In the absence of the parasite, *T. confusum* went extinct in 12 out of 18 mixed-species replicates. Park also censused single-species beetle populations with and without the parasite, from which he concluded that the parasite induced higher mortality in *T. castaneum*, and that this effect was largely responsible for the change in competitive outcomes. This example has become a cornerstone for much of the work on the effects of parasites in communities, illustrating three key concepts that will recur in this book:

(1) Parasites can alter competitive relationships between host species.
(2) By altering interactions such as competition, parasites can play *keystone roles* in ecological communities. That is, the addition of a parasite species can alter the outcome of an interaction, mediating coexistence or exclusion of one or other host species *with knock-on effects throughout the community*.
(3) Alternative host species can act as *reservoirs* for parasite amplification, *enabling maintenance of higher parasite population densities*, and so providing increased opportunities to infect other species in which the parasite is more virulent.

In this chapter we examine the theory and evidence underpinning the first point and begin to explore the ramifications of these effects for community processes (the second and third points). How the effects of parasitism on competition, and on other interactions, translate into community-level processes are explored in more detail throughout the remainder of the book.

2.1.1 Parasitism in modules of competition

In reality, the above example of parasitism in competing flour beetles (Park, 1948) is only one of a great variety of ways in which parasitism can influence competition (Fig. 2.1). We can distinguish between several different processes:

- *Intraspecific competition in the host*: parasites can alter the survival, fecundity or competitive ability of individuals, affecting their interactions with conspecifics and influencing the dynamics of intraspecific competition.
- *Interspecific competition between hosts*: parasites (shared between host species or specialist on one species) can alter interactions between competing host species. Parasites can influence competition through density-mediated effects on host populations (parasite-mediated competition), or modify the competitive abilities of hosts (parasite-modified competition). Both have potential ramifications for host population dynamics and community structure.
- *Apparent competition*: parasites of two or more hosts that do not interact directly can lead to competition-like coupling between the host populations because of the indirect effect of each host acting as a reservoir for infection to the other. This has potentially similar effects on population and community processes to the influence of parasitism on true interspecific competition.
- *Parasite–parasite competition*: parasites that utilise the same host species potentially compete with each other, influencing the structure of parasite communities and parasite evolution.

We can see the differences between these modes of interaction by examining the modules involved (Fig. 2.1); we look at each in more detail in the remainder of this chapter. In Section 2.2, we examine parasitism in intraspecific competition and briefly overview the basic theory that underpins our understanding of parasite–host population dynamics, from which all extensions to the community level ultimately stem. We then progress through structurally more complicated interactions, starting with apparent competition (Section 2.3), in which the hosts do not interact directly, then looking at situations where the hosts do interact directly and in which parasites influence these interactions (parasite-mediated and parasite-modified competition; Sections 2.4 and 2.5).

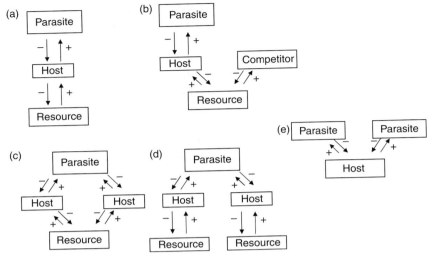

Fig. 2.1. Ways in which parasites can enter into competition modules:
(a) intraspecific competition; (b) interspecific competition (parasite not shared);
(c) interspecific competition (shared parasite); (d) apparent competition;
(e) parasite–parasite competition. Direct (trophic) interactions are shown by
solid arrows; positive (+) or negative (-) effects as indicated.

We then examine some detailed case studies where parasites interact
with potential competitors in natural or managed populations (Section
2.6), before examining competition between parasites in Section 2.7.
From a module perspective, this latter interaction is no more complicated than the host–host competition systems. However, research in this
area has taken a somewhat different approach, because this type of
interaction has immediate evolutionary implications; for example, for
virulence and host range evolution.

2.2 One-host–one-parasite systems

Before we investigate the effects of parasitism on interactions between
competing species, we first need to look at the effects in a single host
species. The mathematical approach underlying most of the community
ecology perspectives on epidemiology originate from Anderson and
May, who included the dynamics of host populations in epidemiological
models of parasitism (Anderson & May, 1981). Before this point, models
had focused on the spread of a parasite in large host populations that
were regulated by some other (unspecified) mechanism to a constant

density (Box 1.2). Because host population density remains constant in these models, they do not tell us what the impact of the parasite is on the host population. For many (largely non–human) host–parasite inter-actions, parasitism may be an important influence on host population dynamics, and similarly, host population dynamics may feed back on parasitism; hence, coupling host and parasite population densities is necessary to fully understand the dynamics of the interaction. Before embarking on the rest of this section, we advise readers new to this area to first review their understanding of three basic concepts in epidemi-ology: the basic reproductive number, R_0; the threshold for disease establishment, N_T; and the force of infection (Box 1.2).

2.2.1 Population dynamics

2.2.1.1 Host–parasite system without host self-regulation

We first extend the equation model in Box 1.2 to include a term for the host population growth. The simplest way to do this is to include constant terms for host birth (a) and death (b) rates (Anderson & May, 1981). In the absence of any intervening forces, such a host population would either grow (if $a > b$) or decline (if $b > a$) exponentially. Let us consider a microparasite infecting a large, randomly mixing (panmictic) host population with continuous breeding. For a parasite with density-dependent transmission (Box 1.4), change in the infected portion (I) of the population is described by:

Gain: transmission of infection

$$\frac{dI}{dt} = \beta SI - (\alpha + b + \gamma)I \qquad (2.1)$$

Rate of change of infected subpopulation

Loss: parasite-induced mortality (α), natural mortality (b) or recovery (γ)

Here, the density of the infected class I is incremented by new infections and reduced by death (parasite-induced or 'natural' – i.e. from causes other than the parasite under consideration) or recovery (see Table 2.1 for definitions).

The density of susceptible hosts S is incremented by recovery from the infectious class and births (from both infected and susceptible hosts,

Table 2.1. *Parameters, variables and functions used in the equations in this chapter*

S	*Variable: susceptible host (prey) class.
I	Variable: infected host (prey) class.
N	Variable: total host population density ($N = S + I$).
a	Host per capita birth rate.
α	Parasite-induced mortality rate.
b, $b(N)$	Host natural death rate (causes other than parasitism); either a constant b or a function of host density $b(N) = b_0 + sN$, where b_0 is background mortality and s is a constant describing the strength of density dependence.
β, β_{ij}	Parasite transmission efficiency per contact. In two-host models, β_{ij} = transmission efficiency from species j to species i.
γ	Recovery rate of infected individuals (per capita return to susceptible class).
r	Intrinsic rate of increase ($r = a - b$).
K	Carrying capacity of the environment for host population with intrinsic density-dependent regulation.
P_i	Population density of parasitoid species i.
c_{ij}	Competition coefficient: c_{ij} describes equivalence of species j in terms of species i; hence c_{ii} = intraspecific competition; c_{ij} = interspecific competititon (impact of species j on species i).

* In Eqn (2.11)–(2.14) there are two host species so parameters and variables take subscripts to denote each host; e.g. N_i = population density of species i; r_i = intrinsic rate of increase of species i.

here assumed to be equally fecund). It is depleted by natural mortality and infection:

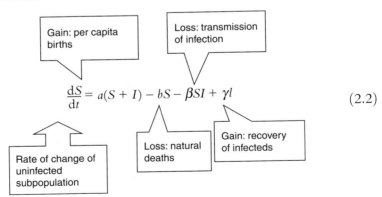

$$\frac{dS}{dt} = a(S + I) - bS - \beta SI + \gamma I \tag{2.2}$$

Because the equation for the infected class is identical to that considered in Box 1.2, we obtain the same conditions for the initial spread of infection (from the requirement $dI/dt > 0$). The basic reproductive number, R_0, must exceed 1, and the host population density must exceed the threshold for establishment, N_T:

$$R_0 = \frac{\beta N}{\alpha + b + \gamma},$$

$$N_T = \frac{\alpha + b + \gamma}{\beta}. \tag{2.3}$$

The total host population density, N, is simply the sum of I and S. By looking at the change in N (summing dI/dt and dS/dt), we can examine the regulatory capacity of the parasite:

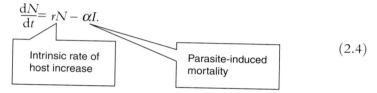

$$\frac{dN}{dt} = rN - \alpha I. \tag{2.4}$$

Intrinsic rate of host increase

Parasite-induced mortality

In this equation we have defined $r = a - b$ (births − deaths), the intrinsic rate of increase of the host. Hence, the population will be regulated ($dN/dt = 0$) by the parasite if pathogenicity exceeds the intrinsic rate of increase of the host ($\alpha > r$). If $\alpha < r$, but the conditions (Eqn 2.3) for parasite spread are still met, host population growth will not be brought to a halt; rather, it will continue exponentially, but at a reduced rate. Solving for $dN/dt = 0$, we find the equilibrium host population size N^* in the presence of the parasite:

$$N^* = \frac{\alpha(\alpha + b + \gamma)}{\beta(\alpha - r)}. \tag{2.5}$$

It can be shown analytically that the equilibrium in this case is stable, so populations will grow or decrease towards this point and, if perturbed from it, they will return there (Table 2.2). By examining how N^* varies with parasite parameters, we can determine which characteristics are important for host population regulation. Firstly, N^* is inversely proportional to the transmission rate β, so higher parasite transmission efficiency will result in a lower disease-regulated population density. The non-linear relationship between N^* and parasite-induced mortality α means that the host population reaches a minimum at an intermediate pathogenicity (Fig. 2.2a). This is because increasing α initially results in a higher death rate, but also reduces the time that individuals spend

Table 2.2. *Terminology used for describing the dynamical behaviour of populations*

Stable equilibrium	Stable state towards which the population will move. The equilibrium is locally stable if the population returns to this state after a slight perturbation. A globally stable equilibrium is one which is moved towards (i.e. 'attracts' the population from any point in state space).
Unstable equilibrium	If the population is precisely at this state, it will remain there, but if it is perturbed, it will move away.
Monotonic damping	One of the ways in which populations move towards equilibrium. Here, the population changes smoothly in the same direction towards equilibrium with no oscillations.
Damped oscillations	Here, the population oscillates towards equilibrium; the oscillations get smaller until the equilibrium is reached.
Undamped (divergent) oscillations	The population oscillates with increasing amplitude; it will never reach equilibrium. In practice, the population will eventually become so small during a trough that it is driven extinct.
Stable limit cycle	The population repeats a trajectory around a point, and continues cycling with constant amplitude and period indefinitely. If perturbed, it returns to this stable trajectory.
Neutral stability	If undisturbed, the population cycles at constant amplitude and period, but if perturbed, it does not return to this course, but continues to cycle in the new trajectory.
Alternative stable states	Multiple stable equilibria in the state space. Each is locally stable, but not globally so. Which equilibrium is reached depends on the current position of the population, and the behaviour of trajectories around the equilibria. A population sufficiently perturbed from one equilibrium may move to a different equilibrium; it may be hard to predict which equilibrium some populations will move towards.
Chaotic	Small differences in initial conditions can give rise to very large differences in trajectory. Consequently, long-range predictions about the population's behaviour are not meaningful.

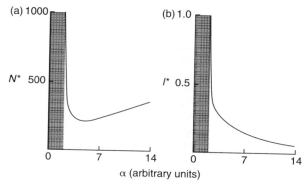

Fig. 2.2. The effect of pathogenicity in a one-host–one-parasite model. Figures plot the effect of increasing pathogenicity (α) on (a) equilibrium host population density (N^*) and (b) equilibrium prevalence of infection (I^*), from Anderson and May's (1981) model for a directly transmitted specialist microparasite and host in the absence of other regulatory factors. The shaded area depicts the region for which pathogenicity is lower than that required for parasite establishment, given the value taken by the other parameters in the example. From Anderson and May (1981).

infectious, reducing the force of infection (Box 1.2) to susceptible hosts. Consequently, the proportion of the population infected (parasite prevalence) also declines with increasing pathogenicity (Fig. 2.2b).

Applications to biological control: the predictions of these models are relevant to the use of specialist parasites in biological control. Parasites can be used in one of two ways: as biopesticides in which the aim is to extinguish local outbreaks of the pest with repeat applications if a further outbreak occurs; or as classical biological control agents that will continue to exist with the pest population, regulating it to lower densities. For this latter class, intermediate levels of pathogenicity are desirable, as the pathogen will be less vulnerable to local extinction (termed 'fade out' or 'burn out') through insufficient transmission; it will also have the largest impact on host (pest) numbers at intermediate pathogenicity (Fig. 2.2). For biopesticides, more virulent parasites can be selected because fade out is an accepted outcome; however, high transmission rates and/ or mechanical delivery systems for assuring a good coverage of infection throughout the pest population are required.

2.2.1.2 *Host–parasite system with density-dependent regulation of the host*

A more realistic model is obtained by including density-dependent regulation of the host (Box 2.1). This adds an additional constraint on parasite persistence. As above, there is a threshold population size for parasite establishment, N_T. However, in the absence of the parasite, hosts with density-dependent regulation are limited to a carrying capacity, K. It follows that the parasite can only persist in a host population if N_T is less than K, otherwise the host population at carrying capacity will not be large enough to permit parasite establishment. In this model, N_T increases with pathogenicity and decreases with transmission efficiency (as before), but is also dependent on the strength of host density dependence (Box 2.1). From the perspective of the host, the form of this new relationship means that hosts with large carrying capacities will be vulnerable to more pathogenic parasites, and to parasites with lower transmission efficiency. From the perspective of the parasite, hosts with strong density dependence result in more stringent conditions for establishment. In general, transmission efficiency must be high enough, and pathogenicity not too high, for successful establishment. As before, parasites of intermediate pathogenicity will have the greatest effect in depressing host population density below the disease-free carrying capacity.

These models make key predictions that have ramifications in a multi-host context:

- Highly pathogenic parasites reach equilibrium at lower prevalences in host populations.
- Parasites with efficient transmission can become established in smaller host populations, or in populations with stronger density-dependent constraints.
- Host population birth and death rates also influence parasite establishment and prevalence at equilibrium.

When parasites are able to infect more than one host species, these restrictions need not apply to both host species. For instance, if the criteria for parasite establishment are fulfilled by host species A, the parasite may also be able to cause outbreaks in another less-suitable species B, in which it could not be sustained independently. In this case A can act as a reservoir for infection to B, potentially with highly detrimental effects. For example, a poxvirus of squirrels is prevalent in English populations of grey squirrels, but is low in pathogenicity, whereas it is highly pathogenic in red squirrels and rapidly burns out in local populations (Box 2.5). Infection by an intestinal helminth cannot be sustained alone in grey partridge populations where it has an $R_0 < 1$; however, $R_0 > 1$ in pheasants may be sufficient to sustain worm populations in both species (Box 2.6).

Box 2.1 A model of parasitism with density-dependent host population growth

Anderson and May (1981; their model F) incorporated a term for density-dependent regulation of the host in the absence of parasitism. Their model, in which density dependence acts on the term for host death rate, can represent intrinsic processes (such as intraspecific competition) or extrinsic processes that lead to density dependence (such as some forms of predation, for instance). The death rate, previously a constant b, becomes a function of host density:

$$b(N) = b_0 + sN. \tag{2.6}$$

Substituting this term for death rate into the equation system above (Eqn 2.1–2.2), we obtain a new system of equations:

$$\frac{dI}{dt} = \beta SI - (\alpha + b_0 + sN + \gamma)I,$$
$$\frac{dS}{dt} = a(S + I) - (b_0 + sN)S - \beta SI + \gamma I. \tag{2.7}$$

As before, $dN/dt = dS/dt + dI/dt$:

$$\frac{dN}{dt} = (a - b_0 - sN)N - \alpha I. \tag{2.8}$$

When the parasite is absent (setting $I = 0$) the host population is regulated to the familiar carrying capacity of the logistic equation:

$$N^* = K = \frac{a - b_0}{s}. \tag{2.9}$$

As before, we can solve for R_0 and N_T:

$$R_0 = \frac{\beta N}{\alpha + b_0 + \gamma + sN}$$
$$N_T = \frac{\alpha + b_0 + \gamma}{\beta - s}. \tag{2.10}$$

From the expression for R_0, we see that if transmission is less than the strength of density dependence ($\beta < s$), then R_0 is definitely < 1,

Box 2.1 (*continued*)

so higher transmission efficiency will be required for parasite establishment in strongly density-dependent populations. Similarly, from
the expression for N_T, higher density dependence s results in a higher
N_T (realised provided $\beta > s$). Further, if N_T exceeds K, the carrying
capacity in the absence of parasitism, the host population will never be
large enough for parasite establishment. The equilibrium population
size in the presence of parasitism (N^*) behaves in a similar way (but is
more complicated) to that above in the absence of density dependence
(Anderson & May, 1981). Host birth and death rates influence N^* and
the equilibrium parasite prevalence; because birth adds new susceptibles to the population, providing more resources for parasitism,
higher birth rates increase parasite prevalence. However, as birth rate
offsets death rate, it reduces the impact of the parasite on the host
population, so that N^* is closer to K at higher birth rates.

Since this pioneering work, several models have been developed that
relax some of these simplifying assumptions or extend the treatment to
other systems. Roberts (1995) and Bonsall (2004) provide accessible
reviews of these; in addition, the volume by Anderson and May
(1991) concerning infectious diseases of humans provides a good
grounding in many of the concepts relevant for one-host–one-parasite
models. The volume edited by Hawkins and Cornell (1999) provides
a comprehensive guide for the application of microparasites and
parasitoids in biological control; Myers and Bazely (2003) provide an
excellent overview of classical biological control for invasive plants.

2.2.2 Competitive release

For host populations constrained by limited resources, parasites can
potentially impact on populations in opposing directions. For instance,
by reducing population density, parasites may actually free-up
resources for the survivors of an epidemic; this has been demonstrated
in mesocosm experiments on infections in larval amphibians. Kiesecker
and Blaustein (1999) stocked experimental ponds at two densities with
pure or mixed tadpole cultures of two species (*Rana cascadae* and *Hyla
regilla*), with or without the fungus *Saprolegnia ferax*. For *R. cascadae*
(but not for *H. regilla*) the fungus altered the effects of intraspecific

competition: tadpoles that survived the infection weighed more at metamorphosis than conspecifics in uninfected cultures, suggesting that surviving tadpoles were 'released' from intraspecific competition by the demise of infected conspecifics. Infection with *S. ferax* also impacted interspecific competition, suggesting that the fungus mediates interspecific interactions as well (see Section 2.4.2; Fig. 2.8).

When parasites are less virulent and do not greatly increase host mortality (at least in the short term), the results of the interaction between parasitism and intraspecific competition are harder to gauge. In this case, competitive release through mortality of conspecifics will be unlikely, but parasitism may still alter competitive outcomes, particularly under stressful conditions. For instance, tadpoles of *R. pipiens* reared at high densities and infected with the trematode *Echinostoma trivolvis* attained lower mass than uninfected conspecifics, possibly due to reduced feeding rates in infected tadpoles (Koprivnikar et al., 2008). Similarly, Bedhomme et al. (2005) found that mosquito larvae *Aedes aegypti* infected with the microsporidian parasite *Vavraia culicis* took disproportionately longer to develop when reared with uninfected conspecifics; hence the fitness costs of harbouring parasites can depend on both absolute density *and* parasite prevalence. The interaction between parasitism and resource availability can therefore have a range of consequences for populations, from reduced to unchanged or even increased population densities, depending on how parasitism affects competitive interactions. For instance, field manipulations of the mosquito *Aedes sierrensis* have shown that more adults emerged from larval populations infected with the ciliate *Lambornella clarki* when food was limiting (Washburn et al., 1991). Some parasites increase resource consumption by infected hosts (for example, the trematode *Posthodiplostomum minimum* in the freshwater snail *Physa acuta* (see Section 7.2.2); and the acanthocephalan *Echinorhynchus truttae* in the freshwater amphipod *Gammarus duebeni celticus* (Box 7.2)); this could translate into resource depletion in food-limited conditions.

The range of outcomes of competition between infected and uninfected hosts is exemplified in studies of parasitoids. Many parasitoids (known as koinobiont parasitoids) only kill the host when they emerge at host pupation; their success is therefore tied to survival of the host throughout larval development. Cameron et al. (2005) found that, in mixed populations, survival to pupation of the moth *Plodia interpunctella* was negatively correlated to the proportion of larvae infected with the parasitoid *Venturia canescens*. In this system it appears that parasitoids reduce the resource requirements of infected hosts (which still reach pupation, but are smaller on average), thus allowing them (and the parasitoid) to survive intense interspecific competition. However, in another moth, the cranberry fruitworm (*Acrobasis vaccinii*), the parasitoid *Phanerotoma franklini* has the reverse effect, increasing survival of healthy hosts in mixed populations. In this system, host larvae engage in contest competition for resources (each larva inhabits and consumes a single berry); parasitised larvae have a reduced growth rate and are less effective in contests, resulting in competitive release for healthy conspecifics (Sisterson & Averill, 2003).

The parasitoid studies make clear that parasites can *modify* the competitive ability of their hosts, in addition to their effects on host population density. Most one-host–one-parasite (or parasitoid) models assume infected hosts are either unaltered in competitive capacity, or do not compete at all (i.e. they are functionally dead). Predicting the outcomes of parasite-modified interactions requires that we allow infected and uninfected hosts to have different parameters for intraspecific competition; models that include this feature (of which there are very few) predict qualitatively different dynamics (e.g. White *et al.*, 2007). In Section 2.5 we examine parasite-modified competition in more detail in regard to its effects in multi-host systems.

2.3 Apparent competition

2.3.1 Baseline theory

The population dynamics of different species that do not compete for resources can become linked through the effects of a shared natural enemy,

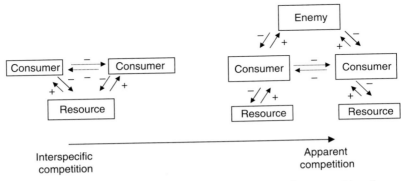

Fig. 2.3. Schematic of the relationship between interspecific competition for resources (exploitative competition) and apparent competition. Positive and negative effects of one species on another are indicated by + and − symbols respectively; solid arrows show direct effects, dashed arrows show indirect effects.

through a process known as apparent competition. Apparent competition, a term coined by Holt (1977), was first considered in the context of predator–prey interactions (Holt, 1977; 1984; 1997). If two prey species are subject to predation by a shared natural enemy, their population dynamics will negatively covary if the population dynamics of each prey species is separately dependent on that of the predator, and vice versa (Fig. 2.3). A number of theoretical papers have also demonstrated that the process can also occur through the action of shared parasites and parasitoids (Table 2.3).

2.3.1.1 Apparent competition mediated by parasitoids

Basic predator and parasitoid models predict that the enemy cannot coexist stably with both prey species: the prey that can support the

Table 2.3. *General models of apparent competition for parasites and parasitoids*

System	Outcomes	Reference
Host–host–microparasite (directly transmitted)	*Repeatable exclusion, coexistence or contingent exclusion in a pattern mirroring outcomes for interspecific competition	(A) Holt and Pickering (1985)
Host–host–microparasite with common pool of free stages	Coexistence of both hosts with parasite not possible; winner is host that can tolerate higher density of free-living stages	Bowers and Begon (1991)
Host–host–parasitoid	Parasitoid mediates exclusion: the winner is the host that can tolerate the higher parasitoid density. Mechanisms for coexistence discussed, including resource limitation of hosts and spatial refuges	Holt and Lawton (1993)
Host–host–parasitoid with spatial structure	Metapopulation with (1) local or (2) global dispersal facilitates coexistence; outcomes depend on fecundity/dispersal or fecundity/resistance tradeoffs	(1) Bonsall and Hassell (2000); (2) King and Hastings (2003)

(cont.)

Table 2.3. (*cont.*)

System	Outcomes	Reference
Host–host–microparasite with intraspecific competition (direct transmission)	Host self-regulation expands the range of conditions leading to coexistence; coexistence may be mediated by resource limitation or parasite; cyclic population dynamics possible	(B) Begon *et al.* (1992)
Host–host–pathogen with intraspecific competition (free-living pathogen stage)	Host self-regulation enhances coexistence; free-living stages increase initial-condition contingency	Begon and Bowers (1994)
Host–host–microparasite with or without intraspecific competition; direct transmission	Counterexamples to (A) and (B) showing that some parameter combinations have no stable equilibria or multiple stable infected equilibria; population cycles possible	Greenman and Hudson (1997)

* This is a non-exhaustive list showing some of the main developments; for mathematical terms used to describe population dynamics, see Table 2.2.

higher density of the natural enemy excludes the other. We can understand this result in terms of the invasion criteria for the prey/host: a prey species i can only increase when rare, provided:

$$r_i > f_i P, \qquad (2.11)$$

where r_i is the prey's intrinsic rate of increase, f_i is the per capita rate of attack on species i and P is the density of predators or parasitoids. The populations will therefore equilibrate when:

$$P_i^* = r_i/f_i. \qquad (2.12)$$

A second prey species j will only be able to invade if its intrinsic rate of increase exceeds the losses due to attack at prevalence P^*. In this case, it will spread until a new equilibrium $P_j^* = r_j/f_j$ is reached. In this case, since $P_j^* > P_i^*$, prey i will decline to zero. In the alternative case, where $P_j^* < P_i^*$, the second prey species cannot increase when rare, so prey

i remains in equilibrium with the parasitoid. This is known as the P^* rule for exclusion by apparent competition; Holt and Lawton (1993) discuss a number of mechanisms that may allow coexistence of hosts with shared parasitoids, and further theory explores some of these possibilities (Table 2.3).

2.3.1.2 Apparent competition mediated by parasites

Apparent competition mediated by parasites was first considered explicitly by Holt and Pickering (1985), who examined two host species sharing one microparasite. Prey population densities respond to parasite population density through the term for parasite-induced mortality; parasite population density is modelled as the populations of infected hosts, so is clearly tied to both host population densities. This coupling provides the prerequisites for apparent competition between the hosts. The model follows the general form developed earlier (Eqn 2.1–2.2), except that now we require equations for the two host species i and j, including terms for intraspecific (β_{ii}) and interspecific (β_{ij}) transmission efficiency (β_{ij}, transmission efficiency to species i from species j; see Table 2.1 for other parameters). Here, I_i and S_i are the infected and susceptible subpopulations for species i, respectively:

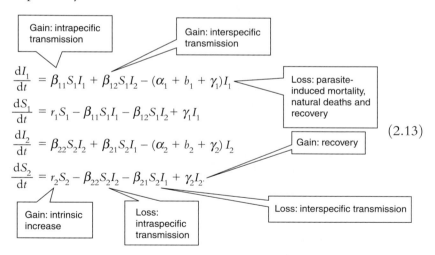

$$\frac{dI_1}{dt} = \beta_{11}S_1I_1 + \beta_{12}S_1I_2 - (\alpha_1 + b_1 + \gamma_1)I_1$$

$$\frac{dS_1}{dt} = r_1S_1 - \beta_{11}S_1I_1 - \beta_{12}S_1I_2 + \gamma_1I_1$$

$$\frac{dI_2}{dt} = \beta_{22}S_2I_2 + \beta_{21}S_2I_1 - (\alpha_2 + b_2 + \gamma_2)I_2 \tag{2.13}$$

$$\frac{dS_2}{dt} = r_2S_2 - \beta_{22}S_2I_2 - \beta_{21}S_2I_1 + \gamma_2I_2.$$

Gain: intrapecific transmission

Gain: interspecific transmission

Loss: parasite-induced mortality, natural deaths and recovery

Gain: recovery

Gain: intrinsic increase

Loss: intraspecific transmission

Loss: interspecific transmission

In this model, the host populations are unregulated in the absence of the parasite (i.e. there is no intra- or interspecific competition, as is definitional for apparent competition). Hence, in the absence of the parasite,

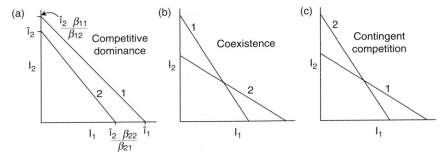

Fig. 2.4. Isocline analysis of apparent competition mediated by parasites. The lines plot the zero-growth isoclines for the infected class for three distinct cases. (a) The isoclines do not overlap, so there is no joint coexistence equilibrium: species 1 excludes species 2 (its isocline lies above that of species 2; meaning that it tolerates a higher population density of infected species 2 than species 2 can tolerate from species 1). (b) The isoclines cross once, and the equilibrium point is stable, meaning that the species will stably coexist at the joint equilibrium values. (c) The joint equilibrium is unstable; a perturbation towards I_1 will exclude species 2, and a perturbation towards I_2 will exclude species 1; in practice this means that whichever species is initially more common will exclude the other; for terminology see Table 2.2. From Holt and Pickering (1985).

the hosts coexist and both populations grow in an unbounded manner. With the parasite present, Holt and Pickering distinguished three outcomes, reminiscent of those for the classic Lotka–Volterra model of interspecific competition:

- *Repeatable exclusion of one host species* (the one most affected by the parasite): this is predicted to occur if interspecific transmission is the same as or greater than intraspecific transmission and if the difference between virulence effects on the two host species is large enough.
- *Host coexistence*: this occurs if interspecific transmission is less than intraspecific transmission, but the two species are similarly affected by the parasite.
- *Exclusion of host species depending on initial conditions*: this occurs when interspecific transmission is higher than intraspecific transmission. Which host species is excluded depends on the sizes of the initial populations. This initial-condition-dependent outcome is analogous to that for two competing species, when interspecific competition exceeds intraspecific competition.

These outcomes can be visualised graphically by plotting the zero-growth isoclines for the infected subpopulation of each host species against that for the other host (Fig. 2.4). These lines plot the density of infected species 2 at which infected species 1 is at equilibrium, and vice versa; these relationships are found by solving the equations above (for $dI/dt = 0$). Further analysis is required to tell whether the equilibria are stable (Table 2.2). For parasite-mediated apparent competition, the isoclines can be placed in three qualitatively different ways, mapping onto the three different outcomes above, paralleling the results for interspecific competition.

The two exclusion results capture the essence of apparent competition: the outcomes look like the result of interspecific competition, but no direct competition is involved. One host species acts as a reservoir for enemies of the other host, and vice versa; the species that can tolerate a higher parasite prevalence excludes the other. However, exclusion of one of the host species is not a certainty – host coexistence is also possible. This is of interest because other models of apparent competition (mediated by predators or parasitoids) do not predict stable coexistence in the absence of other limiting factors (Holt, 1977; Holt and Lawton, 1993; Table 2.3). This arises because for parasitoids or predators, there is no analogue to intra-and interspecific transmission: without population structure or prey preference, the two host species are attacked at random by a common pool of ovipositing parasitoids or predators, and from the P^* rule (Eqn 2.12), whichever one can support the higher density of natural enemies excludes the other.

Inclusion of density-dependent host regulation further expands the range of conditions leading to coexistence, but does not qualitatively alter these outcomes (Begon et al., 1992; Table 2.3).

2.3.2 Implications for biological control

Apparent competition has potentially important implications for biological control, in situations where the biological control agent is not a specialist of the pest. Under these circumstances, the population dynamic effects *of* non-target prey/hosts, and effects *on* non-target prey/hosts, need to be considered. Firstly, the basic models imply that addition of alternative prey/hosts will enhance parasite or predator numbers, which should feed back into enhanced control of the target pest. This underlies the strategy of biological subsidy, in which the addition of alternate non-pest species is used to improve biological control of a target species

(Harmon & Andow, 2004). However, the strategy is not always success-ful (Holt & Lawton, 1993; Holt *et al.*, 1994), and reference to the P^* rule explains why: the species that can support a larger enemy population will win out and exclude the other – so biological subsidies will only succeed if a tolerant species (or one that can be supported at a large population size) provides the subsidy. Furthermore, predator/parasite behaviour and movement patterns can influence outcomes; for example, non-linear (Type II or III) functional responses (e.g. due to satiation or switching between prey (Table 3.2)) can result in reduced impact on the target species in the presence of alternative prey (reviewed in Harmon & Andow, 2004).

Apparent competition also underlies the related strategy of conservation biological control, in which management practices are designed to enhance populations of natural enemies that will contribute to control of a pest species (Meyling & Hajek, 2010). Here, addition of resources or reservoir species to support indigenous parasitoids or parasites may be undertaken. For instance, encouraging patches of nettles *Urtica dioica* supports populations of the nettle aphid *Microlophium carnosum*, which act as a reservoir for the generalist fungal entomopathogen *Pandora neoaphidis*, which is effective against pest aphid species (Pell *et al.*, 2010).

On the other hand, the occurrence of apparent competition can lead to undesired non-target effects: the enemy used to control a pest may also result in reduction or exclusion of species we wish to conserve. This is particularly problematic for the use of parasitoids, which are predicted under simplistic assumptions to result in exclusion of one of the hosts. The incorporation of more realistic assumptions (resource limitation of host species and spatially structured populations, for example) relaxes this outcome and provides some encouragement that parasites might be applied without causing extinction (Begon & Bowers, 1994).

These processes have further implications at the interface of biological control and conservation. The populations of natural enemies, including insect parasitoids and plant pathogens, are often subsidised in high productivity agricultural ecosystems. Spillover to adjoining natural habi-tat fragments which cannot support them at such high densities may result in exacerbated predation/parasite pressure on indigenous prey species (Rand & Louda, 2006). Recent findings suggest that many parasitoids thought of as specialist biological control agents actually utilise a variety of host species (reviewed in Rand & Louda, 2006). For instance, 83% of parasitoids reared from native forest moths in Hawaii were biological control agents (Henneman & Memmott, 2001). For an in-depth discussion of the implications of apparent competition for biological control and conservation, see Holt and Hochberg (2001) and Hawkins and Cornell (1999).

2.3.3 Empirical evidence for apparent competition

There are several examples where apparent competition mediated by parasites might be responsible for population declines in the field (Table 2.4), but it has proved difficult to rule out other factors, such as direct competition or habitat loss, in determining patterns (Hudson & Greenman, 1998). Empirical studies confirm that parasitoids can mediate apparent competition leading to exclusion of one host species. In controlled laboratory studies, Bonsall and Hassell (1997) investigated coexistence between the moths *Plodia interpunctella* and *Ephestia kuehniella* and a shared parasitic wasp, *Veneturia canescens*. The experiment disentangled the two processes of interspecific and apparent competition by preventing two host species from competing directly for resources; a mesh screen separated the moth populations but allowed free passage of the smaller parasitoid. When the moth species were reared separately with the parasitoid, host and parasitoid were maintained over the 15-week course of the experiments, and time series analysis indicated damped oscillations towards a stable equilibrium for both moth species. However, when the three species were housed together, *E. kuehniella* was eventually excluded, with damped oscillations for *P. interpunctella* but diverging oscillations for *E. kuehniella* (Fig. 2.5). As direct competition was prevented, this study conclusively demonstrates that apparent competition can lead to exclusion of one prey species.

Another nicely controlled empirical demonstration of apparent competition involves field manipulation of two planthopper species which feed upon different species of grass, but are infected by two shared species of parasitoids. Manipulation of reservoir availability in the field revealed apparent competition mediated by the parasitoids which had the potential to cause local extinction of one of the planthopper species (Cronin, 2007; Box 2.2). Similarly, parasitoids may themselves be subject to apparent competition if they share the same hyperparasitoid. For example, the parasitoids *Cotesia melitaearum* and *C. glomerata* use different lepidopteran host species and so do not compete directly for resources, but engage in apparent competition mediated by the shared hyperparasitoid *Gelis agilis* (van Nouhuys & Hanski, 2000).

Table 2.4. *Empirical examples of apparent competition, parasite-mediated competition and parasite-modified competition**

Competitors	Parasite	Impact of parasite	Reference
	Microparasites		
Bacterium; *Bordetella*, resistant and susceptible strains	Bacteriophage	Shared	*Apparent competition* (potential direct competition neutralised by high resource levels). The competitive advantage of the resistant strain results from pathogenicity in susceptible strain.	Joo et al. (2006)
Beetles; *Tribolium confusum* and *T. casteneum*	Sporozoan *Adelina tribolii*	Shared	*Parasite-mediated competition.* *T. casteneum* was the superior competitor but the outcome of competition was reversed in the presence of the parasite.	Park (1948)
Amphibians; *Rana cascadae* and *Hyla regilla*	Fungus *Saprolegnia ferax*	Shared	*Parasite-mediated competition.* Parasitism reversed the outcome of competition such that *H. regilla* survival and development were higher in the presence of both competitor and parasite, but were reduced when only the competitor was present.	Kiesecker and Blaustein (1999)
Reptiles; *Anolis wattsi* and *A. gingivinus*	Protist *Plasmodium azurophilum*	Shared, but rare in *A. wattsi*	*Parasite-mediated competition.* In the field, lizards coexist only when the parasite is present. *A. gingivinus* occurs alone when parasite absent.	Schall (1992)

(cont.)

Host	Parasite	Sharing	Description	Reference
Mosquitoes; *Aedes albopictus* and *A. aegypti*	Gregarine *Ascogregarina* sp.	Shared	*Parasite-mediated competition.* Parasite prevalence in *A. albopictus* increased following invasion by *A. aegypti*.	Blackmore *et al.* (1995)
Mammals; red squirrels, *Sciurus vulgaris* and grey squirrels, *S. carolinensis*	Virus Parapox	Shared	*Parasite-mediated competition.* The invasive grey squirrel acts as a reservoir for the virus, which is lethal in the native red squirrel.	Rushton *et al.* (2006)
Isopods; *Porcellio scaber* and *P. laevis*	Virus	Shared	*Parasite-mediated competition.* Parasite prevalence in *P. scaber* higher in the presence of the competitor; competitive strength of infected *P. scaber* reduced.	Grosholz (1992)
Parasitoids				
Planthoppers; *Delphacodes scolochloa* and *Prokelisia crocea*	Parasitoids *Anagrus nigriventris* and *A. columbi*	Both parasitoids shared	*Apparent competition* (planthoppers use different plant species). *P. scholochloa* suffered higher levels of parasitism and decreased population density when *P. crocea* acted as a reservoir for the parasitoids.	Cronin (2007)
Moths; *Plodia interpunctella* and *Ephestia kuehniella*	Parasitoid wasp *Venturia canescens*	Shared	*Apparent competition* (direct competition prevented). Hosts coexisted in the absence of the parasite, but *E. kuehniella* eliminated in presence of competitor and parasitoid.	Bonsall and Hassell (1997)

(cont.)

Table 2.4. (*cont.*)

Competitors	Parasite		Impact of parasite	Reference
Leaf mining beetle *Pentispa fairmairei* and dipteran *Calcomyza* sp.	Parasitoids, several species	Shared	*Apparent competition* (different plants used). Experimental removal in the field of one host species led to a decrease in parasitism in other potential hosts and, in one case, an increase in host abundance.	Morris *et al.* (2004)
Beetles; *Galerucella tenella* and *G. calmariensis*	Parasitoid wasp *Asecodes mento*	Shared	*Apparent competition* (beetles use different host plants). *G. tenella* suffered higher parasitism when in sympatry with *G. calmariensis*. In mixed populations, parasitoid showed selectivity towards *G. tenella*.	Hamback *et al.* (2006)
Primary parasitoids of aphids; *Aphidius ervii* and *A. microlophii*	Secondary parasitoids, several species	Shared	*Apparent competition* (primary parasitoids use different aphid host species). Presence of *A. microlophii* increased parasitism in *A. ervii*.	Morris *et al.* (2001)
Parasitoids; *Cotesia melitaearum* and *C. glomerata*	Hyperparasitoid *Gelis agilis*	Shared	*Apparent competition* (parasitoids use different lepidopteran hosts). Introduction of *C. glomerata* led to decline of *C. melitaearum* populations as a result of increased densities of the hyperparasitoid.	van Nouhuys and Hanski (2000)

(*cont.*)

Leafhoppers; *Erythroneura variabilis* and *E. elegantula*	Parasitoid *Anagrus epos*	Shared	*Apparent competition/parasite-mediated competition.* Interspecific competition weak, but parasitoid densities increased following *E. variabilis* invasion and *E. elegantula* suffers higher parasitism.	Settle and Wilson (1990)
Ladybirds; *Harmonia axyridis* and *Coleomegilla maculata*	Parasitoid wasp *Dinocampus coccinellae*	Shared	*Apparent competition/parasite-mediated competition.* Invasive *H. axyridis* suffered similar parasite attack rates to *C. maculata*. However, parasite less likely to develop in *H. axyridis* which acts as sink for parasitoid. Resource competition not considered in this study, but is likely to occur in the field.	Hoogendoorn and Heimpel (2002)
Lepidopterans; *Pieris napi* and *P. rapae*	Parasitoid *Cotesia glomerata*	Shared	*Parasite-mediated competition.* In southern habitats, the native *P. napi* suffers a higher rate of parasitism than the invader.	Benson *et al.* (2003)
Fungus *Aspergillus niger* and fly *Drosophila melanogaster*	Parasitoid *Asobara tabida*	Not shared	*Parasite-modified competition.* Parasitoid reduces fly feeding.	Rohlfs (2008)
Ants; *Pheidole dentata* and *Solenopsis texana*	Parasitoid phorid fly *Apocephalus* sp.	Not shared	*Parasite-modified competition. P. dentata* dominates confrontation with fire ants in absence of parasite. Outcome reversed in presence of parasite, which induces refuge-seeking and	Feener (1981)

(cont.)

Table 2.4. (*cont.*)

Competitors	Parasite		Impact of parasite	Reference
			prevents defensive behaviour towards fire ants.	
Ants; *Azteca instabilis* and competing ant species	Parasitoid phorid flies *Psuedacteon* sp.	Not shared	*Parasite-modified competition.* Presence of flies decreased abundance of *A. instabilis* at baits, and other ant species were more likely to take over bait when flies present.	Philpott (2005)
Fire ants *Solenopsis geminata* and *S. invicta*	Parasitoid phorid fly *Psuedacteaon browni*	Not shared	*Parasite-modified competition.* The presence of parasitoids decreased stimulated host defensive behaviour which reduced the foraging rates of *S. geminata*.	Morrison (1999)
	Macroparasites			
Amphipods; *Corophium volutator* and *C. arenarium*	Trematode *Microphallus claviformis*	Shared	*Parasite-mediated competition.* Parasite caused mortality of competitively superior *C. volutator*, but not of *C. arenarium*. Parasite may facilitate coexistence.	Jensen *et al.* (1998)
Mammals; sea lions, *Otaria byronia* and fur seals, *Arctocephalus australis*	Hookworm *Unicinaria* sp.	Shared	*Parasite-mediated competition.* Parasite more virulent in sea lions. Parasite burdens higher in sea lions sympatric with fur seals. Infection may mediate displacement of sea lions.	Georgenascimento and Marin (1992)

(*cont.*)

Host	Parasite		Description	Reference
Reptiles; geckos *Lepidodactylus lugubris* and *Hemidactylus frenatus*	Cestode *Cylindrotaenia* sp.	Shared	*Parasite-mediated competition.* In the field, cestode prevalence in *L. lugubris* higher when sympatric with introduced *H. frenatus*. However, experimental manipulation found no increase in parasite prevalence in mixed-species populations.	Hanley *et al.* (1995); Hanley *et al.* (1998)
Birds; ring-necked pheasant, *Phasianus colchicus* and grey partridge, *Perdix perdix*	Nematode *Heterakis gallinarum*	Shared	*Parasite-mediated competition.* Pheasants act as a reservoir for *Heterakis* infection, which is more virulent in partridges.	Tompkins *et al.* (2000a); Tompkins *et al.* (2000b)
White-tailed deer *Odocoileus virginianus* and moose, *Alces alces*	Nematode *Parelaphostrongylus tenuis*	Shared	*Parasite-mediated competition.* Parasite is more virulent in moose (the stronger competitor). Moose density was found to be inversely proportional to deer density and moose densities were lower in areas of high parasite prevalence in deer.	Whitlaw & Lankester (1994)
Fruit flies; *Drosophila putrida* and *D. falleni*	Nematode *Howardula aoronymphium*	Shared	*Parasite-mediated competition.* Parasite decreases competitive superiority of *D putrida*. In the field, relative abundance of *D. putrida* lower when parasitoid present.	Jaemike (1995)

* In many cases, more than one of these mechanisms is in operation.

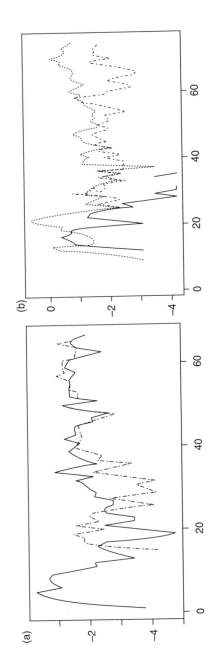

Fig. 2.5. Bonsall and Hassell's empirical demonstration of parasitoid-mediated apparent competition. Figures plot Log$_{10}$ population densities for each species against time (in weeks): (a) *Ephestia kuehniella* (solid line) is maintained after introduction of the parasitoid *Venturia canescens* (dashed line); (b) *E. kuehniella* is eliminated when cultured with *Plodia interpunctella* (dotted line) after introduction of *V. canescens*. From Bonsall and Hassell (1997).

Many real systems will tend to lie somewhere on a continuum with varying relative strengths of indirect effect mediated by the natural enemy and direct competition. Although the original theoretical formulation of apparent competition assumes no direct competition between the prey species (Holt, 1977), the process can nevertheless be important in modules with direct competition (Begon & Bowers, 1995). Deconstructing systems into direct and indirect effects helps to clarify the problem (Hatcher *et al.*, 2006): two competing species that share a parasite interact directly via use of the shared resource and indirectly via effects of the parasite on the other's population density. In some cases (when resources are not depleted), the system is dominated by parasite-mediated effects and resembles pure apparent competition. In contrast, when competition is severe because resources are strongly limiting, the module is dominated by resource-mediated effects (Fig. 2.7).

Box 2.2 Apparent competition between planthoppers

In prairies in North Dakota, there are many ponds or potholes surrounded by stands of sprangletop grass (*Scolochloa festucacea*). At higher elevations, but contiguous with sprangletop, are patches of prairie cordgrass (*Spartina pectinata*). Two species of planthopper are the dominant herbivores, *Delphacodes scolochloa*, which utilises sprangletop, and *Prokelisia crocea*, which feeds on cordgrass. These planthoppers provide an opportunity to look at the effects of apparent competition as they are specialists of their host plants, and hence do not engage in direct competition. However, both species are attacked by the egg parasitoids *Anagrus nigriventris* and *A. columbi*.

Apparent competition is asymmetric and operates through effects on host density. In natural planthopper populations, parasite prevalence was similar in the two host species. However, field manipulation of planthopper densities revealed that apparent competition in this system was asymmetric. A five-fold increase in the density of *D. scolochloa* led to a two-fold increase in parasitism of *P. crocea* as *D. scolochloa* acted as a reservoir for the parasitoids. As a result, *P. crocea* density declined over two generations. In contrast, a six-fold increase in the density of *P. crocea* did not affect the density or parasite prevalence of *D. scolochloa* (Fig. 2.6).

Box 2.2 *(continued)*

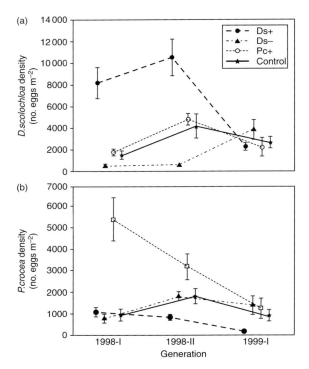

Fig. 2.6. Mean densities of *Delphacodes scolochloa* (a) and *Prokelisia crocea* (b) in response to manipulation of planthopper densities. Ds+ is the addition of *D. scolochloa*, Ds− is the removal of *D. scolochloa* and Pc+ is the addition of *P. crocea* to individual prairie potholes. Control potholes are unmanipulated. From Cronin (2007).

Apparent competition causes extinction at the patch scale. An elegant manipulation of reservoir proximity demonstrated that apparent competition led to extinction of *P. crocea* at the patch scale. Experimental cordgrass patches containing low or high densities of *P. crocea* eggs were placed in the field in one of three treatments: in isolation; close to other cordgrass (where nearby *P. crocea* provided a source of parasitoids); or close to sprangle top (*D. scolochloa* providing a source of parasitoids). Over two generations, proximity to *D. scolochloa* populations increased parasitoid prevalence and the likelihood of extinction in *P. crocea*.

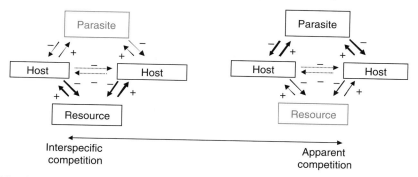

Fig. 2.7. Schematic of the relationship between interspecific competition and apparent competition. The strength of direct interactions is depicted by thickness of solid arrows; less influential components in the module are shown in grey. In either case, the direct interactions result in the same pattern and sign of indirect interactions (dashed arrows) between the host (or prey) species.

2.4 Parasite-mediated competition

Whenever two species compete and a parasite infects one or both species, there is the potential for parasite-mediated or parasite-modified competition. This can result in similar outcomes to that of apparent competition, including increasing the range of parameters that lead to exclusion of one host species, or, alternatively, enhancing the likelihood of coexistence. We can distinguish between parasite-mediated competition, in which parasites alter the competitive relationship by virtue of their effects on host densities (through parasite-induced mortality or fecundity reduction) and parasite-modified competition (in which the competitive abilities of individuals are changed by parasitic infection (Section 2.4.2)). Most models have concentrated on the former, although many empirical examples show evidence for the latter process. Unlike the case for apparent competition, both specialist and shared parasites can, in theory, mediate competition because only one of the two competing 'host' species need be affected by parasitism to influence the outcome. Like apparent competition, the parasite population must respond numerically to its host species, and vice versa, for these effects to occur.

2.4.1 Specialist parasite-mediated competition

Host–host competition can be mediated by shared parasites, and also by specialist parasites that infect only one of the competing species. Anderson and May (1986) developed a number of simple community models for parasitism, including the case for two species that compete intra- and interspecifically, one of which is also host to a microparasite. This specialist parasite system demonstrates many of the salient features of parasite-mediated competition, even though the parasite is not shared (Fig. 2.1b). In the absence of the parasite, the standard three outcomes for interspecific competition apply: the superior competitor wins; there is stable coexistence; or the outcome depends on initial conditions. If the parasite infects the inferior competitor, there is no change in outcome. However, if it infects the superior competitor, it can enable the otherwise excluded competitor to persist.

Few examples where only one of an interacting species pair is infected have been studied in the context of parasite-mediated competition (Table 2.4). A possible example conforming to this model involves malarial parasites, which are generally species-specific to *Anolis* lizards in the Caribbean (Schall, 1992). *Anolis gingivinus* is a stronger competitor than *A. wattsi*, and sympatric populations are found only when the malarial parasite *Plasmodium azurophilum* is present. This parasite rarely infects *A. wattsi*, but can be common in *A. gingivinus*, leading Schall *et al.* to suggest that the parasite permits coexistence of the two competing species and hence appears to be responsible for maintaining species diversity.

2.4.2 Shared parasite-mediated competition

The models discussed above demonstrate how shared parasites can cause apparent competition and how specialist parasitism can interact with competition to change coexistence outcomes for competing species. However, there are not many empirical examples that meet these strict assumptions. A more general framework for parasitism–competition modules requires that we include all three forces at work: intraspecific competition so that hosts are self-limited; interspecific competition between hosts; and a parasite capable of infecting both hosts, the population dynamics of which are coupled to both of its host populations. These extensions were developed for microparasite, macroparasite and parasitoid models during the 1990s (Table 2.5), and a baseline model for a microparasite is discussed in Box 2.3. The general conclusion from these models is that whilst intraspecific competition expands the range of conditions that lead to host–host–parasite coexistence (Table 2.5), the addition of interspecific competition tends to oppose this, reducing the parameter space for coexistence. However, a class of cases can be demonstrated where the parasite enables coexistence of hosts that would otherwise not coexist; this can occur when interspecific competition is strong but the parasite has a greater detrimental effect on the superior competitor (Box 2.3).

Table 2.5. *General models of parasite-mediated competition, parasite shared. For dynamical terminology, see Table 2.2*

System	Outcomes	Reference
Host–host–macroparasite (directly transmitted); parasite affects mortality and fecundity	Parasite may enhance or disrupt host coexistence. Infected coexistence possible when superior competitor more affected by parasite. Increased aggregation of parasites within hosts reduces regulatory effect of parasite.	Yan (1996)
Host–host–microparasite (directly transmitted)	Repeatable exclusion, contingent exclusion, uninfected or infected coexistence possible. Coexistence mediated by resource limitation and/or interspecific parasite transmission.	(C) Bowers and Turner (1997)
Host–host–microparasite (directly transmitted)	As (C), but counters that infected coexistence does not have single, stable equilibrium for some parameter combinations (i.e. there are alternative stable states).	Greenman and Hudson (1999)
Host–host–macroparasite (common pool of free-living stages)	Interspecific competition reinforces apparent competition to expand parameter range for exclusion. Parasite-induced effects on fecundity can lead to population cycles.	Greenman and Hudson (2000)

2.4.2.1 Keystone effects

These models (Table 2.5, Box 2.3) show that parasites can theoretically enhance or reverse the effects of direct competition between the hosts. The outcome depends on which species can tolerate or support a higher prevalence of infection: if the parasite is more virulent in the inferior competitor, it can lead to elimination of that species; if it is more virulent in the superior competitor, it can reverse competitive outcomes, enabling

Box 2.3 A model of parasite-mediated competition

Examination of the interaction between parasitism and 'full' competition between hosts requires that we add interspecific competition into the models discussed above for apparent and intraspecific competition (e.g. Begon *et al.*, 1992; Section 2.3). Hence, for a directly transmitted microparasite of two continuously breeding hosts we obtain (parameters as defined in Table 2.2):

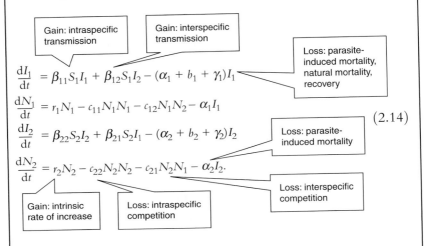

$$\frac{dI_1}{dt} = \beta_{11}S_1I_1 + \beta_{12}S_1I_2 - (\alpha_1 + b_1 + \gamma_1)I_1$$

$$\frac{dN_1}{dt} = r_1N_1 - c_{11}N_1N_1 - c_{12}N_1N_2 - \alpha_1I_1$$

$$\frac{dI_2}{dt} = \beta_{22}S_2I_2 + \beta_{21}S_2I_1 - (\alpha_2 + b_2 + \gamma_2)I_2$$

$$\frac{dN_2}{dt} = r_2N_2 - c_{22}N_2N_2 - c_{21}N_2N_1 - \alpha_2I_2.$$

(2.14)

Labels in figure: Gain: intraspecific transmission; Gain: interspecific transmission; Loss: parasite-induced mortality, natural mortality, recovery; Loss: parasite-induced mortality; Loss: interspecific competition; Gain: intrinsic rate of increase; Loss: intraspecific competition.

Bowers and Turner (1997) characterised the outcomes for this model in terms of the ability of one host species to withstand invasion by the other, depending on the strength of interspecific competition (tending to reduce host coexistence), intraspecific competition (limiting population sizes and therefore reducing resources for parasite establishment) and parasite infection (where the relative frequency of intra- versus interspecific transmission and the relative pathogenicity to each host will determine outcomes). Infected coexistence can arise when the two hosts can coexist stably without the parasite (a case of resource-mediated coexistence (Begon & Bowers, 1995)), or when they would not. In this latter case of parasite-mediated coexistence, if the inferior competitor suffers less from the effects of parasitism, it can coexist with the superior competitor (similar outcomes apply for directly transmitted macroparasites (Yan, 1996)).

Mathematical complexity: the full model of parasitism and competition is the analytic equivalent of the first experiments on parasite-mediated competition (the *Tribolium* beetles of Park, 1948;

Box 2.3 (*continued*)

Section 2.1), and it is perhaps surprising that it took nearly 50 years for the theory to catch up with the data. One reason for this is that as the equation systems become more complicated, they become less analytically tractable. There are no standard techniques for solving equations with more than four variables, so full characterisation of such systems is not usually possible. Bowers and Turner (1997) were unable to prove the uniqueness or stability of the joint coexistence equilibria (see Table 2.2 for terminology), and indeed further analysis has shown that, for some combinations of parameter values, alternative stable states may arise (Greenman & Hudson, 1999). Similarly, Yan (1996) did not provide a complete analytical account for the equivalent macroparasite model. It is not known whether the 'exceptional' parameter combinations that yield multiple equilibria are relevant to any real biological systems.

the inferior competitor (which might otherwise have been excluded) to persist, or even excluding the superior competitor. From a community ecology perspective, these results suggest that parasites can act as keystone species, their presence altering the likelihood of coexistence or species exclusion and hence affecting community structure. We discuss some quantified examples of keystone parasite–competition interactions in natural and managed populations in Section 2.6. In addition to its interaction with competition, parasitism can have keystone effects in a variety of community modules, including predation (Chapter 3) and intraguild predation (IGP) (Chapter 4), and one of our chief aims in this book is to examine the broader consequences of these keystone roles for community structure and ecosystem processes (Chapters 6–8).

2.4.2.2 Parasites as biological weapons

The keystone effect of parasitism in competitive interactions has led to the suggestion that, from an evolutionary standpoint, shared parasites can potentially be used as biological weapons enabling dominance of one host species over a competing susceptible species (Price *et al.*, 1986). Models of plant populations with spatially local pathogen transmission show that reduced resistance can be favoured in the inferior competitor as this enhances the negative impact on the dominant competitor (Brown & Hastings, 2003). This type of relationship may be an important driving factor in some biological invasions, where the invader harbours a parasite to which native species are vulnerable; spatial eco-logical models of this phenomenon demonstrate a wave of disease spreading through the landscape ahead of a wave of replacement (Bell *et al.*, 2009). The biological weapons process acting in the short term appears likely to underlie a number of animal and plant invasions (Chapter 6), including the replacement of native red squirrels by grey squirrels in England (Section 2.6.1). However, in the longer term, the evolutionary stability of harbouring a deleterious parasite that inflicts damage on a competitor is sensitive to many factors. Whether this strategy is adaptive for the host depends on a fine balance between within-population and between-population selective pressures (Wodarz & Sasaki, 2004). The direction of selection depends on the costs of main-taining immunity, or delaying recovery, compared to the benefits of killing or debilitating competitors. The magnitude of costs and benefits will therefore depend, among other things, on relative virulence in the two species, immune mechanisms and their costs, and within- and between-species contact rates. Contact rates determine parasite trans-mission dynamics, and will depend on population structure and the rates at which each species recolonise or go extinct in local patches. Kuo *et al.* (2008) developed a theory showing that the evolutionary stable strategy for carrying parasites as a biological weapon varies depending on whether the population is panmictic (where biological weapons are favoured) or structured as a metapopulation (where bio-logical weapons can be disadvantageous under high relative local extinction rates that lead to strong localisation of parasites).

2.4.2.3 Empirical examples

In contrast to the case of specialist parasites of competing species, there are many examples in the literature of shared parasites altering the outcome of competition (Table 2.4). However, it should be noted that in many empirical examples, it is not always clear whether effects are mediated via density (as assumed here in the models of parasite-mediated competition) or trait changes (i.e. are examples of parasite-modified competition (Section 2.5)). Density effects have, however, been dem-onstrated and quantified in a series of mesocosm experiments involving larval amphibians where a combination of parasitism, intraspecific and interspecific competition appears to determine outcomes for the frogs *Rana cascadae* and *Hyla regilla* (Kiesecker & Blaustein, 1999). In this system, the pathogenic water mould *Saprolegnia ferax*, which infects both

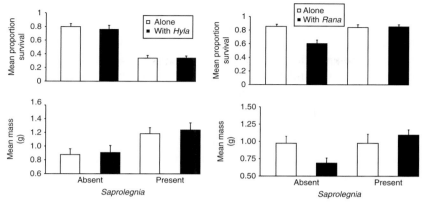

Fig. 2.8. Intraspecific and interspecific effects of parasitism on larval amphibians. Survival (top) and mass at metamorphosis (bottom) are shown for *Rana cascadae* (panels on left) and *Hyla regilla* (right). Data are plotted for each species cultured alone or with the competing species, with or without the fungal pathogen *Saprolegnia ferax*. *R. cascadae* is not impacted significantly by *H. regilla* alone, but is strongly impacted in the presence of *S. ferax*; *H. regilla* is impacted by *R. cascadae* in the absence of *S. ferax* but released from interspecific competition by the effects of *S. ferax* on *R. cascadae*. For *R. cascadae*, survivors of infection are also released from intraspecific competition (see bottom-left panel). From Kiesecker and Blaustein (1999).

species, interacts with intraspecific competition (Section 2.2); it also reverses the outcome of interspecific competition in experimental ponds stocked with both species. In the absence of the fungus, *R. cascadae* tadpoles had strong negative effects on the growth, development and survival of *H. regilla*, but there were no such effects when the fungus was present (Fig. 2.8). Microparasite-mediated competition is also implicated in natural populations (Section 2.6).

Macroparasites may also mediate host–host competition. Laboratory experiments revealed that the shared trematode parasite *Microphallus claviformis* promotes coexistence of two competing species of amphipod, *Corophium volutator* and *C. arenarium* (Jensen *et al.*, 1998). This parasite has an indirect life cycle with a bird definitive host and two intermediate hosts – a snail and an amphipod. *C. volutator* and *C. arenarium* often occupy different habitats in the intertidal zone, and *C. volutator* is competitively superior. However, laboratory experiments revealed parasite-mediated competition; whilst *M. claviformis* showed no host preference, it caused higher mortality in the competitively superior *C. volutator*. In addition to parasite-mediated competition, predation-mediated competition has also been suggested to play a role in coexistence in this system, as *C. volutator* spends more time in surface activity where it is at risk of predation. Other systems in which macroparasites mediate interspecific competition are discussed in Section 2.6.

2.5 Parasite-modified competition

Parasites may influence competitive interactions by another means, by *modifying* the competitive ability of hosts (Fig. 2.9) and thereby altering the dominance relationship between competing species (or between competing conspecifics (Section 2.2.2)). Classically, parasite-mediated apparent competition acts via long-term density effects (Holt, 1977), and the general models considered here (Sections 2.3–2.4) reflect that mechanism.

The most widely applicable general population dynamic models for interspecific competition (Box 2.3, for example) do not incorporate parasite-modified competition because parasitism only influences outcomes via its effects on mortality (a density-mediated effect). However, in a community module context, parasite-induced modification of

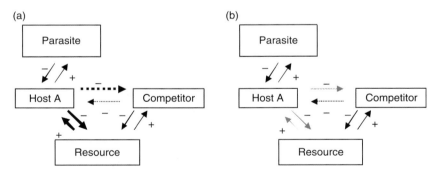

Fig. 2.9. Relationship between parasite-mediated and parasite-modified competition. A specialist parasite mediates competition via its influence on host population density, with indirect effects on the competing species (see Fig. 2.1b). The parasite may also modify competition by (a) enhancing (denoted by thicker, bolder arrows) or (b) reducing the host's impact on the shared resource (thinner, paler arrows), altering the indirect effects between the competitors (dashed arrows); in (a), the host has a net negative impact on the competing species; in (b), the direction of impact is reversed. In both cases, the effect of the host on its competitor is now the combination of density and trait-mediated indirect effects.

competitive ability is an example of a trait-mediated indirect effect (TMIE) (Bolker et al., 2003; Werner & Peacor, 2003). A model of parasite-modified competition should, strictly speaking, incorporate competition parameters that depend upon infection status. Parasite modification of competitive ability might seem an inconsequential distinction from parasite-mediated competition, but it is potentially important as such effects can produce qualitatively different population dynamics (White et al., 2007). Furthermore, because TMIEs operate on a shorter timescale than demographic effects, they have the potential for large impacts on population dynamics (Box 1.3).

Several clear examples of parasite-modified competitive ability have been reported in the context of intraspecific competition (Section 2.2.2). For instance, parasitoids often reduce the growth rate of their host, which can either reduce (Sisterson & Averill, 2003) or enhance (Cameron et al., 2005) host competitive ability in comparison to uninfected conspecifics. Infection-induced effects on growth also modify competitive interactions between plants; infected hosts have a lower growth rate, which tends to reduce their competitive ability where space is the limiting resource (Alexander & Holt, 1998). Native grasses infected with barley yellow dwarf viruses (BYDVs) suffer reduced competitive ability against invasive species, implicating the virus in their replacement across US prairies (Boxes 5.3 and 6.4). Many plant–endophyte interactions show the reverse pattern: fungal endophytes that form symbioses with their hosts enhance the ability of the host in intra- and interspecific competition, with knock-on effects for plant and arthropod community composition (Section 5.5).

There are also convincing examples of parasite modification of competition in animal hosts. One interesting system involves interking-dom competition between *Drosophila melanogaster* and the filamentous fungus *Aspergillus niger* that is modified by the parasitoid *Asobara tabida* (Rohlfs, 2008). The fly and the fungus compete for access to dead organic matter. The fly larvae suppress fungal growth by disrupting hyphae formation, whilst the fungus secretes metabolites that are toxic to the fly. The parasitoid was found to exert a negative impact on *Drosophila* feeding ability, reducing the impact of single parasitised larvae on reproduction of the fungus, but this effect disappeared at high larval densities. Hence the competitor *A. niger* also mediated an Allee effect on parasitised *Drosophila*: larvae at high densities were disproportionately successful as foraging *en masse* they disrupted the growth of fungal filaments.

Another well-studied association that provides an insight into the role of parasite-modified competition is that of ant communities and their phorid fly parasitoids. Parasitoid phorid flies tend to be very host-specific and induce behavioural defences in the host that reduce the likelihood of infection. A series of studies reveal that these defences have a knock-on effect on host competitive ability, influencing both resource exploitation and antagonistic interactions with competing species (Box 2.4).

Box 2.4 Parasite-modified competition: phorid flies and competing ants

Phorid flies are a diverse dipteran family, many of which parasitise social insects either as nest symbionts or as parasitoids. Parasitoid phorids attack adult worker ants. The larva develops in the head capsule of the host, which eventually drops off. These flies are host-specific and so generally attack only one of a competing pair of ant species. A series of studies, some prompted by the need for biological control of invasive ants, indicate that phorid flies can modify ant–ant competition and that, in some cases, the impact of the parasitoid reverses the outcome of competition.

The first evidence for parasite-modified competition by phorid flies was presented by Feener (1981), who demonstrated that the presence of *Apocephalus* sp. induces refuge-seeking behaviour in the host *Pheidole dentata*, thus preventing defensive behaviour towards the competing fire ant *Solenopsis texana* and reversing the outcome of competition. Whilst Feener's study found that the phorid flies mediated aggressive encounters between competing ant species, other studies have shown that phorid flies also mediate exploitation competition. Morrison (1999) investigated the impact of a native parasitoid *Psuedacteon browni* on competition between its host fire ant *Solenopsis geminata* and the invasive *S. invicta*. Aggressive behaviour between *S. geminata* and *S. invicta* was not influenced by the parasite; when the invader was present, *S. geminata* ignored attacking phorid flies. However, defensive behaviour (the adoption of a stationary, curled posture) in the presence of phorids led to reduced foraging rates of *S. geminata* of up to 50%.

The presence of phorid flies can influence ant dominance hierarchies and community structure, not least because the majority of phorid parasitoids attack ant species that are dominant competitors. Feener (2000) suggested that those characteristics that allow an ant species to

Box 2.4 (*continued*)

dominate resources also make it a conspicuous evolutionary target for host switching and colonisation by these specialist parasitoids. Dominant competitors such as the well-known invasive fire ants tend to have larger colonies, strong pheromonal recruitment trails and large numbers of visible workers. These traits allow them to dominate resources, but may also provide olfactory and visual cues that are exploited by parasitoids. For example, a study of interactions between two genera of *Linepithema* found that the production of alarm cues in response to a competing species had the additional effect of attracting and increasing attacks by the phorid fly *Pseudacteon pusillus* (Orr *et al.*, 2003).

Release from attack by phorid fly parasitoids (enemy release is discussed in detail in Chapter 6) may be a factor in the success of fire ant invasions (Feener, 2000). In their native South American range, *Solenopsis wagneri* coexists with other competing species. However, invasive populations in North America occur at higher densities and lead to competitive exclusion of native species. In the invasive range, host-specific native parasitoids may be absent. Hence invasive fire ants may escape the direct effects of parasitoid attack as well as the indirect effects of the parasitoid on competition, leading to competitive dominance in the invasive range.

2.6 Examples from conservation and management

The effects of parasites on competitive interactions can have implications both for the conservation of natural populations and for the management of species of economic importance. This is illustrated by four comprehensively studied systems where the consequences of parasite-mediated, -modified and apparent competition have been analysed for natural (red squirrels; moose) and managed (grey partridge; red grouse) populations.

2.6.1 Red squirrels, grey squirrels and poxvirus

The Eurasian red squirrel *Sciurus vulgaris*, a native to the United Kingdom, is being replaced by the introduced grey squirrel *S. carolinensis*. English populations of the grey squirrel harbour a virus (squirrel pox virus: SQPV), which causes asymptomatic infections in greys but is highly pathogenic in reds. Mathematic models indicate that it interacts with direct competition, driving the replacement of red squirrels by grey squirrels (Box 2.5).

Box 2.5 The ecological replacement of red squirrels in England

Red squirrel (*Sciurus vulgaris*) populations have declined and been eliminated over much of the United Kingdom during the twentieth century, along with the spread of the grey squirrel (*S. carolinensis*), a species native to the United States that was introduced on a number of occasions from the 1920s to the 1970s. Red squirrels are now considered to be endangered in Britain. Competition appears to be partly responsible for the decline of reds: in mixed or broadleaf woodland, reds and greys compete for food and possibly nesting sites (Gurnell *et al.*, 2004), and greys attain a higher population density than reds.

English populations of the grey squirrel harbour a virus (SQPV (Thomas *et al.*, 2003), previously considered to be a parapoxvirus). The infection causes no discernable pathology in greys but is highly virulent in reds, with an incubation period of 2–3 weeks, resulting in an estimated 75% mortality within a few days of the appearance of secondary lesions (Tompkins *et al.*, 2002). Grey squirrels have also been introduced to Northern Italy, Ireland and Scotland, but these populations are not infected with SQPV; expansion of greys in these regions is proceeding at a slower rate (Tattoni *et al.*, 2006). The speed of decline of red squirrels throughout England, and the pathogenicity of SQPV to red squirrels, has led to the suggestion that SQPV plays a role in driving the decline of reds through parasite-mediated competition. Rushton *et al.* (2000) used a spatially explicit individual-based model (in which the interactions between individual hosts are simulated as probabilistic events) to model the decline and extinction of reds across sites in Norfolk from 1966–1980, for which extensive contemporary data have been collated (Reynolds, 1985). Inclusion of SQPV infection dynamics was necessary to attain the observed rate of extinction of reds across the Norfolk habitat patches (Fig. 2.10).

Tompkins *et al.* (2003) developed an SIR model of the system (with equations for the susceptible, infectious and recovered classes of grey squirrels and susceptible and infectious classes for reds). They extended this model spatially (by simulating the dynamics of populations arranged in a square grid with dispersal between neighbouring patches), and when parameterised it produced an excellent fit to the Norfolk data (Fig. 2.11). The similarity of outcomes from these two

Box 2.5 (*continued*)

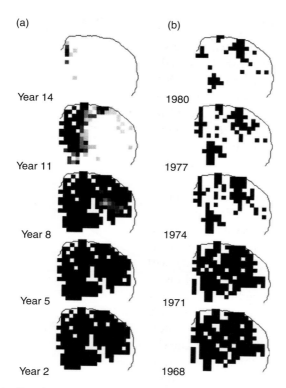

(a)   (b)

Year 14   1980

Year 11   1977

Year 8   1974

Year 5   1971

Year 2   1968

Fig. 2.10. Predicted (a) and observed (b) distribution of red squirrels in Norfolk in the period 1968–1980. Intensity of shading indicates probability that red squirrels inhabited 5 km grid squares (black = 100%, white = 0%). From Rushton *et al.* (2000).

very different modelling approaches provides convincing evidence that SQPV was involved in driving the Norfolk invasion. Moreover, because the Tompkins *et al.* model is generic and can be applied to other populations, it provides theoretical ground for the suggestion that SQPV might be implicated in the replacement of reds in other parts of the United Kingdom. Recent empirical data is also strongly suggestive of a role for SQPV in red squirrel decline. Rushton *et al.* (2006) compared the rate of spread of greys in Cumbria and Norfolk (where greys are SQPV-positive) to those in Scotland and Italy (where greys do not carry the virus); they found invasion rates 17–25 times faster in localities with infected greys. These authors have applied individual-based models (IBMs)

Box 2.5 (*continued*)

to examine the ongoing expansion of grey squirrels in Cumbria (Rushton *et al.*, 2006) and Scotland (Gurnell *et al.*, 2006), with the aim of assessing management and control strategies for the conservation of red squirrels in these areas (Box 6.3).

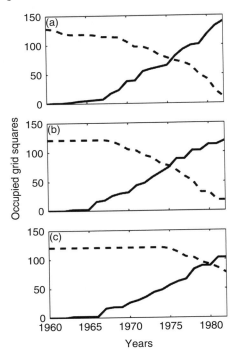

Fig. 2.11. Predicted and observed decline of red squirrels in Norfolk from 1960 to 1982. Figures plot the number of 5 km grid squares occupied by red (dashed lines) and grey squirrels (solid lines). (a) Predicted outcomes with competition and parasitism; (b) observed outcome; (c) predicted outcome with competition alone. From Tompkins *et al.* (2003).

Grey squirrels as a reservoir for disease. The squirrel models (Box 2.5) demonstrate an important point: pathogen prevalence need not be high for it nevertheless to drive the dynamics of populations. Indeed, seroprevalence of SQPV in greys sampled from the United Kingdom was 61%, whereas less than 3% of red squirrels had antibodies to the infection (Sainsbury *et al.*, 2000). Prevalence is low in reds because the virus is virulent and leads rapidly to the death of most infected individuals, whereas for greys, the infection is benign and a high prevalence can be maintained. This has consequences for the spread and persistence of the disease: the virus is too virulent to be supported in red squirrel populations alone, as it rapidly spreads through populations causing their decline and local extinction. Grey squirrel populations provide a reservoir for successive infection of newly encountered red squirrel populations. This suggests that SQPV may 'burn out' in coniferous forest, where greys cannot colonise, providing a potential refuge for red squirrels (Box 6.3). Reservoir hosts are an important feature in parasite-mediated and apparent competition; we return to this concept in the examples that follow.

2.6.2 Grey partridge, pheasants and nematodes

The caecal nematode (*Heterakis gallinarum*) may be partially responsible for exclusion of the native UK grey partridge (*Perdix perdix*) by the ring-necked pheasant (*Phasianus colchicus*). There has been a marked decline in grey partridge numbers in the United Kingdom over the last 50 years. At the same time, there has been an increase in the release of reared pheasants for shooting; as pheasants are a good host for the parasite, the impact of parasite-mediated competition may have increased in parallel for partridges (Box 2.6).

Box 2.6 Caecal nematodes in grey partridges and ring-necked pheasants

The intestinal nematode *Heterakis gallinarum* infects a variety of gamebird species, with measured effects on host survival and condition in the grey partridge and, to a lesser extent, the ring-necked pheasant. Birds acquire the infection through ingestion of nematode eggs as they forage (there are no intermediate hosts). The decline in grey partridge populations in England over the last 50 years and the failure of populations to recover has been attributed, in part, to macroparasite-mediated competition (Tompkins *et al.*, 2000a; 2000b). Several lines of evidence support the hypothesis of parasite-mediated or apparent competition, but its precise role in the decline of the grey partridge remains unclear.

Preliminary evidence for apparent competition: Tompkins *et al.* (1999) exposed caged birds to *Heterakis* infection. For pheasants, 12/12 became infected and had higher worm burdens and worm growth rate than the 11/12 partridges that became infected.

Box 2.6 (*continued*)

An inverse relationship between condition (measured as breast weight) and worm burden at the end of the study was found for partridges but not for pheasants, suggesting that pheasants are the main source of infection to partridges in the wild and that partridges suffer more from the effects of infection. Further support was provided by a correlative field study (Tompkins et al., 2000a) showing that worm burden in partridges on game-bird estates was positively related to prevalence in pheasants the previous year.

Model parameterisation and predictions: Tompkins et al. (2000b) used the above data and additional laboratory experiments to parameterise a model of the *Heterakis* system. The data were used to estimate the basic reproductive number (R_0; see Box 1.2) for the parasite. R_0 for *Heterakis* in the pheasant alone was 1.23, suggesting that this host could maintain populations of the parasite in the absence of partridges. However, R_0 in the partridge was only 0.057; as a minimum $R_0 = 1$ is required for parasite maintenance, the parasite would be lost from pure partridge populations. Simulations revealed that the parasite would be maintained in mixed pheasant/partridge populations with the eventual exclusion of the partridge (Fig. 2.12a). Further data suggest, however, that apparent competition may be insufficient to cause exclusion of the partridge (Tompkins et al., 2001). Taking account of possible negative impacts of infection for pheasants, the parameterised model predicted coexistence of pheasants and partridges (Fig. 2.12b).

However, another experimental infection study found smaller negative effects of infection on partridges and, in contrast to the data above, breast mass (an index of body condition) was positively correlated to worm burden (Sage et al., 2002). This puzzling result could suggest that breast mass in fact measures the costs of mounting an immune response to infection, with birds that successfully clear the infection able to channel fewer resources into growth.

Translating models into predictions for real populations: in reality, parameter values are likely to vary between populations or locations, as will the strength of resource competition, which was not considered in the model. What is perhaps more important than the precise values of the parameters is their relative effect. The 20-fold

Box 2.6 (*continued*)

(a)

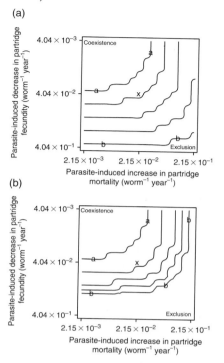

(b)

Fig. 2.12. Partridge exclusion–coexistence boundaries under different scenarios for the effect of *Heterakis gallinarum* on pheasant fitness. (a) Effect of fecundity reduction (infected pheasant fecundity is reduced successively along line a–b); (b) effect of increased mortality (infected pheasant mortality is increased successively along line a–b). Crosses indicated predicted outcome at the empirical estimates. From Tompkins *et al.* (2001).

difference in R_0 in pheasants and partridges is unlikely to be entirely due to experimental artefact, hence the prediction that pheasants are a reservoir for *Heterakis* infection to partridges is probably robust. Depending on the precise values of the parameters, the resulting apparent competition could be stronger (leading to partridge exclusion) or weaker (perhaps contributing to their decline but not causing exclusion). The nematode is also suspected to impact on fecundity of both partridges and pheasants in the field; models incorporating this effect produce sustained population cycles that would further complicate interpretation of population data (Greenman & Hudson, 2000).

2.6.3 White-tailed deer, moose and brainworm

A disease of longstanding conservation interest in the United States is that of brainworm (*Parelaphostrongylus tenuis*), a meningeal nematode with an indirect life cycle that can infect a variety of ungulates (Box 2.7). In parts of the United States and Canada, widespread declines and failures of reintroduction for moose (*Alces alces*) have been attributed to brainworm because it is largely benign in white-tailed deer (*Odocoileus virginianus*), but frequently fatal or debilitating in other cervids (Anderson, 1972). White-tailed deer populations have increased dramatically in North America since 1900, concomitant with habitat changes, the decline of hunting and top predators and the eradication of one of its probable regulators (the primary screwworm *Cochliomyia hominivorax*, a blowfly; once a major pest of livestock in the southern United States). The rise in deer populations is also implicated in the emergence of a range of tick-borne diseases in humans (Paddock & Yabsley, 2007), including Lyme disease (Box 7.1). Models of the deer–moose–brainworm system demonstrate that parasite-mediated competition can occur, but may or may not lead to exclusion of moose (Box 2.7). This example echoes the findings above for partridges: accurate parameter estimates are necessary in order to predict population outcomes when systems lie near the boundary between states of coexistence and exclusion.

Box 2.7 Brainworm in white-tailed deer and moose

Brain worm (*Parelaphostrongylus tenuis*) is a nematode inhabiting the meninges between the brain and skull of the definitive host, the white-tailed deer (*Odocoileus virginianus*). Presumably as a result of a long co-evolutionary history, brainworms cause little pathology in white-tailed deer but are highly pathogenic to other ungulates, including moose, elk, other deer species, goats, sheep, cattle and llamas. The neurological damage caused by brainworm in these unsuitable hosts can result in lameness, blindness, paralysis and death, and the nematode itself usually does not complete its life cycle in such hosts. In white-tailed deer, adult helminths produce eggs that develop into first-stage larvae; these are then passed from the body in the faeces. The larvae can then infect the intermediate host (a variety of land snails and slugs). Later-stage larvae infect white-tailed deer or other ungulates when infected intermediate hosts are incidentally consumed during grazing; parasite-induced behavioural changes may make snails more vulnerable to predation (Fig. 2.13).

Box 2.7 (*continued*)

Fig. 2.13. Life cycle of brainworm in white-tailed deer. Eggs laid in the brain of the host are transported in the blood to the lungs, coughed up and swallowed and then excreted. The larvae emerge and penetrate the gastropod intermediate host. Transmission of the L3 larva to the definitive host occurs when an infected gastropod is eaten.

Empirical evidence for parasite-mediated competition: a series of field studies found patterns of brainworm infection in accord with those predicted under apparent competition. Moose (*A. alces*) density was found to be inversely proportional to white-tailed deer density (Whitlaw & Lankester, 1994), but this could reflect direct or parasite-mediated competition. Further support for parasite-mediated competition comes from observations that the frequency of moose symptomatic for the parasite increased in areas of high deer density (Dumont & Crete, 1996) and that moose densities were lower in areas of high parasite prevalence in deer (Whitlaw & Lankester, 1994).

Predicting population outcomes: Schmitz and Nudds (1994) developed a detailed model of this system, deriving equations for the population dynamics of white-tailed deer, moose, gastropod intermediate hosts and the three parasite life stages. The model had three basic outcomes reminiscent of the generic models for parasite-mediated competition (Section 2.4): (1) competitive exclusion of deer (when parasitism is infrequent but competitive differences are large); (2) exclusion of moose (when parasite virulence and transmission are high and competitive differences are small); (3) stable or

Box 2.7 (*continued*)

unstable coexistence of moose and deer (for other combinations of these factors). Parameterised versions of the model using data from the literature predicted parasite-mediated exclusion of moose, stable coexistence or even competitive exclusion of deer. Outcomes were particularly sensitive to coefficients of interspecific competition, parasite-induced mortality rate and rate of increase for intermediate hosts (all parameters that were difficult to estimate in the field). Hence, exclusion of moose was theoretically possible, but on the basis of available data, coexistence was equally likely, and more robust parameter estimates are required to predict the outcomes of this interaction.

This might be encouraging news for moose conservation; Schmitz and Nudds concluded that reintroductions of moose into areas occupied by white-tailed deer would not necessarily fail as a result of the parasite. However, reintroductions of moose, elk and caribou into areas occupied by white-tailed deer have frequently been unsuccessful. Concerns remain that white-tailed deer will spread to the region west of the Great Plains in the United States, threatening populations of many susceptible ungulate species in habitat also suitable for deer and intermediate hosts.

2.6.4 Red grouse, deer, mountain hare, sheep and louping ill virus

Apparent and parasite-mediated competition can also be mediated by the vectors of parasites. In this case the parasite can be a specialist or generalist with a generalist vector; vector hosts provide a reservoir for vector amplification to the detriment of those species affected by the pathogen. This is the case for red grouse (*Lagopus lagopus scoticus*), which suffer substantial mortality from infection with louping ill virus (LIV). This pathogen, which can infect a range of domestic and wild vertebrate hosts, is transmitted by the sheep tick (*Ixodes ricinus*). Sheep ticks feed on a range of vertebrate species, but can only complete their life cycle on larger mammals. Models of this system show that a combination of two or more vector and viral hosts together are more likely to maintain infection; for instance, the virus will not persist in grouse populations alone as the tick vector cannot persist (Box 2.8).

Box 2.8 Louping ill virus in red grouse, deer, mountain hare and sheep

The LIV system: Louping ill virus (LIV) can infect many vertebrates, including sheep, pig, cattle, red grouse (*Lagopus lagopus scoticus*), red deer (*Cervus elaphus*), mountain hares (*Lepus timidus*), roe deer (*Capreolus capreolus*) and short-tailed voles (*Microtus agrestis*), and is transmitted by the sheep tick (*Ixodes ricinus*). Only two of these hosts, sheep and red grouse, produce a viraemia sufficient for feeding ticks to acquire the virus, but mountain hares allow a low level of non-viraemic transmission through cofeeding of ticks on infected hosts. Immature tick stages feed on a number of vertebrates, but adults generally feed only on a larger mammal, such as sheep, red deer and mountain hares. LIV causes substantial mortality in red grouse and sheep, with economic implications for management of upland areas of the United Kingdom. Treatment of sheep with acaricide and vaccination against LIV are generally used to control infection in domestic stock, and this can reduce LIV prevalence in grouse as sheep are effectively removed from the system as a major reservoir for both virus and ticks (Laurenson *et al.*, 2007). With sheep removed from the system, at least three host species are relevant on managed gamebird estates: red deer amplify the tick but do not amplify the virus; mountain hares amplify both tick and virus at low rates; and grouse amplify the virus but not the tick (Fig. 2.14).

Models and predictions: a combination of models and empirical data have been used to investigate this system. Norman *et al.* (1999)

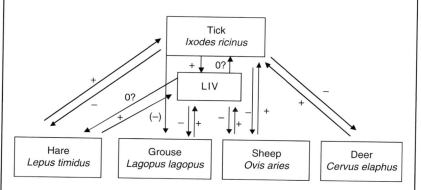

Fig. 2.14. Community module for LIV in grouse moorland. The sign of direct interactions is shown where known or suspected. The virus is placed below its tick vector for convenience.

Box 2.8 (*continued*)

developed a model involving red grouse, ticks, virus and hares in which non-viraemic transmission was not included. Because hares amplify only ticks, and grouse amplify only virus, both hares and grouse are necessary for LIV persistence; hares at low-to-moderate densities inflict apparent competition on grouse via the tick. Gilbert *et al.* (2001) extended the model to include tick amplification in red deer and non-viraemic transmission in hares. Analysis of isoclines (Fig. 2.15) plotting the threshold densities for virus and tick persistence revealed how LIV can persist in the system. In grouse–deer systems, both hosts are required for LIV persistence, provided deer densities are neither too low (resulting in insufficient ticks) or too high (such that infected ticks are more likely to bite deer than grouse, thus reducing LIV transmission; Fig. 2.15a). In grouse–hare systems, LIV can persist in the absence of grouse, if hare densities are high enough to support LIV through cofeeding on infected hares (Fig. 2.15b). In grouse–hare–deer systems, LIV is predicted to persist unless both hare and grouse are at very low densities (Fig. 2.15c).

LIV management in sheep: analysis of LIV infection in sheep from managed moorland in Lancashire suggests that, even with LIV vaccination and use of acaricides, leaving sheep out of the models may be too simplistic (Laurenson *et al.*, 2007). Vaccination and acaricide use on sheep are of interest to gamebird managers and sheep farmers as these techniques theoretically remove sheep as a reservoir for both virus and vector. However, strict adherence to both regimens is necessary. Laurenson *et al.* found that after eight years of treatment practice, very few farms had eradicated the disease and that great spatial variation in infection rates probably reflected different management practices. Eradication of LIV is particularly difficult as grouse, and the many wild mammal species that support either ticks or virus, cannot be effectively treated.

Conservation issues: an analysis of the effect of mountain hare culling on LIV prevalence in grouse on shooting estates in Morayshire, Scotland led to some heated debate on control of LIV (Laurenson *et al.*, 2003). Culling of wild hares led to sharp declines in tick and LIV prevalence compared to estates where culling was not introduced. The suggestion that mountain hare populations, which are themselves of conservation interest in the United Kingdom, should be reduced to lower LIV impact on gamebird populations is

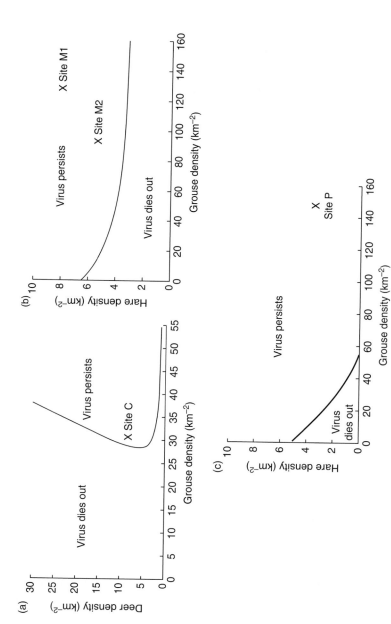

Fig. 2.15. Isoclines for persistence of LIV in communities consisting of (a) grouse and deer; (b) grouse and hare; and (c) grouse, hare and deer. The lines plot the joint densities for which $R_0 = 1$ (in (c) deer are at a constant, reflecting field densities); sites used to parameterise or test the model are indicated with an X. From Gilbert *et al.* (2001).

Box 2.8 (*continued*)

controversial (Cope *et al.*, 2004; Laurenson *et al.*, 2004). However, Laurenson *et al.* point out that if grouse shooting becomes uneconomical owing to the reservoir effect of mountain hares, it will no longer be economical to maintain heather moorland that provides a habitat suitable for grouse *or* hares.

Reservoir and dilution effects in multi-host systems. In the LIV system, several different hosts act as reservoirs, boosting the population densities of either the pathogen, the vector or both. From the perspective of the grouse, hares and deer are reservoirs for the vector, and hares are also a reservoir for the virus. In contrast to the reservoir effect, a dilution effect can occur whereby the addition of more host species reduces the prevalence or likely persistence of a disease (Section 7.3.2). This also occurs in the LIV system: high densities of deer cause a dilution effect because infected ticks are more likely to bite deer, which do not transmit the infection, rather than grouse. This illustrates a point we will return to: parasite establishment in host communities is likely to depend on biodiversity. A dilution effect of biodiversity has been reported in other vector-borne diseases, including two human diseases: Lyme disease (Box 7.1) and West Nile virus (Box 8.8). The interplay of dilution and reservoir effects not only has consequences for parasite establishment and maintenance in host communities, it can feed back on community structure as a whole (Sections 7.3 and 7.4).

2.7 Competition between parasites

So far we have concentrated on how parasites affect interactions between competitors (or between 'apparent' competitors). There is another case to consider: when parasites themselves are the competitors. Multiple parasite, pathogen or parasitoid species that attack a single host species, and even the same individuals within a population, are themselves community modules (Fig. 2.1.e), and coinfection is common in natural populations (Table 2.6). The population dynamics of these interactions are complicated and in many respects bear similarity to those for modules of IGP (Chapter 4). Theory in this area has developed from two viewpoints: firstly, population dynamic outcomes for parasite–parasite modules, mainly focusing on criteria for coexistence (Box 2.9); and

Table 2.6. *Empirical examples in which two or more species of parasite compete for resources within a single host individual*

Competing parasites	Host	Outcomes	Reference
Trematode *Schistosoma mansoni*	Mollusc *Biomphalaria glabrata*	*Virulent strain competitively superior.* Parasite reproduction was higher in mixed strain infections, which were more virulent.	Davies *et al.* (2002)
Protist *Plasmodium chabaudi*	Mice *Mus musculus*	*Virulent strain competitively superior.* More virulent strains of malaria had a competitive advantage in mixed infections and the outcome of competition determined the relative transmission success of the strains.	de Roode *et al.* (2005a); de Roode *et al.* (2005b)
Three strains of the bacterium *Pasteuria ramosa*	Cladoceran *Daphnia magna*	*Virulent strain competitively superior.* Virulence resembled that of the more virulent competitor, which also produced most transmission stages if simultaneous infections or virulent strain first to infect. However, if the less virulent strain was first to infect, then it produced most transmission stages.	Ben-Ami *et al.* (2008)
Tick *Ixodes ricinus* and nematode *Heligmosomoides polygyrus*	Mouse *Apodemus flavicollis*	*Asymmetric competition.* Field survey and experimental manipulation revealed an increase in tick numbers in mice with low nematode burdens.	Ferrari *et al.* (2009)
Acanthocephalans *Pomphorhynchus laevis* and *Acanthocephalus clavula*	Brown trout *Salmo trutta*	*Asymmetric competition, reduction in niche width.* Infection intensity was lower for both species in coinfected hosts, and *P. laevis* showed reduced niche width when in competition with *A. clavula.*	Byrne *et al.* (2003)
Baculovirus, different strains of nucleopolyhedrovirus	Lepidopteran *Mamestra brassicae*	Persistent, non-lethal baculovirus infection triggered into an overt lethal infection in response to infection of the host by a second baculovirus.	Burden *et al.* (2003)

(cont.)

Table 2.6. (cont.)

Competing parasites	Host	Outcomes	Reference
Bacteria *Photorhabdus* and *Xenorhabdus* spp.	Lepidopteran *Galleria mellonella*	*Interference competition.* Toxins produced by the bacteria that kill competing strains. In multiple infections, the toxin-producing strain excludes the other. When both strains produced toxins, coexistence and reduced virulence were observed.	Massey *et al.* (2004)
Gut bacteria *Escherichia coli* and pathogenic *Salmonella typhimurium*	Mice *Mus musculus*	*Resident parasites protected the host from invasion by new pathogen.* The presence of virulent or avirulent *E. coli* resident bacteria delayed establishment of pathogenic bacteria in the gut of germ-free mice. This effect was not seen in normal mice, indicating that the resident gut flora exerted the same protective effect.	Hudault *et al.* (2001)
Resident gut bacteria and invasive pathogenic bacterium *Serratia marcescens*	Locust *Schistocerca gregaria*	*Resident parasite diversity protected the host from invasion by new pathogen. S. marcescens* prevalence and burden were negatively correlated with the diversity of resident bacteria.	Dillon *et al.* (2005)
Lice *Columbicola columbae* and *Campanulotes compare*	Rock doves *Columba livia*	*Host-mediated competition. C. compar* is the stronger competitor and competition is mediated by host defences. When host defences were blocked, the competitive impact of *C. compar* increased.	Bush and Malenke (2008)
Parasitoid wasp *Cotesia plutellae* and bacterial pathogen *Bacillus thuringiensis*	Lepidopteran *Plutella xylostella*	*Host-mediated competition.* The outcome of competition between parasitoid and pathogen depended upon host resistance and the phenotype of the host. In susceptible hosts, the pathogen was the superior competitor. In resistant hosts, parasitoids were freed from interspecific competition.	Chilcutt and Tabashnik (1997)

Box 2.9 Parasite–parasite–host population models

A comprehensive comparison of population dynamic models for two parasites sharing a host species can be found in Begon and Bowers (1995); for a useful update with simplified models, see Holt and Dobson (2006).

Microparasite–microparasite, a simple model: Holt and Dobson (2006) discuss a number of models in which two microparasites compete for a host with fixed population density. These models, like most others, assume no coinfection or a very brief window of coinfection before one parasite supplants the other. In the simplest case, coexistence for the two parasites is not stable: the parasite that can persist at the lowest density of susceptible hosts will exclude the other. This parallels basic resource–consumer models where a single species (the one that can persist at the lowest level of resource) is expected to exclude all others competing for a single limiting resource. However, the mechanisms that make coexistence possible in resource–consumer models can also be applied to parasite–parasite competition. For instance, if the two parasites utilise more than one host species they can coexist if they each specialise to some degree on different hosts (an example of niche differentiation); this arises if each parasite has a higher transmission efficiency to, or lower rate of clearance from, a different host. Similarly, if intraspecific transmission is higher than interspecific transmission for each parasite, this increases opportunities for coexistence (recall the similar coexistence results for apparent and parasite-mediated competition models; Sections 2.3 and 2.4). A number of other regulating factors, including frequency-dependent mortality (e.g. from predation) and spatial subsidies (maintenance of an inferior competitor through spillover from another source (Section 8.2)) enhance the likelihood of parasite coexistence.

Microparasite–microparasite, with coinfection: a full analysis of two parasites competing for a host with host density as a dynamic variable regulated by the parasites is developed by Hochberg and Holt (1990). Coinfection is assumed possible, but one parasite quickly usurps the other. Predicted outcomes confirm the reasoning above, that parasite–parasite coexistence requires some form of niche differentiation. For instance, if one pathogen is better at infecting healthy hosts and the other better at infecting hosts harbouring the first pathogen, coexistence can occur if the advantage of the latter pathogen in usurping the former is not too great. This type of model is very similar in structure to intraguild predation (IGP), because the

Box 2.9 (*continued*)

usurper effectively consumes its competitor (Chapter 4). The predicted requirements for coexistence are also similar: coexistence requires that the 'predator' is less efficient at competing for the shared resource. Interestingly, the predicted impact of coexisting parasites on host density is intermediate to that for the single host–parasite associations, suggesting that from a biological control perspective, it would be better to utilise only the parasite that most efficiently exploits healthy hosts.

Macroparasite–macroparasite: Dobson (1985) develops models for macroparasites sharing a host with long-term coinfection permitted (as is frequently found for macroparasites such as helminths). Niche differentiation in the form of spatial aggregation within hosts, and also an aggregated distribution among hosts, enables coexistence. Interference between the parasites also influences the outcomes, in some cases resulting in coexistence where elimination of one parasite species was predicted under exploitation competition alone. At its extreme, interference competition will produce systems similar to those of IGP (Chapter 4).

Parasitoid–pathogen: a useful discussion of models and an empirical model system demonstrating the predicted patterns is given by Begon *et al.* (1999). Hochberg *et al.* (1990) examine a parasitoid and microparasite competing for a host with host density as a dynamic variable. Three-way coexistence can occur with constant, cyclic or chaotic populations, or one parasite can exclude the other (see Table 2.2 for terminology). Coexistence again requires a balance between efficient search and transmission to new hosts (the parasitoid might be expected to have an advantage here), and ability to infect and usurp other parasites from already-infected hosts (where many fungal pathogens have an intrinsic advantage). As for the macroparasite case, aggregation (in the form of aggregated parasitoid attacks) enhances coexistence. As above for the microparasite model, priority effects on coinfection are important in determining outcomes, and as before, there is an element of IGP in the models; we discuss this and its implications for biological control in more detail in Section 4.4.

secondly, evolutionary outcomes of within-host competition between parasite strains (focusing on how this influences the evolution of virulence; see Section 8.3). These are burgeoning research fields worthy of greater depth than afforded here, not least because of the ramifications for human health (see

Read & Taylor, 2001). Below we review a selection of empirical examples from a community ecology perspective. Competition between parasites can involve direct competition for resources, or apparent competition mediated by the host's defences/immune system or via behavioural manipulation of the host (Tompkins *et al.*, 2010). As we found for studies of host–host competition, it is often difficult to discriminate the relative importance of these mechanisms in empirical systems of parasite–parasite competition.

The theories discussed in Box 2.9 suggests that persistence of two parasite species sharing a host species will not occur unless there is some form of niche differentiation between the parasites. This can take the form of differential specialisation on body tissues or life stages of the host, aggregated distribution of parasites among hosts or parasite spillover from other sources (Box 2.10). Coinfection is common in nature, and in many cases it is clear that a degree of niche differentiation does exist (Table 2.6). Indeed, most empirical studies focus on cases of coinfection and its consequences for parasite virulence and transmission, as we discuss below.

2.7.1 Competition for resources

Competition can regulate parasite populations within a host, as the presence of one parasite may protect a host from infection with a second parasite if the resident out-competes it. For example, Dillon *et al.* (2005) compared the establishment success of a pathogenic bacterium (*Serratia marcescens*) in wild-type and germ-free locusts (*Schistocerca gregaria*) and found that the natural gut biota protected the host from establishment of *S. marcescens*. By exposing germ-free locusts to varying numbers of bacterial strains, the authors also found a negative correlation between *S. marcescens* density and the diversity of the resident community, indicating that, in accord with general ecological theory (Elton, 1958), diverse parasite communities are more resistant to invasion by new species. At the host population level, Kennedy and colleagues also looked at parasite diversity, comparing helminth species richness in fish at the regional (UK) and local (population) level (Kennedy & Guegan, 1994). For indigenous fish they found a curvilinear species–area curve consistent with predictions of community saturation, whereas the relationship for invasive fish (which have had less time to acquire parasites) was linear. However, other studies suggest that these curves may not be a result of saturation, but could also reflect the local availability of parasite species (Poulin, 1997; Rohde, 1998), whilst a meta-analysis of helminth diversity in mammals has revealed a strong influence of host species on parasite diversity, which may reflect the host's habitat or immune response (Bordes & Morand, 2008).

Parasite–parasite competition can also have evolutionary outcomes, driving changes in parasite virulence. In a single parasite infection, virulence results from the tradeoff between maximising host growth and survival (thereby ensuring good resource availability for the parasite) and maximising parasite reproduction and transmission (often resulting in virulence). However, in dual infections, a more virulent parasite strain or species may have competitive advantage as it may kill the host or exclude a competitor before the competitor has achieved transmission (Frank, 1996). de Roode *et al.* (2005a) tested this empirically with different strains of *Plasmodium chabaudi* in mice and found that the virulent strains had a competitive advantage in mixed-strain infections, resulting in higher transmission success to the mosquito. Burden *et al.* (2003) found that the virulence of a baculovirus infecting the cabbage moth *Mamestra brassicae* was altered dramatically by the presence of a second baculovirus. *M. brassicae* populations in the field harbour a persistent, non–lethal baculovirus. However, this was triggered into an overt lethal infection in response to infection of the host by a second baculovirus.

Parasites which occur in the same host individual can also show interference competition. For example, Massey *et al.* (2004) found that competition between toxin-producing strains of bacteria led to reduced virulence in the lepidopteran host *Galleria mellonella*, which could be attributed to interference competition. The different bacterial strains produced biocides to target interspecific competitors. Thus, in caterpillars that harboured multiple infections of bacteria that could kill each other, coexistence of the bacteria occurred and the host experienced lower virulence than in single-strain infection. Brown *et al.* (2009) reviewed interference competition between parasites and found a range of outcomes for the host. As illustrated by the *Galleria* example, competition led to reduced virulence by reducing overall microbial densities. However, for some systems the interference competition itself may damage the host either through toxins or manipulation of host immunity. Interference between coinfecting and competing parasites can also be considered as an example of intraguild competition (Box 2.9; Chapter 4).

2.7.2 Apparent and host-mediated competition

So far in this section we have considered resource-driven competition (although the influence of the host cannot be discounted). Other studies have demonstrated the role of host defences in the outcome of competition between parasites (Pedersen & Fenton, 2007; Tompkins *et al.*, 2010). In mammalian hosts, acquired immunity to one parasite species may have negative effects on a second species, whilst positive effects can result from immuno-suppression and from tradeoffs arising between the immune responses to different parasites. The impact of acquired immunity on coinfection was studied by Telfer *et al.* (2008), who carried out a long-term study of microparasite communities in natural vole (*Microtus agrestis*) populations. Coinfection was found to change the susceptibility to and infection duration for a number of parasites. For example, the presence of the tick-borne protist *Babesia microti* reduced the duration of infection by the flea-transmitted bacterium *Bartonella taylorii*. This negative interaction could result from competition for resources (red blood cells). Alternatively, clearance of *B. taylorii* may be enhanced as a result of up-regulation of the host's immune response by *B. microti*.

In a study of gut helminth communities in the rabbit *Oryctolagus cuniculus*, Lello *et al.* (2004) found a network of positive and negative interactions between the parasites. By considering the direction of impact and the biology and location of the host, the authors found evidence both for direct competition between parasites and for immune-mediated interactions (Box 2.10).

In mammalian hosts, a tradeoff is seen between the different immune responses induced by microparasites (which induce Th_1-type helper cells) and macroparasites (which induce Th_2 cells). Graham (2008) conducted a meta-analysis of microparasite and helminth coinfection levels in mice. She found that regulation of microparasite abundance resulted from competition with helminths for resources, as well as from helminth-mediated immune responses of the host. In situations where helminths induced anaemia, a reduction in red-blood-cell-dependent microparasites was observed, indicating competition for resources. However, in the absence of resource limitation (i.e. for microparasites that were not red-blood-cell-dependent), an increase in microparasite abundance was observed for coinfected hosts. The increase in microparasites was mediated by helminth-suppressed production of microparasite-specific cytokines by the Th_2 cells. Hence, microparasite abundance was regulated by both competition and host-mediated competition resulting from helminth-induced tradeoffs in the host immune response. Similarly complex patterns of interaction occur between parasites and pests of plants, mediated by host-plant chemical defence systems (Section 5.3).

Box 2.10 Interspecific competition between gut helminths in rabbits

Five species of helminths are common in populations of the wild rabbit *Oryctolagus cuniculus* in the United Kingdom. The stomach harbours the nematode *Graphidium strigosum*; in the small intestine are found the nematode *Trichostrongylus retortaeformis* and the cestodes *Mosgovoyia pectinata* and *Cittotaenia denticulata*; the large intestine harbours the nematode *Passalurus ambiguus*. Lello *et al.* (2004) studied the distribution of helminths in the gut, using general linear models to examine and predict the strength and direction of parasite–parasite interactions. They found a network of positive and negative interactions between the parasites, which resulted from direct and host-mediated competition (Fig. 2.16). To identify the possible mechanisms of interaction, they took into account the biology and location of these parasites along the gut. They speculated that interactions between parasites from the same region of the gut, or between an upstream and downstream parasite, may result from direct competition for resources, whereas the impact of a downstream parasite on one upstream is more likely to be mediated by the host's immune system. For example, numbers of *T. retortaeformis* were predicted to increase in response to a positive downstream influence from *G. strigosum*, probably due to modulation of

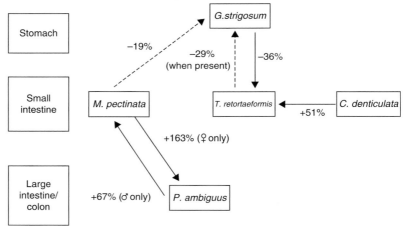

Fig. 2.16. The relative positions of rabbit gut helminths in the host gut and the directions and strength of parasite–parasite interactions. From Lello *et al.* (2004).

Box 2.10 (*continued*)

host immunity by *G. strigosum*. Conversely, acquired immunity in response to *T. retortaeformis* infection may underpin the predicted decrease of coinfecting *G. strigosum*.

The authors went on to investigate the likely impact of parasite–parasite interactions on the outcome of a vaccination programme against one of three interacting species. Their models predicted that vaccination against a single parasite species could release other parasites from the regulatory effects of competition, resulting in the unexpected increase in the abundance of the (previously low virulence) non-target species. Hence a better understanding of parasite community relationships should be a priority for parasite control strategies for both wild and managed populations.

In addition to the immune defences, the host's behavioural responses to infection can also mediate competition between parasites. For example, Bush and Malenke (2008) investigated competition between two louse species infecting the rock dove *Columba livia*. The louse *Campanulotes compar* is found on the body, while *Columbicola columbae* is found on the wings of the host. *C. compar* was found to be the stronger competitor, and its regulatory impact on *C. columbae* was strengthened when the doves were prevented from preening, indicating that competition between the two species of lice was mediated by host defences.

2.7.3 Coinfections and trait-mediated indirect effects

Parasite–parasite competition may result from density-dependent effects which may be mediated by the host, as illustrated above. In addition, the trait-mediated indirect effects (TMIEs) of parasites on their hosts can determine the outcome of parasite–parasite competition.

Many trophically transmitted parasites manipulate their host's anti-predator behaviour to enhance their transmission through predation by the definitive host (Thomas *et al.*, 2005b; Table 2.7). The outcome of coinfections involving manipulative parasites reflects the life cycle of the different parasites and the (similar or conflicting) selective pressures on the different parasite species. For example, if two species of parasite share the same definitive host, then the second parasite may benefit from manipulation of the intermediate host by the first. However, if predation of the host leads to death of the second parasite through dead-end

Table 2.7. *Empirical examples in which hosts are infected by two or more species of parasite, one or both of which manipulate the host's behaviour*

Competing parasites	Host	Outcomes	Reference
Trematode *Maritrema* sp. and acanthocephalan *Profilicollis* sp.	Shared intermediate crab host *Macrophthalmus hirtipes* and definitive bird (gull) hosts	*Synergistic manipulation.* The two parasite species were correlated among crab hosts. Serotonin concentrations in coinfected crabs correlated negatively with helminth abundance, suggesting synergistic manipulation of host behaviour.	Poulin et al. (2003)
		Reduced fecundity. Profilicollis egg production as an adult is reduced as a result of intraspecific and interspecific competition that was experienced in the crab intermediate host.	Fredensborg and Poulin (2005)
Trematodes *Microphallus papillorobustus* and *M. subdolum*	Parasites use the same second intermediate amphipod host *Gammarus insensibilis* and the same definitive (aquatic bird) hosts	*Hitchhiking. M. subdolum* cercariae preferentially infect *M. papillorobustus*-infected *G. insensibilis* and benefit from manipulation of the host by *M. papillorobustus*.	Thomas et al. (1997)
Trematodes *Microphallus papillorobustus, M. hoffmanni* and *Levinseniella tridigitata*	Parasites use the same second intermediate amphipod host *Gammarus aequicauda* and definitive (aquatic bird) hosts	*Lucky passenger. M. hoffmanni* and *Levinseniella tridigitata* benefit from manipulation of the host by *M. papillorobustus*, but do not preferentially infect these hosts.	Thomas et al. (1998)
Trematodes *Coitocaecum parvum* and *Microphallus* sp.	Shared intermediate amphipod host *Paracalliope fluviatilis*, different definitive hosts	*Avoidance of costly host manipulation.* In coinfected hosts, *C. parvum* is more likely to undergo progenesis and reach maturity in	Lagrue and Poulin (2008)

(cont.)

Table 2.7. (cont.)

Competing parasites	Host	Outcomes	Reference
		the intermediate host, thereby reducing the chances of dead-end transmission to a non-host (bird) predator as a result of host manipulation by *Microphallus*.	
Trematode *Microphallus papillorobustus* and nematode *Gammarinema gammari*	Shared amphipod host *Gammarus insensibilis*, but only the trematode is trophically transmitted (to aquatic bird definitive hosts)	*Sabotage and avoidance.* Trematode-infected *G. insensibilis* that did not display manipulated behaviour harboured higher nematode burdens, a result suggesting sabotage of manipulation by the nematode (thereby keeping its host alive). However, experimental infection/removal of the nematodes did not influence manipulation by the trematode. Male gammarids infected by *M. papillorobustus* harboured fewer nematodes. Possible avoidance of coinfection, or decreased exposure owing to behavioural manipulation.	Thomas *et al.* (2002); Fauchier and Thomas (2001)
Acanthocephalan *Polymorphus minutus* and microsporidium *Dictyocoela* sp.	Shared amphipod host *Gammarus roeseli*, but only *P. minutus* is trophically transmitted (to aquatic birds)	*Sabotage.* Dual-infected hosts showed lowered manipulation by the acanthocephalan, increasing the chances of vertical transmission by the microsporidium.	Haine *et al.* (2005)

transmission, then the second parasite should be selected to prevent or avoid manipulation of the host by its competitor. A series of studies involving crustacean hosts illustrate the various outcomes for coinfection involving manipulative parasites (Box 2.11).

Box 2.11 Coinfections in amphipods involving manipulative parasites

Trematode and acanthocephalan parasites have indirect life cycles that involve a definitive host and one or more intermediate hosts (Fig. 2.17). Parasite transmission to the definitive host occurs when it preys upon an infected intermediate host, and these parasites often enhance their transmission by manipulating host antipredator behaviour (Chapter 3). For parasites which infect the same host, selection to maximise transmission has led to a range of outcomes from synergistic to antagonistic effects.

Synergistic effects. Coinfection of hosts might be advantageous to manipulative species that share the same host if they act synergistically. For example, the acanthocephalan *Profilicollis* spp. and the trematode *Maritrema* both use the crab *Macrophthalmus hirtipes* as an intermediate host and are trophically transmitted to definitive shorebird hosts. Poulin *et al.* (2003) found a positive association between these two species, suggesting preferential coinfection. Manipulation of the host is mediated through reduced serotonin levels, and Poulin *et al.*'s study found that serotonin levels in the host brain were correlated with parasite burden in coinfected (but not singly infected) hosts, suggesting a synergistic effect on host antipredator behaviour.

Hitchikers. Coinfection may be advantageous to a trophically transmitted parasite that does not itself affect host behaviour but benefits from manipulation induced by another species of parasite. For example, the trematode *Microphallus papillorobustus* has two intermediate hosts. Eggs which are released in the faeces of the definitive bird host are ingested by snails (*Hydrobia* sp.) in which cercariae develop. These cercariae leave the snail and penetrate the second intermediate host, the amphipod *Gammarus insensibilis*. The parasite encysts in the brain and induces positive phototaxis and negative geotaxis, thereby increasing vulnerability of the gammarid host to predation by the definitive host. Interestingly, *G. insensibilis* also acts as a host for a trematode, *Maretrima subdolum* (later re-identified as *Microphallidae* sp.), which does not manipulate host

Box 2.11 (*continued*)

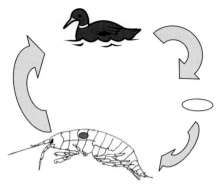

Fig. 2.17. Stylised life cycle of an acanthocephalan parasite. Adult worms reproduce in the definitive vertebrate host. Eggs are released via the faeces and ingested by the intermediate arthropod host. Transmission to the next definitive host occurs when the intermediate host is preyed upon by a suitable definitive host.

behaviour, but shares the same definitive host (Mouritsen, 2001). Thomas *et al.* (1997) found *M. subdolum* increased its transmission by hitchhiking in *M. papillorobustus*-infected hosts. In the wild, they found that coinfection was more frequent than should be predicted if infection of the host is random. Furthermore, *M. subdolum* cercariae were found to swim higher in the water column than *M. papillorobustus* cercariae, a behaviour that is likely to bring them into contact with manipulated *M. papillorobustus*-infected gammarids.

Two other trematodes, *M. hoffmanni* and *Levinseniella tridigitata*, also benefit from manipulation of the *G. insensibilis* host by *M. papillorobustus*. However, there was no evidence that these passengers preferentially attacked *M. papillorobustus*-infected hosts, a situation Thomas *et al.* (1998) term 'lucky passengers'.

Avoidance of costly manipulation. In contrast to the above situations, one parasite might be selected to prevent or avoid manipulation of the host by its competitor if trophic transmission of one parasite in fact disrupts the life cycle of the second parasite. The amphipod *Paracalliope fluviatilis* is host to two species of trematode; *Microphallus* sp. uses aquatic birds as definitive hosts whilst *Coitocaecum parvum* has a fish definitive host. *Microphallus* manipulates host behaviour to enhance predation by birds. However, bird predation leads to the death of *C. parvum*. In response to coinfection (and the risk of

Box 2.11 (*continued*)

transmission to the wrong host), *C. parvum* changes its life history from its normal three-host life cycle to protogenesis, reaching maturity in the amphipod host and producing eggs which are released at host death. This shift in life history results in rapid egg production before the manipulated host is eaten by the predator and dead-end transmission occurs (Lagrue & Poulin, 2008).

Sabotage. A coinfecting parasite may also sabotage the manipulative effect of its competitor to avoid dead-end transmission. Populations of the amphipod *Gammarus roeseli* in France are host to the acanthocephalan parasite *Polymorphus minutus*, as well as to a microsporidian parasite, *Dictyocoela* sp. These two parasites have contrasting life cycles and transmission strategies. *P. minutus* is trophically transmitted and induces negative geotaxis in its amphipod host, which increases the risk of predation by the definitive bird host. However, the microsporidian is vertically transmitted from generation to generation of hosts and predation of its host leads to the death of the parasite. Haine *et al.* (2005) found that coinfected hosts showed reduced manipulation by the acanthocephalan, suggesting that manipulation by the acanthocephalan had been sabotaged by the competing parasite.

2.8 Conclusions

Parasites can potentially interact with hosts in a great variety of ways as mediators, modifiers or ameliorators of competition. In many cases it can be hard to determine which processes apply, and indeed several different processes (such as a combination of apparent, parasite-mediated and parasite-modified competition) occur together in many systems. Given this complexity of interactions, if we suspect two species compete and a parasite is also involved, what outcomes might we expect? Fig. 2.18 summarises the questions we need answers to in order to predict the outcome. Key processes influencing the outcome include:

- The transmission dynamics of the parasite: is intraspecific transmission stronger than interspecific transmission, and if so, for which host species?
- The importance of competition: is intraspecific competition strong, limiting hosts to small carrying capacities? Is interspecific competition strong, and if so, which species is the superior competitor?

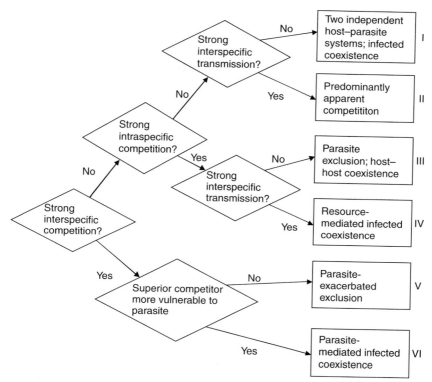

Fig. 2.18. Schematic of outcomes for parasite-mediated competition models (see Box 2.3). This is a qualitative classification: precise outcomes depend on the parameters, including pathogenicity and host reproductive rates; for macroparasites, degree of aggregation within hosts; for parasitoids, search efficiency, host preference, dispersal and distribution of host populations. For parasites (but not parasitoids), when competition of any form between hosts is not strong, both host populations are at large carrying capacities and can maintain the parasite independently; if interspecific transmission is rare, the host–parasite systems are effectively uncoupled (outcome I), but if interspecific transmission is more common, the system resembles that of apparent competition (II). If interspecific competition is weak but intraspecific competition strong, the parasite may be excluded (if interspecific transmission is rare, each host is at too low a carrying capacity for parasite establishment in the separate populations; case III). However, if interspecific transmission is more frequent, the two hosts can together support the parasite (IV). If interspecific competition is strong, outcomes depend on whether the parasite is most pathogenic in the inferior competitor (in which case it is excluded by the combined forces of parasitism and competition; V), or of sufficiently greater virulence in the dominant competitor (in which case the parasite may mediate coexistence; VI).

- The relative pathogenicity of the parasite in each host: is the parasite more pathogenic in the dominant competitor?

Fig. 2.18 shows how parasite-mediated coexistence and exclusion processes are related, but it is an over-simplification of the processes involved. For instance, we need to overlay the possibility of parasite-modified competition into this figure; it is difficult to say how this will influence outcomes as we do not yet possess a good body of theory to predict how trait- and density-mediated effects interact. In addition, different parasites may be competing for or within hosts; outcomes are again difficult to generalise, and theory linking within-host dynamics to outcomes at the scale of populations is in its infancy. Some important consequences for communities and ecosystems arise from some of these predictions:

- communities of host species may support parasites where individual host species would not (case IV in Fig. 2.18);
- parasites may speed up exclusion of a host species in competition with another, less susceptible species (case V);
- parasites may enable coexistence of an inferior competitor if they are more deleterious to a superior competitor.

In the rest of this book, we examine the consequences of these and other impacts of parasitism for populations, communities and ecosystems.

There is a further problem with the decision-tree approach; it assumes that we already understand the composition of – and interactions within – the communities under study. Parasites are ubiquitous in ecological communities, and we cannot easily tease out how they have influenced interactions. In Park's (1948) classic experiments (Section 2.1), parasite presence was the norm. Park suspected that his flour beetle populations were infected by *Adelina tribolii* in part because of his prior research on the beetles and on the parasite. In this four-year study, during which 211 *Tribolium* populations were subject to a complete census every month (an endeavour that Park estimated took 40 hours every week), Park went to considerable lengths to produce parasite-free colonies. In order to obtain the appropriate comparisons, 'sterile' colonies were obtained by isolating eggs from the beetles and washing with a solution of mercuric chloride; multiple sterile rinses were followed by cultivation in a heat-sterilised medium. Sterile colonies were kept in a separate incubator, and counts were performed with separate laboratory instruments wearing separate clothing. Despite these precautions, several of these colonies became reinfected before the experiments were due to end. In many ways, it is unfair to conclude that the beetle (*T. castaneum*) that persisted in the absence of the parasite was the 'dominant competitor', because beyond Park's meticulous manipulations, these beetles probably rarely interact in the absence of parasitism. To his credit, Park did not refer to *T. castaneum* as dominant over *T. confusum*; it is only in hindsight, with another 60 years of development of ecological concepts, that we tend to think of the interactions as nested this way. What we need, especially where natural communities are concerned, is to work back from the other direction: we have observed 'outcomes', and we have to tease out what processes occurred to produce them; parasitism is as much a fundamental unit of interaction as is competition.

3 · *Parasites and predators*

3.1 Introduction

In the preceding chapter we examined the effects of parasites on inter-actions between members of the same trophic level (intra- and interspecific competition and apparent competition). We can regard the indirect effects in these systems as horizontal ramifications of parasitism; however, parasites can also have vertical ramifications, affecting the interactions between species at different trophic levels. In this chapter, we look at examples of vertical ramifications, examining the role of parasites in predator–prey interactions. Parasites can enter into predation modules in a variety of ways, and their consequences for population dynamics and community structure will depend on their position in the module. We can distinguish modules involving parasites of the prey species, or of the predator species, or of both (Fig. 3.1). These modules may be further differentiated depending on the degree of predator specialisation. For instance, a specialist predator with population dynamics tightly coupled to that of its prey might be expected to be more sensitive to the impact of parasitism on the prey (Section 3.1). Nevertheless, as we examine in Section 3.2, generalist predators may be influenced by, and influence, parasitism of the prey.

3.1.1 Overview of predation modules

We can get some qualitative understanding of the likely impact of parasites by comparing the topologies of different predation–parasitism modules. Parasites that infect the prey species (Fig. 3.1a) are in essence competing with predators; thus the addition of parasites turns this into a competition module (with the prey/host species the resource). We may anticipate here that interspecific competition theory should apply; predator and parasite may act together to reduce the resource (i.e. host species), and there may be resource-related restrictions on predator–parasite coexistence (Sections 3.2 and 3.3). This interaction may be further complicated if parasites are consumed and killed when an infected host is eaten, in which case this is

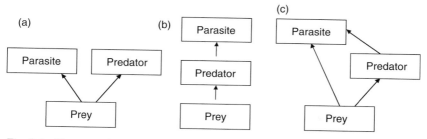

Fig. 3.1. How parasites enter into predation modules: (a) parasite of prey; (b) parasite of predator; (c) parasite of predator and prey.

effectively an intraguild predation (IGP) module (Chapter 4). Parasites that infect only the predator (Fig. 3.1b) are effectively consumers of the predator, turning this module into a linear food chain (Section 7.2). We may anticipate here that the effect of parasites might be similar to that of a top predator, perhaps regulating intermediate predator (i.e. host) levels and in some cases initiating trophic cascades to lower trophic levels, releasing the prey from predation. Many parasites infect both predator and prey (Fig. 3.1c); these are parasite species with complex life histories whereby the prey is the intermediate host and the predator is the definitive host. Parasite transmission from prey to predator depends on infected prey being eaten (a process known as trophic transmission). Here, the parasite and predator may be in competition for uninfected prey, but the parasite benefits if infected prey are killed and eaten (provided this occurs at the right stage in the parasite's life cycle). Arguably, this module can also be seen as a form of IGP, in which the parasite is the stronger intraguild predator (Chapter 4). Importantly, when predator and prey are both hosts, parasite transmission from prey to predator depends on infected prey being eaten. In this case it will be in the parasite's evolutionary interests to increase the chances that infected prey are eaten, and there are numerous empirical examples where infected individuals are more prone to predation than uninfected conspecifics (Section 3.5.1). Although some changes in host behaviour may be a by-product of infection (Section 3.2.2), for others there is strong evidence for adaptive manipulation by the parasite (Section 3.5.1).

In this chapter we build up from structurally simpler modules to more complicated cases, and then examine some of the applied problems relating to the interaction between parasitism and predation. In Section 3.2 we examine theory and empirical examples of parasitism of prey with a specialist predator, before moving on to the parallel case for generalist predators

(Section 3.3). In Section 3.4 we examine parasites of predators, before moving on to parasites of both predators and prey (Section 3.5). Applications of this body of work are relevant to population management through predator control (which may interact with parasitism (Section 3.6.1)), biological control (for instance, using pathogens to control top predators (Section 3.6.2)) and the harvesting of infected populations (Section 3.6.3).

3.2 Parasites of prey with specialist predators

Specialist and generalist predators differ in fundamental ways with regard to the population dynamics and outcomes of their community modules. This is because specialist predators have population dynamics tied to that of their prey species. At the other extreme, a broadly generalist predator utilises many different prey, and its population dynamics can be assumed to be completely decoupled from any one prey species.

3.2.1 Baseline theory

The theory of predator–prey relationships with infected prey is quite well developed. Most of the models have their origins in the models of Anderson and May (1986), a version of which we present here to give an idea of some of the more important components that need to be considered. As before, for models of competition and parasitism (see, for example, Section 2.3.1), for microparasite systems the prey population is comprised of susceptible (S) and infected (I) classes (for definitions, see Table 3.1):

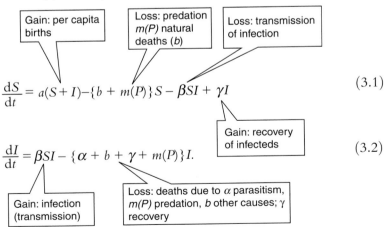

Gain: per capita births

Loss: predation $m(P)$ natural deaths (b)

Loss: transmission of infection

$$\frac{dS}{dt} = a(S+I) - \{b + m(P)\}S - \beta SI + \gamma I \qquad (3.1)$$

Gain: recovery of infecteds

$$\frac{dI}{dt} = \beta SI - \{\alpha + b + \gamma + m(P)\}I. \qquad (3.2)$$

Gain: infection (transmission)

Loss: deaths due to α parasitism, $m(P)$ predation, b other causes; γ recovery

These equations model a directly transmitted pathogen with density-dependent transmission (at rate β per contact between an infected and

Table 3.1. *Parameters, variables and functions used in the equations in this chapter*

S	Variable: susceptible host (prey) class
I	Variable: infected host (prey) class
P	Variable: predator class
a	Prey per capita birth rate
α	Parasite-induced mortality rate
b	Prey natural death rate (causes other than parasitism and predation)
β	Parasite transmission efficiency per contact
γ	Recovery rate of infected individuals (per capita return to susceptible class)
$m(P)$	Mortality rate of prey as a function of predator density (to model a linear predator functional response, m is a constant, so $m(P)=mP$; to model a non-linear functional response, m is a more complicated function of P)
δ	Predator per capita reproduction rate
d	Predator per capita mortality rate

susceptible prey). In this simple model, the host is unregulated in the absence of predator and parasite. Once an individual becomes infected it becomes immediately infectious, but recovers (without becoming immune) at a rate γ; hence, this is an SIS (susceptible–infectious–susceptible) model (Box 1.4). For systems with a specialist predator, we need to include an equation for the predator population P coupled to prey population density:

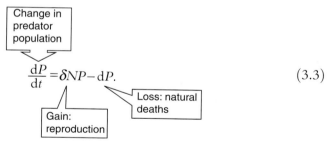

$$\frac{dP}{dt} = \delta NP - dP. \tag{3.3}$$

Here, predator reproduction at per capita rate δ is proportional to total prey density N. This is equivalent to assuming that predators catch prey at a rate proportional to prey density, and each are converted with constant conversion efficiency into predator offspring. Hence, this model assumes a linear (Type I) functional response (see Table 3.2). Functional response shape is important in determining the dynamics of predation–parasitism modules; we consider other functional responses below (Section 3.3.2 and Tables 3.3 and 3.4).

Table 3.2. *Common forms of predatory functional responses*

Type	Relationship between predation intensity and prey density	Mechanism underlying relationship
Type I	Linear	Predation proportional to predator–prey contact rate; contact proportional to prey density; short handling time and no predator satiation within prey density range
Type II	Saturating (convex) curve	At high prey densities, predator becomes satiated, or long prey handling times restrict rate of prey consumption
Type III	Sigmoid (S-shaped) curve	Density-dependent predator preference (switching behaviour) or learning (which reduces handling time or increases capture rate) such that rare prey types are under-represented in predator diet, with saturation at high densities

Selective predation for or against infected prey is another important factor; in Eqn (3.3) predators are assumed to attack prey regardless of infection status. Selective predation for infected prey (also known as parasite-induced vulnerability to predation (PIVP)) is considered towards the end of this section. Together, these equations form a modified Lotka–Volterra predator–prey system with subclasses for susceptible and infected prey. As before (Box 1.2; Section 2.2.1), we can examine population outcome with reference to basic reproductive numbers (R_0 for the parasite, and R_P for the predator – an approach exemplified in Hethcote *et al.* (2004)) and thresholds for establishment (N_T for the parasite, and N_P for the predator; see below).

3.2.1.1 Thresholds for parasite and predator invasion

Two thresholds in prey population size are key to understanding the outcomes for this and related models. Firstly, as parasite transmission is density dependent, there is a threshold for parasite establishment (Box 1.2); if the prey population falls below this the parasite cannot be maintained. We can derive this threshold from the equation for the

Table 3.3. *General models of parasite–host systems involving a specialist predator of the host*

System	Outcome	Reference
Lotka–Volterra SI model (no intrinsic prey regulation, no recovery or immunity); Type I functional response; density-dependent transmission	Two establishment thresholds: for parasite and predator. Coexistence possible if both thresholds met. Predator and parasite jointly regulate prey to stable equilibrium or stable limit cycles.	Anderson and May (1986)
Logistic SIR model (density-dependent prey population, frequency-dependent transmission with recovery); Type I or II functional response, selective predation on infected subclass	Parasitism enhances predator persistence via selective predation; predation can exacerbate parasite extinction by reducing prey population below disease threshold, or via selective predation on infected subclass. Coexistence with alternative stable states possible.	Hethcote *et al.*(2004)
Logistic SI model (density-dependent prey population, density-dependent transmission). Predator consumes lethal toxin when it consumes infected prey (e.g. botulism); Type II functional response	Stable coexistence in parts of parameter space; unstable in others. Toxins exacerbate likelihood of predator extinction. Harvesting of prey population prevents predator extinction by reducing infection prevalence.	Chattopadhyay and Bairagi (2001); Chattopadhyay *et al.* (2003)
Logistic SIS model (density-dependent prey population, density-dependent transmission), Type II functional response	Characterises equilibria and conditions leading to stable limit cycles; discusses differences between SI versus SIS models and different models for transmission and predation functional response	Haque and Venturino (2006)

This is a selection covering the main varieties of model; references to others, often of a highly technical nature, can be found in the more recent papers above. Functional response: Type I: linear relationship between predation and prey density; Type II: saturating relationship (see Table 3.1). Model structure (see Box 1.4): SI: transition between susceptible and infectious classes only; SIS: susceptible–infectious–susceptible; SIR: susceptible–infectious–immune.

dynamics of the infected subclass (Eqn 3.2); for initial parasite invasion we require $dI/dt > 0$:

$$N_T(P) = \frac{\alpha + b + \gamma + m(P)}{\beta}. \qquad (3.4)$$

This threshold is of the same form as that we derived earlier (e.g. Box 1.2), except that now mortality due to predation is included, and is a function of predator density. Indeed, the previous models have been used to reason about the effect of predation (Anderson & May, 1981; Holt & Dobson, 2006), with predation assumed as one of the factors included in 'natural mortality' b. Alternatively, many predation models dispense with b, assuming that other sources of mortality can be subsumed within the expression for predation (for instance, Packer *et al.*, 2003).

As predator dynamics are density dependent on the prey, predator persistence is also subject to a minimum prey population density (by similar reasoning, $dP/dt > 0$ requires $N_p > d/\delta$). If the prey population falls below this threshold, the predator will go deterministically extinct.

In the model above (Eqn 3.1–3.3), the relative magnitude of these two thresholds determines the community outcomes. If the predator can establish in a prey population smaller than the parasite threshold N_T, the parasite is excluded, resulting in a predator–prey module. Alternatively, if the parasite can persist in a population smaller than the predator threshold N_P, the predator is excluded and we obtain a pathogen–host module. If both thresholds are met, the predator and parasite can both become established and can potentially regulate the prey to a joint equilibrium lying above the two thresholds. However, three-way coexistence is not always possible, and the stability of the equilibrium and form of population dynamics predicted varies depending on the model assumptions. Some interesting cases are discussed below and summarised in Fig. 3.2 (for a reminder of the terms used to describe dynamical properties of these systems, see Table 2.2).

3.2.1.2 *Without recovery, parasite and predator cannot coexist*

In the system above, if infected individuals do not recover ($\gamma = 0$), three-way coexistence of parasite, predator and prey is not possible (Anderson & May, 1986). This is because infected individuals neither recover nor reproduce; from a demographic perspective they are effectively dead, so the parasite is acting as a predator. Hence, the pathogen and predator are

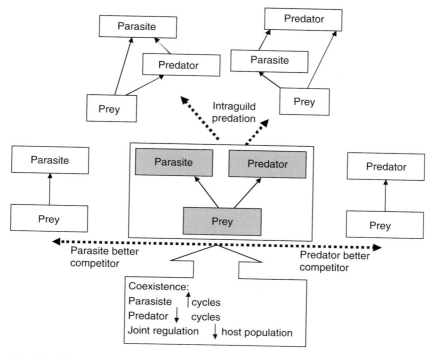

Fig. 3.2. Schematic of outcomes for parasite–predator–prey module, with a parasite of the prey. The full module (shaded) can devolve to a parasite–host module (left), predator–prey (right) or intraguild predation (top), depending on the parameters for the system. If three-way coexistence can occur, parasites may introduce cyclic population dynamics, predators may reduce cycling and the host population may be regulated to a lower level than in the absence of one of its consumers.

utilising the same resource (the prey) in the same way and so experience strong interspecific competition; competitive exclusion of the species that utilises resources less efficiently (the one with the higher threshold for establishment) is the result. The inclusion of density dependence on the prey or biased predation on infected prey does not qualitatively alter this outcome (Anderson & May, 1986; Table 3.3). If infected prey are allowed to recover or reproduce, three-way coexistence becomes possible (Anderson & May, 1986; Hethcote *et al.*, 2004; Table 3.3). In this case, the pathogen is no longer analogous to a standard predator as its 'prey' continue to contribute to population growth before they die; this temporal decoupling allows coexistence for some of the parameter range. In this case, parasite and predator jointly regulate the prey to an equilibrium N^*, and coexistence is predicted provided N^* lies above the establishment thresholds for both the parasite and predator.

3.2.1.3 Parasite-induced vulnerability to predation

In many real systems, predators attack infected prey more or less frequently than healthy prey. In particular, parasitism may weaken prey, making them easier to catch (parasite-induced vulnerability to predation: PIVP), and there is evidence that predators utilise chemical cues released by infected individuals for prey location (Hudson et al., 1992a; Section 3.3.2). If the predator selectively preys on infected individuals, parasitism can enhance the range of conditions, allowing predator persistence (Table 3.3). At the extreme, if predators utilise only infected individuals, the predator population can only be maintained in infected populations (Anderson & May, 1986). More generally, if specialisation is less complete but based on infected prey providing an 'easy' resource for predators to exploit, parasitism may allow persistence of a predator which would otherwise go extinct in the absence of infection (Hethcote et al., 2004). Conversely, PIVP has consequences for parasite populations, because it increases the threshold for disease establishment. We can see this if we modify the rate equation for the infected subclass and then solve it to find N_T:

$$\frac{dI}{dt} = \beta SI - \{\alpha + b + \gamma + mv(P)\}I$$
$$\Rightarrow N_T = \frac{\alpha + b + \gamma + mv(P)}{\beta}. \tag{3.5}$$

Here, we have adapted Eqn (3.2) with a constant v expressing the relative vulnerability of infected individuals to predation. If v is greater than 1, the host suffers PIVP; from Eqn (3.5), as v increases, so does N_T.

When predators focus their attacks on infected individuals, we might expect a greater-than additive effect of parasitism and predation together in regulating population numbers. Such a synergistic interaction has been demonstrated theoretically for snowshoe hares (*Lepus americanus*), helminth-infected individuals of which are subject to strong PIVP (Box 3.1).

3.2.1.4 Increasing oscillations and other dynamics

Anderson and May (1986) note that for several of these models, if three-way coexistence is possible, a stable equilibrium is attained or, in some cases, the populations oscillate under stable limit cycles (Table 2.2). Limit cycles are not predicted to occur in this type of predator–prey interaction in the absence of parasitism (recall that these simple models involve a linear (Type I) functional response by the predator); hence the parasite can be seen as 'destabilising' the predator–prey interaction (in terms of the population dynamics). More detailed analyses of more complex models reveal further population dynamic complications (Table 3.3), including regions of parameter space where addition of a parasite leads to undamped or divergent oscillations leading to eventual extinction of one or more players (Chattopadhyay et al., 2003; Roy & Holt, 2008), or alternative stable states with population trajectories potentially moving between them (e.g. Hethcote et al., 2004).

3.2.1.5 Immunity

Anderson and May (1986) also examine the effect of including acquired immunity in the host (Table 3.3). In their brief treatment of outcomes, immunity does not appear to alter outcomes qualitatively, but we note here that more recent models for specialist and generalist predators show that immunity can dramatically alter the relationship between predation and parasitism (Section 3.6.1). We suspect that similar models for specialist predators will demonstrate similar behaviours, so if acquired immunity is likely to be important in a system, the reader should examine this recent body of work before making specific predictions about the effect of predators on particular parasite–host interactions.

In summary the theoretical consensus is that predators and parasites of prey interact to a greater or lesser extent as competitors exploiting a shared resource. Parasitism can sometimes destabilise an existing predator–prey interaction, inducing or increasing the amplitude of population oscillations. In many circumstances, the introduction of a parasite is predicted to reduce specialist predator populations because prey are regulated to a lower equilibrium density by the joint effects of parasitism and predation. In some cases, parasites may drive specialist predators extinct if they bring the host population below the threshold population density for maintenance of the predator. Equally, predators can interfere with parasite population dynamics, regulating host density below that required for parasite establishment (in which case the parasite is extirpated), or reducing parasite prevalence through selective predation on infected prey. Finally, we note that these models predict that parasites and predators together will generally limit host populations to a lower density than either one alone, although the effect is not necessarily additive. Precise outcomes will depend on the details of the interaction. One reason for this is that in many cases predation–parasitism modules can resemble intraguild predation (IGP): if the predator destroys the parasite when consuming infected prey, the predator is the dominant intraguild (IG) predator; if the parasite goes on to infect the predator when infected prey are consumed, the parasite is the stronger IG predator. Indeed, similar theoretical outcomes to those above are obtained in models of a pathogen and specialist parasitoid that share an arthropod host species (Hochberg et al., 1990), another system that might be regarded as an example of IGP. As we explore in Chapter 4, IGP can result in additive, greater-than-additive and less-than-additive regulation of the shared resource (Section 4.4).

3.2.2 Empirical examples

Well-characterised empirical examples of parasites of prey with a specialist predator are rare, although it should be noted that many similar cases fall into the category of trophic transmission (Section 3.5), where the parasite also infects the predator. There is considerable anecdotal evidence that predators of rabbits (*Oryctolagus cuniculus*) were adversely affected by the introduction of the Myxoma virus in the United Kingdom and Australia in the 1950s, although in the United Kingdom most eventually recovered to some degree by adapting to other prey (Sumption & Flowerdew, 1985; see also Section 7.2.1). More recently, the viral disease rabbit haemorrhagic disease (RHD) has been introduced into Australia and Europe, again initiating predator population declines. In particular, the rapid spread of RHD throughout Spain in 1988 led to further population declines of two endangered predators of the Iberian peninsula. The Iberian lynx (*Lynx pardinus*; Box 8.10) and the Spanish imperial eagle (*Aquila adalberti*) are particularly vulnerable as they feed almost exclusively on rabbits (which are native to Spain) and appear unable to adapt to other prey species (Ferrer & Negro, 2004). Indeed, vaccination of rabbits in this region has been proposed in order to conserve lynx and eagle populations. Where predators can switch to other prey species, knock-on effects on these other prey can occur. For instance, hunting-bag records for mainland Spain suggest that red-legged partridge (*Alectoris rufa*) populations declined following the arrival of RHD (Moleon *et al.*, 2008). The most likely explanation for this is that partridges suffered increased predation from predators such as foxes once the supply of rabbits diminished (a process known as hyperpredation). Partridge populations recovered more rapidly than those of rabbits, possibly indicating reduced predation pressure in the longer term as the reduction in rabbit populations led to reduced predator densities (Moleon *et al.*, 2008).

In some ecosystems, a single species may be sufficiently numerically dominant that many predators rely on it as a resource even though they are not strictly specialists. A well-quantified system of this type is that of the snowshoe hare (*Lepus americanus*), its nematode parasites and its guild of predators (Box 3.1). Population cycles of snowshoe hares are thought to be driven by resource level and predation, but parasitism may interact with predation to increase the tendency to cycle. Murray *et al.* (1997) found that hares treated with an anthelminthic had 2.4-fold increased survival over that of controls, in six populations where predation was estimated to be responsible for 95% of the mortalities. The helminths had no other measurable effects on host fecundity or direct mortality; PIVP (Section 3.2.1) appears to be solely responsible for the effects of parasitism in this system.

In Ives and Murray's (1997) model, predation and parasitism act synergistically to drive prey oscillations, setting up time-lagged predator and parasite cycles (Box 3.1). As infection makes hares more vulnerable to predation (PIVP), increasing parasite burdens reinforce the effects of increasing predator density, resulting in sustained oscillations of all the players. Hence, in this system parasites tend to destabilise predator–prey dynamics as including parasites shifts the damped oscillations of predator–prey dynamics towards long-term cycles or diverging oscillations, mirroring the results of Anderson and May (1986). It is interesting to note that inclusion of a Type II (saturating) functional response by the predator in this parameterised model was not sufficient to invoke cycles in the absence of parasitism (this is not always the case; see Hethcote *et al.*, 2004); in the next section we see that models for generalist predators show that non-linear functional responses in combination with parasitism may underpin a range of cyclic, unstable and even chaotic population dynamics.

Box 3.1 Snowshoe hares, helminth parasites and predator population cycles

Snowshoe hare (*Lepus americanus*) populations throughout the North American boreal forest have cyclic fluctuations over a period of 8–11 years. Historically, these cycles have been attributed to time-lags in hare numerical response (via effects on mortality and natality) to predation and food supply. However, empirical work suggests that parasitism from a variety of nematode species can strongly influence predation rates (Murray *et al.*, 1997). Ives and Murray (1997) showed that parasitism could interact multiplicatively with predation to increase the amplitude of population cycles, using a parameterised model where sustained cycles did not occur in the absence of parasitism.

Snowshoe hares are subject to several avian and mammalian predators. Although no single species can be regarded as a specialist predator of hares, the population dynamics of this predator guild are nonetheless strongly coupled to those of the hares. To reflect this, Ives and Murray modelled a hypothetical predator population (possessing the 'average' properties of the hare predator guild) with a population growth rate dependent on both hare density (a specialist predator, as in Anderson and May's 1986 models), and on intrinsic (predator) density (to reflect competition among predators for other prey species). The helminths were also modelled as a single population, using the standard negative binomial approach to model parasite aggregation within hosts (Box 1.4). Parasites were assumed to be benign apart from their effect on predation, which was assumed to increase linearly with burden. The resulting equations for host, parasite and predator population dynamics were parameterised using estimates from the empirical literature. For the basic model (assuming a Type I linear predator functional response), predator and prey settled to a stable equilibrium in the absence of parasites, or with parasites held at a constant density (Fig. 3.3a, Fig. 3.3b). If parasite population dynamics were included, sustained oscillations in predator, prey (Fig. 3.3c) and parasite (Fig. 3.3d) occurred. A Type II (saturating) predator functional response (probably more realistic for this class of predators), and density dependence in helminth reproduction (likely but not quantified for this system) produced qualitatively similar results.

Box 3.1 (*continued*)

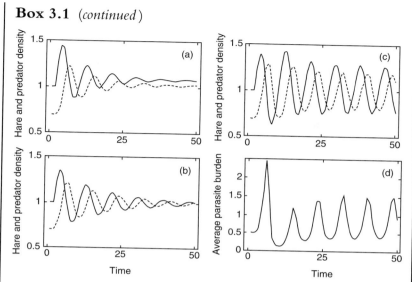

Fig. 3.3. Example population trajectories for the snowshoe hare system (a) in the absence and (b)–(d) in the presence of parasites (based on the model equations parameterised with realistic values). When the parasite population is assumed constant, the system returns to a stable equilibrium (b), but when parasite population dynamics are included, sustained cycles occur (c). Population densities of hares (solid lines) and predators (dashed lines) shown in (a)–(c); (d) shows average parasite burden for the model in (c). From Ives and Murray (1997).

The snowshoe hare system (Box 3.1) emphasises that PIVP can have important consequences for host population dynamics and stability, producing qualitatively different outcomes to those for the combined effects of non-selective predators and parasites. As parasitic infections frequently have debilitating effects on the host, PIVP is likely to be widespread. PIVP has also been demonstrated in field studies of Townsend's voles (*Microtus townsendii*). Predation by birds of prey and mustelids is the major cause of mortality for Townsend's voles. In a field manipulation, Steen *et al.* (2002) compared the predation rates of control voles and voles that had been treated with anthelminthics to rid them of botflies (*Cuterebra grisea*) and helminths. They found a 17% reduction in the monthly survival of untreated voles in comparison with parasite-free voles, indicating that PIVP may be an important influence on population regulation by predators. However, parasitism can also decrease vulnerability to predation in the host. Sticklebacks infected with the microsporidian parasite *Glugea anomala* show increased antipredator behaviour, foraging at greater distances from a potential predator than do uninfected individuals (Milinski, 1985). *G. anomala* has a simple life cycle, being transmitted between stickleback hosts. Hence the decreased vulnerability to predation may result from parasitic manipulation of the host to avoid host and parasite death. In addition, in some circumstances predators may be selected to avoid consumption of infected hosts (Lafferty, 1992).

3.3 Parasites of prey with generalist predators

3.3.1 Baseline theory

A basic model for microparasites of prey with a generalist predator takes a similar form to that for a specialist predator, except that predator dynamics are decoupled from those of the prey. At its simplest, we can make the assumption that predators are regulated to a carrying capacity by other forces; hence we can represent the predator population as a constant:

$$\frac{dS}{dt} = a(S + I) - \{b + m(P)\}S - \beta SI + \gamma I, \tag{3.6}$$

$$\frac{dI}{dt} = \beta SI - \{\alpha + b + \gamma + m(P)\}I, \tag{3.7}$$

with P = predator density (constant).

The equations for the susceptible and infected prey subclasses are the same as before, with three sources of mortality (predation, parasite-induced and other 'natural' sources – Table 3.1). Infected individuals recover at rate γ, but do not become immune (so this is an SIS model – Box 1.4). As before, this basic model assumes that predators do not preferentially select infected or uninfected prey, but we can easily modify the equations (see Eqn 3.5, for example) to reflect PIVP. Some models (e.g. Roy & Holt, 2008) even assume that parasite-induced mortality itself is negligible, with the effect of the parasite entirely mediated through increased predation on the infected class. This may be realistic for some systems (for example, in snowshoe hares (Box 3.1)). Because predator density is constant, if the predator has a Type I (linear) functional response as assumed in the above model, then the rate of predation on each prey subclass will be linearly related to its density. We can solve for the threshold population size for parasite establishment as

Table 3.4. *General models of parasite–host systems with a generalist predator of the host*

System	Outcomes	Reference
SIS model; no intrinsic host self-regulation, density-dependent transmission, constant predator density; Type I functional response; selective predation, macroparasites also modelled	Predation may increase or decrease total population size; predation reduces Prev* and sometimes I*, stronger predation on infected class enhances this; aggregated distribution for macroparasites exacerbates effect.	Packer *et al.* (2003); Ostfeld and Holt (2004)
As above; mortality from predation subsumed within general mortality effects	As above; reduced predation can increase opportunities for parasite spillover to other species via its effect on I*.	Holt and Dobson (2006)
SIR model with acquired immunity; host population size fixed or without intrinsic regulation, density-dependent transmission	Hump-shaped relationship between predation pressure and Prev* or I*. Prev* and I* maximised at intermediate predation levels for a broad range of conditions.	Holt and Roy (2007)
As SIR above; parasite with no mortality effects other than selective predation on infected class; extensions for intrinsic host density-dependence or extrinsic resource dependence, frequency-dependent transmission; Type II functional response	Confirms hump-shaped relationship above for a broad range of models. Dynamics characterised: stable or unstable equilibria, persistent cycles or divergent oscillations, alternative stable states possible.	Roy and Holt (2008)

Functional response: Type I: linear relationship between predation and prey density; Type II: saturating relationship (see Table 3.1). Model structure (see Box 1.4): SI: transition between susceptible and infectious classes only; SIS: susceptible–infectious–susceptible; SIR: susceptible–infectious–immune. Prev*: equilibrium parasite prevalence $= I^*/(S^* + I^*)$; I*: equilibrium frequency of infected individuals. Terms for dynamical behaviour are described in Table 2.2.

before, setting $dI/dt = 0$; since these equations are essentially the same as those assuming a specialist predator, we recover the same threshold:

$$N_T(P) = \frac{\alpha + b + \gamma + m(P)}{\beta}. \tag{3.8}$$

There is no analogous threshold for predator establishment (as the predator is a generalist), so the population outcomes will be either: (1) the parasite cannot establish (N^* in the presence of predator $< N_T$) and a predator–prey system results; or (2) the parasite can be maintained ($N^* > N_T$) and a parasite–predator–prey system is attained. If the latter holds, a stable equilibrium is reached:

$$S^* = \frac{d}{\beta}, I^* = \frac{rd}{\beta(d-e)}. \tag{3.9}$$

In these equations, we have combined some parameters for convenience: d represents the total losses from the infected class ($d = \alpha + \beta + \gamma + m$ (P)); r is the intrinsic rate of increase of susceptibles ($r = a - b - m(P)$); e is the contribution to the susceptible class from infected individuals (in this formulation, $e = \gamma + a$).

From the equation for S^*, we see that increased predation on the infected subclass (which increases d) will always increase S^*, so the absolute population density of healthy hosts will increase with increasing predation pressure (Packer et al., 2003). By similar reasoning, this model predicts that increased predation will always decrease the prevalence of infection (prevalence = $I/(I+S)$). Together, these predictions suggest that predation on infected populations may actually keep them 'healthy', reducing parasite prevalence in the population (see Section 3.6.1 for applications of this concept).

Because predators are held constant, we cannot get oscillations in their density. We might therefore expect such a generalist predator to simply reduce population densities with little impact on the dynamics predicted for host–parasite interactions; but the models predict that we can get qualitatively different behaviours in modules when all three players are present (Table 3.4). For instance, Packer et al. (2003) showed that even in the simple SIS model above, with a Type I (linear) predator functional response, predation can potentially increase total population size; hence parasitism and predation have clearly non-additive impacts. This outcome occurs when the reduction in infectives owing to predation is insufficient to match the increase in susceptibles discussed above. When non-linear predator functional responses are included, a range of population dynamic outcomes, including sustained oscillations, are possible (Table 3.2); we examine some of these dynamic possibilities in more detail in the following section, with reference to specific models of natural systems.

3.3.2 Empirical examples

There is correlative evidence that disease in prey species can impact on generalist predators in similar ways to its effect on specialists (Section 3.3.1) if (1) the host forms a significant part of the predator's diet (as is the case for the snowshoe hare predator guild – for example, see Box 3.1); or (2) the parasite is itself a generalist such that it can infect most species in a predator's diet. The latter situation arose in the case of rinderpest, a morbillivirus with a broad host range affecting many ungulate species, including the large East African grassland herd species, buffalo and wildebeest (see Sections 6.2 and 7.2.1). To our knowledge, data are not available for the effect of rinderpest on predator populations when it became prevalent in native African ungulates in the 1890s. However, its eradication from the Serengeti ecosystem in the 1960s was associated with increased populations of lions and hyenas following increases in the ungulate host populations (Section 7.2.1). Below, we focus on quantified examples where the influence of both parasite and predator on host population regulation has been examined.

3.3.2.1 Predators can reduce cyclic dynamics in host–parasite systems

Hudson and colleagues have combined empirical and theoretical approaches to examine the interplay of parasitism and predation in red grouse (Box 3.2). Like many ground-nesting birds, grouse are vulnerable to generalist predators when incubating eggs. Red grouse are also subject to a parasitic nematode, which is implicated as the driving force behind significant population oscillations. When the effect of predation is incorporated into the grouse–parasite system, a reduction in the amplitude of cycles was predicted. This occurs because predation reduces parasite prevalence, reducing the regulatory impact of the parasite on its host. Field data support this prediction: in gamebird estates with less gamekeeper control of predators, parasite prevalence was significantly reduced (Box 3.2).

An interesting point raised by the work discussed in Box 3.2 is that predators may reduce infection prevalence by selective predation on infected hosts; predator control programmes may therefore have counter-intuitive consequences for managed prey populations (Section 3.6). In the grouse system, predators are also predicted to reduce population oscillations, in contrast to the prediction for snowshoe hares (Box 3.1),

Box 3.2 Red grouse, nematode parasites, predators and gamekeepers

The population biology of red grouse (*Lagopus lagopus scoticus*) and its caecal nematode *Trichostrongylus tenuis* has been studied extensively in upland estates in the northern United Kingdom (Hudson *et al.*, 1992b; 1998). Grouse populations undergo significant fluctuations in density, with a period of 4–8 years that varies between populations. Worm burden in adults also fluctuates, with intensity highest during years in which grouse numbers fall. Population-dynamic models of this system exhibit cycles driven by the time-lagged effect of parasitism on host populations that occur when parasites reduce host fecundity more than they increase mortality (Dobson & Hudson, 1992).

Effect of predation – gamekeepers: grouse are also subject to predation by generalists such as foxes and birds of prey; on managed estates, this is kept in check by the action of gamekeepers. There is evidence for parasite-induced vulnerability to predation (PIVP) (Section 3.2.1); birds killed by predation had significantly higher adult worm burdens, and the proportion of birds with high worm burdens was higher on game estates with less gamekeeper control of predators (Hudson *et al.*, 1992a).

Incorporating predation: Hudson *et al.* (1992a) modified their original host–parasite model to include the effect of generalist predators (modelled as a constant population density) with a Type I (linear) functional response to prey density. Analysis of the model revealed that increased predation reduced mean parasite burden in the grouse population, reducing the regulatory role of the parasite and resulting in a reduction in the amplitude of grouse population oscillations. Predator selection of infected prey further enhanced this effect, but was not required to produce it.

Here, predators do appear to temper the effects of parasites: in this system, parasites alone are responsible theoretically for generating cycles (via their delayed density-dependent effects on host fecundity), and predation reduces the influence of parasitism by one of two means. Selective predators lead to a direct reduction in average worm burden by removing heavily infected hosts; non-selective predators reduce host densities, which results in reduced parasite transmission and consequently reduced parasite prevalence.

Box 3.2 (*continued*)

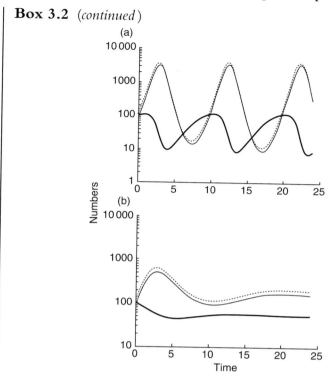

Fig. 3.4. Example population trajectories for the red grouse–nematode system (a) in the absence and (b) in the presence of predators (based on the model equations parameterised with realistic values). Addition of predators dampens the host–parasite population cycles. Horizontal axes: time in years; vertical axes: population numbers; thick line: grouse; thin line: mean adult parasite burden; dashed line: mean total parasite burden (adults plus larvae). From Hudson *et al.* (1992a).

where parasitism and predation acting together are predicted to increase population oscillations. There are a number of possible explanations for this discrepancy, based on the different perspectives taken: Hudson *et al.* add predators to a host–parasite system where the parasites have a strong regulatory influence on the host even in the absence of predation (Hudson *et al.*, 1998); Ives and Murray add otherwise benign parasites to a predator–prey system where cycles in prey numbers were thought to be largely driven by predation and resource limitation. Importantly, the models differ in their treatment of predator dynamics; Ives and Murray include predator dynamics coupled to the prey, whereas Hudson *et al.* assume predator densities are constant. Predator and parasite have multiplicative effects on host density in Ives and Murray's model via PIVP (Section 3.2.1); although PIVP is present in Hudson *et al.*'s model, it cannot feed back on predator density. These differences underline the importance of assumptions about parasite life history and predator numerical and functional responses in determining outcomes.

3.3.2.2 *Predators and parasites may have complementary regulatory roles*

The different regulatory roles of parasites and predators can lead to complicated population dynamic trajectories, especially when predators have non-linear functional responses. This could explain the erratic population dynamics of defoliating forest Lepidoptera, characterised by oscillations at low population densities followed by sporadic outbreaks (Box 3.3). Recent work by Dwyer *et al.* (2004) shows how pathogens and generalist predators may combine to regulate these insect populations at different ends of the density scale. Previously, pathogens have been implicated as regulatory factors, but parasite–host models are not good predictors of the variable time between outbreaks. Predator–prey models also fail to describe the dynamics well, predicting longer cycle times between outbreaks. However, combined predator and parasite attack with additional stochasticity to reflect environmental fluctuations can provide a close fit to the data (Box 3.3).

The work discussed in Box 3.3 suggests that gypsy moth populations are regulated by different forces at each of the population scales: at low densities, predators are most important, and at high densities, when predation has saturated, disease becomes important. In this model, predation and parasitism do not interact synergistically (compare the case for snowshoe hares, where parasitism enhances the effect of predation), but they produce qualitatively different population dynamics when combined than when interacting with the host/prey alone. In fact, Dwyer *et al.*'s analysis suggests that predation and parasitism combined can make it difficult for host/prey populations to persist: for biologically feasible parameters, the lower equilibrium can be unstable; this means that any small perturbations in the moth's population density could lead to its extinction as population oscillations take it too close to zero density.

3.3.2.3 *The impact of predator functional response*

The above studies suggest that predator functional response is important in determining the dynamics of and outcomes for parasitism–predation

Box 3.3 Pathogen and predator regulation of forest Lepidoptera

Dwyer *et al.* (2004) show how two enemies can have a differential impact depending on prey density, resulting in complex dynamics for the host. Their model concerns the gypsy moth *Lymantria dispar*, one of several defoliating forest Lepidoptera, which has outbreaks every 8–11 years. In the extensive deciduous forests of the north-eastern United States, outbreak populations may be several hundred times larger than those at 'normal' low densities, and can lead to forest defoliation on a massive scale. *L. dispar* is host to specialist baculovirus pathogens which cause a fatal infection in larvae when they consume foliage contaminated with infectious cadavers. The moths are also subject to a variety of generalist predators including birds, spiders, small mammals and parasitoids.

Dwyer *et al.* assumed a constant predator density and incorporated a non-linear predator functional response. Specifically, predation was modelled to reflect prey switching (i.e. a Type III functional response), with attacks rising steeply with increasing moth population densities to reflect specialisation on newly abundant prey, but saturating at high densities when predators become overwhelmed by over-abundant resources. Variability in host susceptibility to baculovirus was also an important feature contributing to the dynamics. A discrete generation model was used with overwintering of the host egg stage and a term for overwinter survival of the pathogen in the environment; these features have a tendency to produce cyclical dynamics because they automatically introduce time delays.

When the parameterised host–pathogen model was run without predation, it generated limit cycles with similar periodicity but greater regularity than the empirical data. However, when predation was included, a better fit to the data was obtained, particularly when a degree of stochasticity was added to reflect chance environmental influences (Fig. 3.4). The addition of predation creates a low-density equilibrium where the predator regulates the host and the pathogen is relatively unimportant, in addition to the high-density equilibrium found in the pathogen–host model; adding a small amount of environmental stochasticity drives the trajectory unpredictably between the two. For slightly different parameters, the lower equilibrium becomes unstable and chaotic dynamics are observed.

Box 3.3 (*continued*)

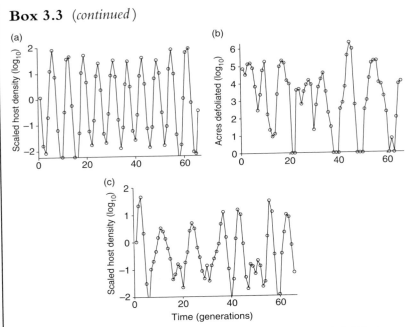

Fig. 3.5. Comparison of model output and data for gypsy moth defoliation, New Hampshire, USA. Graphs show (a) example simulation of host–pathogen model; (b) empirical time series; (c) example simulation of model combining host, pathogen and predation. From Dwyer *et al.* (2004).

interactions. A detailed analysis of the effect of functional response on the interplay between predation and parasitism is presented by Hall *et al.* (2005) for a microparasite–zooplankton–fish system (Box 3.4).

The work discussed in Box 3.4 shows that predators with a Type II (saturating) functional response can destabilise host–parasite systems, resulting in complex population dynamics. The combined effects of parasitism and predation frequently result in diverging oscillations that ultimately drive the host and/or parasite extinct. The dynamics depend on predation rate and predator selectivity, and also productivity (determining density dependence for the host), as well as parasite transmission and virulence. Saturating functional responses have this effect because at low prey densities, the relative impact of predation is higher (compared to a linear response). This has a destabilising effect: regulation at high

---

Box 3.4 *Daphnia*, predators and pathogen outbreaks

Daphnia dentifera, a dominant zooplankton grazer in many northern temperate lakes in the United States, is subject to virulent infections by several microparasites, including the yeast *Metschnikowia bicuspidata* and the bacterium *Spirobacillus cienkowskii*. In many midwestern lakes, *Daphnia* is subject to predation by bluegill sunfish *Lepomis macrochirus*, which have a Type II functional response, saturating at intermediate *Daphnia* densities. Bluegills locate prey using predominantly visual cues and show strong selective predation for infected *Daphnia* (i.e. there is parasite-induced vulnerability to predation – see Section 3.2.1); infected individuals are opaque and more visible than translucent uninfected conspecifics (Duffy *et al.*, 2005; Duffy & Sivars-Becker, 2007).

Hall *et al.* (2005) modelled this system with an SIS model similar to the baseline example in this section, modified to include density-dependent host reproduction and alternative predator functional responses. With a Type I (linear) functional response, infected and susceptible classes settle to a stable equilibrium dependent on productivity (i.e. the degree of density dependence in the host), predation rate and selectivity (provided the host population threshold N_T for parasite establishment is met). With a Type II (saturating) functional response, strongly selective predators that preferentially prey on infected hosts introduce an Allee effect on the parasite, driving it extinct if it falls below a threshold parasite population density. Invading parasites not only require a minimum *susceptible host* population density for invasion (i.e. N_T), they must also exceed a critical propagule size of *infected* hosts. Moderately selective predation interacts with parasitism and can result in damped or divergent population oscillations depending on the intensity of predation (Fig. 3.6). At intermediate predation intensities, neither parasite nor predator has a dominant regulatory influence, and the resulting divergent oscillations ultimately drive the parasite extinct. In contrast, if the predator is non-selective (or prefers uninfected hosts), the combined forces of parasitism and predation acting together may drive the host extinct.

Hall *et al.* suggest that the dynamics for strongly selective predators might explain empirical observations in the *Daphnia* system of epidemic invasions with very sudden crashes and apparent extinction after only one or two oscillations. Epidemics occur in response to seasonal decreases in fish predation, but die out after only a single oscillation

Box 3.4 (*continued*)

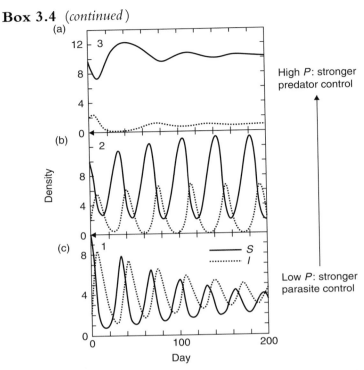

Fig. 3.6. Example population dynamics for the *Daphnia* parasite–predator model in which infected hosts are moderately more vulnerable to predation. Panels show dynamics at three different predation intensities. At high predation intensity (a) cycles are damped and infection is held to a low level; moderate predation intensity (b) results in divergent oscillations and ultimately parasite and/or host extinction; low predation intensity (c) results in damped oscillations to a stable coexistence equilibrium. Graphs plot the population densities of susceptible (solid lines) and infected (dashed lines) hosts. From Hall *et al.* (2005).

(Duffy *et al.*, 2005). This could reflect dynamics leading the parasite population below the Allee critical point, from which it declines to extinction. If predation is sufficiently intense and selective, the parasite would be unable to reinvade host populations that year if it did not attain the critical propagule size. There may be other explanations for curtailment of disease outbreaks, including differential scaling of host and parasite vital rates with temperature (for instance, if parasite physiology scales more steeply with temperature than predator physiology, predation may limit epidemics in

Box 3.4 (*continued*)

cooler seasons (Hall *et al.*, 2006)) or the presence of other *Daphnia* species that do not transmit the infection as efficiently (causing a 'dilution effect' – see Section 7.3; Hall *et al.*, 2009). Rapid evolution of susceptibility to infection in *Daphnia* clones may also explain epidemic curtailment (Box 3.5).

prey densities is relatively weak, and at low densities the prey cannot recover from the pressure of predation.

3.3.3 Evolutionary dynamics and predation

A complicating issue with these analyses is that evolution in the host can occur on similar timescales to that of the ecological dynamics, and this may also drive population dynamics and coexistence outcomes. This may have important ramifications for our interpretation of empirical data on the strength of interaction between species. For instance, in experimental rotifer–algae and phage–bacteria cultures, predator and pathogen populations can exhibit large amplitude fluctuations, whereas prey populations remain essentially unchanged. During these experiments, resistant strains are selected and largely replace the more susceptible ones, so although the strength of the predator–prey or parasite–host relationship is not apparent from changes in prey density, the interaction is nonetheless strong (Yoshida *et al.*, 2007).

Rapid evolution does not appear to be confined to the hosts and parasites with the shortest generation times, such as those in Yoshida *et al.*'s studies. This process, in combination with selective predation on infected hosts, has also been implicated as a factor driving the dynamics of *Daphnia* and its pathogens (Boxes 3.4 and 3.5).

The interaction between predation and parasitism can have long-term ramifications. Empirical studies demonstrate that parasitism and predation can impose opposing selective pressures, resulting in the maintenance of polymorphism in an aphid/parasitoid/predator system (Losey *et al.*, 1997). Field and laboratory manipulations showed that red morphs of the pea aphid (*Acyrthosiphon pisum*) suffer higher parasitism (from the parasitoid *Aphidius ervi*), whereas green morphs suffer higher predation (from the ladybird *Coccinella septempunctata*). Simple models demonstrated that biased, density-dependent parasitism and/or predation were sufficient to maintain the colour morphism in the population (Losey *et al.*, 1997).

Box 3.5 Rapid evolution in host–parasite systems

Several processes may contribute to patterns of pathogen infection in the freshwater zooplankton *Daphnia dentifera*, which are characterised by sporadic outbreaks followed by variable periods in which infections are rare or absent. An alternative (or indeed complementary) explanation to the predator model described in Box 3.4 is that selection of *Daphnia* clones occurs sufficiently quickly for it to influence seasonal population dynamics (Duffy & Sivars-Becker, 2007).

Cross-infection experiments between strains of *Daphnia* and the fungal parasite *Metschnikowia bicuspidata* collected from different lakes showed that *Daphnia* populations differed significantly in their susceptibility to the parasite, but parasite strains did not differ in their infectivity. *Daphnia* collected from lakes that had recently had a *Metschnikowia* outbreak were significantly less susceptible and had lower variation in susceptibility than samples from other lakes. There was no such evidence for rapid evolution by the parasite, suggesting that it could be out-manoeuvred over the course of a season by its host (Duffy & Sivars-Becker, 2007). Simulations using realistic parameter values indicate that rapid host evolution in

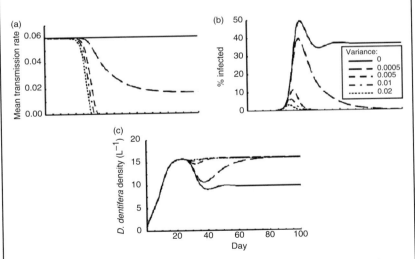

Fig. 3.7. Simulation results for the effect of population variance on *Daphnia* susceptibility to *Metschnikowia*. Plots show how (a) mean transmission rate, (b) infection prevalence and (c) host population density change with time (days) over the course of an outbreak, plotted for different population variances in host resistance (as labelled). From Duffy and Sivars-Becker (2007).

Box 3.5 (*continued*)

combination with selective predation on infected *Daphnia* might terminate pathogen outbreaks after a single peak in infected individuals (Duffy & Sivars-Becker, 2007; Duffy & Hall, 2008; Fig. 3.7). If resistance entails a cost to the host, it will be selected against in the absence of parasitism, allowing parasite reinvasion in subsequent years and setting in motion the dynamics for another outbreak.

These models demonstrate that if the host has sufficient genetic diversity, selection for resistance occurs over the course of a season, resulting in reduced pathogen transmission, reduced prevalence and eventual extinction of the pathogen. Consequently, host population size returns towards pre-epidemic sizes and from this demographic perspective (that is, ignoring the underlying genetic changes), the influence of the parasite is rather short-lived.

Long-term data from natural populations reveal a contrasting pattern in a UK freshwater lake, where an invasion by a novel pathogen appears to have permanently shifted growth patterns of the host. The growth rate of perch *Perca fluviatilis* in Windermere, Cumbria appears to be under opposing selective pressures from predation by pike *Esox lucius* (driving growth rate up), and infection with an unidentified pathogen (driving growth rate down) (Edeline *et al.*, 2008). After an outbreak of the pathogen in the 1970s, the distribution of perch size classes shifted downwards. This pathogen-driven selection keeps perch in classes where they are more vulnerable to predation by pike, thus shifting the trophic balance in the lake. Because fewer perch reach the largest size classes, the strength of competition between adult perch and juvenile pike is also reduced by parasitism. These long-term changes can thus have consequences for food web structure and stability (Chapter 7). From an evolutionary perspective, mortality imposed by predation, in particular PIVP, can influence the evolution of parasite traits such as virulence. Classical models of virulence evolution show that virulence evolves to higher levels with increasing host background mortality (i.e. 'natural' mortality b in the earlier equations) and with increasing predation (Williams & Day, 2001). However, if predators select infected prey, this relationship can theoretically be reversed (Choo *et al.*, 2003). The precise outcome depends on whether the predator is a specialist or a generalist, underlining the need for detailed knowledge of the biology of systems before specific predictions can be made.

3.4 Parasites of predators

3.4.1 Baseline theory

In this structure, we only need to consider the case of a specialist predator interacting with parasitism, because generalist predator populations will not be coupled to the density of the focal prey/host species. Lotka–Volterra predator–prey models can again be modified to incorporate a parasite infecting the predator:

Rate of change of prey | Gain: intrinsic rate of increase | Loss: predation

$$\frac{dN}{dt} = rN - m(P)N \tag{3.10}$$

$$\frac{dS}{dt} = \delta NS - bS - \beta SI + \gamma I \tag{3.11}$$

$$\frac{dI}{dt} = \beta SI - (\alpha + b + \gamma)I. \tag{3.12}$$

Rate of change of infected predators | Gain: infection | Loss: α parasite-induced, b natural mortality, γ recovery

Here, N represents the prey population, while the predator population is composed of susceptible S and infectious I subclasses (other parameters and variables as in Table 3.1). In this model, only uninfected predators reproduce, but both classes capture prey at the same rate (hence prey loss due to predation occurs at rate m, a function of P, the total predator density). Predator breeding gains are therefore proportional to prey density (δ represents capture and conversion of prey into predators as in Eqn (3.3)); this assumes a Type I (linear) functional response as in the other simple models we have discussed. The above equations (based on Anderson & May, 1986, but allowing for recovery from infection) can be modified to take account of reduced predation or some breeding by the infected class. By solving $dI/dt = 0$, we can again find the critical predator population density for parasite establishment (N_T); and by solving $dP/dt = 0$, we can find the critical prey population density for predator establishment (N_P). Clearly, if the predator is unable to invade, the parasite will also become extinct, so both these thresholds must be met for parasite establishment. If the predator persists, two outcomes are possible: either the parasite fails to establish (if the predator is at an equilibrium density lower than the threshold for parasite invasion), or all three species coexist (Anderson & May, 1986; Table 3.5). As is the case for the parasites-of-prey models, previously stable predator–prey dynamics may become oscillatory when a parasite of the predator is introduced (Anderson & May, 1986). Under some circumstances, parasites of herbivores (which, as consumers of basal resources, bear similarity to predators) can induce chaotic dynamics (for instance, if they impact on host reproductive success (Grenfell, 1992; Table 3.5)). However, adding parasites can sometimes stabilise otherwise oscillatory predator–prey relationships (Table 3.5). This occurs when the predator has a saturating (Type II) functional response to prey density with an inherent tendency to cyclic dynamics. By reducing predator numbers, a parasite of the predator reduces the regulatory role of the predator; this increases prey numbers such that density dependence in the prey population plays a stronger role in regulation (Hilker & Schmitz, 2008). Hence, the addition of parasites is predicted to switch control from top-down (predators) to bottom-up (density-dependent resource competition) modes. Moose (*Alces alces*) in a community where canine parvovirus (CPV) invaded the resident wolf population appear to show exactly this type of pattern (Box 3.6).

When parasites infect predators, they effectively become 'top predators' (or 'top parasites' (Grenfell, 1992)), and the relevant community module takes the form of a parasite–predator–prey food chain. Grenfell (1992) concluded that the theoretical dynamics of parasite–herbivore–plant systems was, in many ways, similar to that of classical predator–prey systems, but there were also differences. These chiefly stem from the non-lethal effects of parasites on hosts, such as effects on fecundity, which introduce time delays into the dynamics contributing to cyclic or chaotic dynamics. Despite these differences, from a community ecology perspective we would tentatively expect aspects of food chain theory to apply; for instance, we might expect to see trophic cascades induced by the parasite, and indeed there is mounting empirical evidence for parasite-induced trophic cascades (Section 7.2).

3.4.2 Empirical examples

There are few well-quantified examples in the literature of the population effects of parasites of predators on predator and prey population dynamics. However, there are several convincing but unquantified accounts of population and community-level effects, particularly for parasites of grazers, in which knock-on effects at other trophic levels

Table 3.5. *General models of parasite–predator–prey interactions with a parasite of the predator*

System	Outcome	Reference
SIS model; no intrinsic prey regulation, density-dependent transmission; Type I functional response; extensions tabulated	Threshold for parasite establishment. If met, then stable or cyclic coexistence, otherwise reverts to predator–prey system.	Anderson and May (1986)
Herbivore–macroparasite model. Host population dynamics linked to basal resource; Type II functional response (grazing); parasite has free-living stages ingested when grazing; parasite affects fecundity and mortality	Coexistence results in stable or unstable equilibrium with stable, cyclic or chaotic dynamics; increased parasitism reduces herbivore population, reducing grazing pressure on basal resource.	Grenfell (1992)
SIRS; Lotka–Volterra model; density-dependent transmission, density-dependent prey population; Type I functional response	Population dynamics and outcomes characterised; predator extinction, predator–prey, or predator–prey–parasite outcomes; parasite can increase population density of predators.	Auger *et al.* (2009)
SIS model; frequency-dependent transmission, density-dependent prey population growth; Type II functional response	Addition of parasite to oscillatory predator–prey system reduces oscillations (in the region of unstable equilibrium with stable limit cycles). Parasite reduces predator population or may drive it extinct.	Hilker and Schmitz (2008)

This is a selection covering the main varieties of models; references to others, often of a highly technical nature, can be found in the more recent papers above. Functional response: Type I: linear relationship between predation and prey density; Type II: saturating relationship (see Box 1.4). Model structure (see Table 3.2). SIS: transition between susceptible and infectious classes with recovery to susceptible class; SIRS: susceptible–infectious–immune, with loss of immunity to re-enter susceptible class.

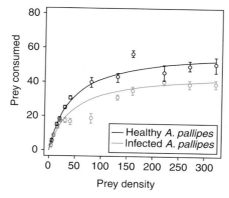

Fig. 3.8. The effect of parasitism on the host's predatory functional response. Relationship between the number of prey eaten and prey density, the 'functional response' after 12 hours, for unparasitised *Austropotamobius pallipes* and those parasitised with *Thelohania contejeani* (Dunn, Haddaway & Wilcox unpublished data).

are clear. These parasite-induced trophic cascades are discussed in the context of effects on community structure in Section 7.2. In order to induce such trophic cascades, the pathogen must decrease predation either by density- (mortality) or trait- (attack rate) mediated effects on the predator, thereby releasing prey from predation. We review examples below where these effects have been demonstrated.

A recent study has revealed trait-mediated effects (decreased predation rate) of parasitism in the white-clawed crayfish *Austropotamobius pallipes*, although the likely impact on predator–prey dynamics has not been explored theoretically. In freshwater streams and lakes in the United Kingdom, the amphipod *Gammarus pulex* is an important food item in the diet of the white-clawed crayfish. Laboratory studies have demonstrated a reduction of more than 30% in the predatory impact of the crayfish on *G. pulex* as a result of infection by the microsporidian parasite *Thelohania contejeani*. Dunn, Haddaway and Wilcox (unpublished) measured the impact of parasitism on the predatory functional response (Table 3.2) of the host. Both infected and uninfected individuals displayed a typical Type II functional response; prey intake increased with prey density, reaching a plateau as prey handling time limits prey intake. However, infected individuals showed an increased handling time and decreased prey intake (Fig. 3.8). Parasite prevalence varies between field populations and can reach >25%; hence, parasite-modified predation may have a strong impact on predator–prey dynamics. Other studies show that parasitism can increase predation rates of infected amphipod hosts (Dick *et al.*, 2010; Box 7.2) and increase or decrease grazing rates of molluscs (Section 7.2.2).

A recent study of the effects of a disease-induced population crash in wolves (*Canis lupus*) provides evidence for density-mediated effects of parasitism on predators, with long-lasting effects on the community (Box 3.6).

Box 3.6 Wolves, moose and canine parvovirus

In the early 1980s, canine parvovirus (CPV) was accidentally introduced to wolves (*Canis lupus*) in the Isle Royale National Park, USA. CPV spread rapidly throughout the wolf population, causing it to crash between 1980 and 1982 (Peterson *et al.*, 1998). Since then, wolf population density has remained at substantially lower levels, and the density of moose (*Alces alces*) has shown erratic fluctuations with increased variance (Fig. 3.9). CPV was no longer detectable after 1990, but wolf and moose populations show no signs of recovery to their pre-CPV levels (Vucetich & Peterson, 2004).

Wilmers *et al.* (2006) used linear regression models to analyse data for the Isle Royale populations, concluding that the outbreak of CPV has shifted moose regulation from top-down (wolf) to bottom-up control, principally through climatic effects on moose and their primary winter food resource, balsam fir (*Abies balsamea*).

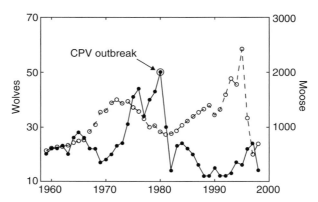

Fig. 3.9. Wolf and moose population dynamics in Isle Royale, 1959–1998. Solid line: wolves, dashed line: moose. From Wilmers *et al.* (2006).

Box 3.6 (*continued*)

Prior to CPV onset in 1980, wolf population density explained 38% of the variation in moose population growth rate, and bottom-up factors (balsam fir abundance and climatic conditions) explained 13%. After 1980 top-down effects explained only 1% of the variation, whereas bottom-up effects explained 28%. The effect of climate appears to have changed in a way consistent with its indirect effects propagating through other trophic levels of changing importance: prior to CPV, climate exerted a weak, one-year lagged effect on moose population growth rate (this would be expected if its effects were propagated via its impact on the predator), but after CPV it exerted a stronger, two-year lagged effect (suggestive of climatic effects propagating via plant growth reflected as resource availability the following year).

An important point from the example discussed in Box 3.6 is that the effects of the pathogen can outlive the epidemic: although CPV has been absent from Isle Royale for 20 years, wolf and moose populations remain in the post-epidemic pattern, possibly because genetic bottlenecks in the wolf population were generated by the crash, and these prevent recovery of the wolf population, which may now have reset to a new lower equilibrium population density (Wilmers *et al.*, 2006). The empirical and theoretical examples above suggest that in some systems parasites of predators may be an effective means of limiting the impact of predators; the potential application of this to biological control of predators is discussed in Section 3.6.2.

3.5 Parasites of predators and prey

In many cases, parasites infect both the prey species and (one or more species of) its predators. It is not particularly helpful to present a general baseline theory for this category because systems differ in a variety of key features which require different model structures; even 'baseline' models become quite complex as (at the very least) infected and susceptible classes must be considered for both predator and prey species. Instead, we start by examining empirical examples representing the different types of interaction and go on to consider the theoretical predictions for these interactions.

In a community context, the parasite-induced behavioural changes that occur in trophic transmission can be regarded as trait-mediated indirect effects (TMIEs) (Werner & Peacor, 2003) between predator and prey that result in increased exposure of hosts to predation (Mouritsen & Poulin, 2005). This type of interaction is of interest because it acts on a shorter timescale than population–density mediated effects, and so TMIEs have the potential to propagate rapidly to different trophic levels (Hatcher et al., 2006). We are still lacking a general theoretical framework for understanding TMIEs in parasite community modules, but several models of specific systems involving TMIEs are considered in this section and demonstrate the potential power of this type of effect on population dynamics and community outcomes.

3.5.1 Patterns and evolution of manipulation and trophic transmission

There are numerous examples of parasites which manipulate the behaviour of their host and thereby increase opportunities for transmission to the next host individual, as we discuss below (see also Box 4.3 and Box 7.2). Parasitic manipulation refers to changes in host behaviour that are adaptive to the parasite, as opposed to 'boring by-products' of host morbidity (Poulin, 1995; Lefevre et al., 2008). Parasite manipulation of host behaviour represents a trait-mediated indirect effect (TMIE) and has evolved many times across a range of parasite and host taxa (Moore, 2002; Thomas et al., 2005a; Lefevre et al., 2008). Parasitic manipulation frequently involves increased vulnerability to predation by the next host in a parasite's life cycle, but can also affect predation by non-host predators. Dobson (1988) distinguishes four different types of manipulation which reflect the life cycle of the parasite.

Parasites with direct life cycles (transmitted directly from one definitive host to the next, of the same species) may influence host behaviour or physiology to increase the chance of intraspecific transmission (e.g. sneezing of cold viruses). This constitutes a one-parasite–one-host relationship, and the manipulation does not directly affect predation. Therefore we do not consider this further here.

Parasites (usually microparasites) with indirect life cycles involving vector transmission; here, the vector (usually a biting insect) can be considered a 'predator' on the larger (usually bird or mammalian) 'prey'. The vector does not kill and consume its 'prey', but transmission can nonetheless be considered 'trophic' and the parasite may manipulate either or both the vector and vertebrate host to enhance transmission. For example, mice infected with malaria (*Plasmodium berghei*, *P. chabaudi* and *P. yoelii*) showed lethargy and reduced anti-mosquito behaviour. The change in behaviour is not simply a by-product of infection, but occurs only during periods of peak gametocyte infectivity, leading to increased biting by the mosquito vector (Day & Edman, 1983). Malaria also influences the biting behaviour of its vector, with mosquitoes showing reduced biting behaviour during the early stages of *P. yoelii* infection, but increased biting persistence when infective sporozoites have developed (Anderson *et al.*, 1999).

Macroparasites with indirect life cycles involving free-living larvae are not trophically transmitted as the larvae emerge from the host before actively seeking out the next host. Nonetheless, manipulation can alter predator–prey dynamics or even lead to novel predator–prey relationships. For example, the hairworm *Parasgordius tricuspidatus* causes its terrestrial cricket (*Nemobius sylvestris*) host to jump into water, where the free-living stage of the worm exits the host and swims away to seek a mate (Thomas *et al.*, 2002b). Host manipulation by hairworms has been shown to provide novel opportunities for predation of terrestrial crickets by fish (Sato *et al.*, 2008). This allochthonous resource input may play an important role in food web dynamics and energy flow; a study of streams in Japan revealed that trout (*Salvelinus leucomaenis japonicus*) ingested a greater mass of crickets (*Diestrammena* sp.) than of any other prey species.

Finally, *parasites (usually macroparasites) with indirect life cycles that utilise both members of a classical predator–prey relationship as hosts* provide numerous examples of parasite-modified predation (e.g. Lefevre *et al.*, 2008). These parasites manipulate the antipredator behaviour of the prey intermediate hosts, thereby facilitating parasite transmission to the predator definitive host. Parasite-induced vulnerability to predation (PIVP) in these systems is not simply a by-product of infection, but results from parasite-specific changes in host behaviour that coincide with the transmission stage of the parasite life cycle (Poulin, 1995). For example, the trematode parasite *Euhaplorchis californiensis* induces behavioural changes in its intermediate fish host that increase the risk of predation, and hence provide increased opportunities for trophic transmission (Lafferty & Morris, 1996). *E. californiensis* has an indirect life cycle. Several species of bird act as definitive hosts. The first intermediate snail (*Cerithidea californica*) host becomes infected after ingesting the trematode eggs. The snail host suffers castration by the parasite and produces cercariae that leave and then penetrate the second intermediate killifish (*Fundulus parvipinnis*) host, encysting in the brain. The life cycle is completed when the killifish is preyed upon by a definitive bird host. Lafferty and Morris

(1996) found that parasitised killifish exhibited more conspicuous behaviour, showing an increase in surfacing, shimmying and other behaviours that would make them more visible to predators. The authors went on to measure the effect of these TMIEs on predation and found that a relatively small (four-fold) increase in conspicuous behaviour led to a 30-fold increase in predation in the field. Such TMIEs benefit the parasite as they increase opportunities for trophic transmission to the definitive host.

Parasitic manipulation of host behaviour may also result in increased vulnerability to non-host predators. For example, the trematode *Curtuteria australis* uses the oyster catcher (*Haematopus* sp.) as its definitive host. Eggs passed out in the faeces hatch into miracidia that go on to infect the first intermediate host, a whelk. Cercariae are released from the whelk and go on to penetrate cockles *Austrovenus stutchburyi*, developing into metacercariae in the host foot. The life cycle is completed when an infected cockle is preyed upon by the definitive host. Parasite manipulation leads to reduced burrowing activity by the intermediate cockle host, and so infected cockles suffer an increase in predation by the definitive host. However, the increased surfacing behaviour of infected cockles also makes them vulnerable to a second, non-host predator, the fish *Notolabrus celidotus*, which causes non-lethal predation by cropping the foot of exposed cockles (Mouritsen & Poulin, 2003).

Parasite-modified changes in predator–prey dynamics can have ramifications throughout the community, affecting community structure and productivity. These wider community effects are considered in Chapter 7.

3.5.2 Theoretical impacts on populations and communities

A treatment of population dynamic outcomes for trophically transmitted parasites and their hosts follows the same principles as applied earlier; we derive R_0 for the parasite and the minimum population sizes N_T for parasite establishment. Most theoretical studies have concentrated on how parasite-induced changes in host traits influence R_0, N_T and population dynamic outcomes. Dobson (1988) derives these expressions for the three indirect life cycle cases discussed above; see Dobson and Keymer (1985) for a full treatment of the classic parasite–predator–prey case, and Aron and May (1982) for vector transmission.

For macroparasites that are transmitted when the definitive host eats the intermediate host, there are two thresholds (one for each host); prey and predator must exceed their respective thresholds for parasite establishment. These thresholds are dependent (among other parameters) on the level of parasite-induced vulnerability to predation (PIVP); specifically, as PIVP increases, R_0 increases and the threshold population densities decrease (Dobson, 1988). However, the predator threshold decreases more rapidly with manipulation than the prey threshold, suggesting that relatively small increases in PIVP could enable a parasite to establish when predators are rare or only occasionally visit habitat occupied by the prey (Dobson, 1988).

In the case of parasites with vector transmission, there is a minimum threshold vector population size and a maximum prey (vertebrate host) population size; if prey become too numerous, the chances of an individual vector biting both an infected and an uninfected host become too small for parasite maintenance (Aron & May, 1982); this 'dilution effect' reflects similar findings for louping ill in red grouse (Box 2.8). In this case, parasite manipulation (increasing biting rates or effective transmission per bite) increases R_0, reduces the minimum vector population threshold and increases the maximum prey population threshold – again, expanding the range of population sizes in which the parasite can be maintained.

Parasitic manipulation can also be beneficial from the predator's perspective, in some cases enabling a predator to persist that would otherwise go extinct (Hadeler & Freedman, 1989; Freedman, 1990). This occurs because PIVP makes prey more available to predators (reducing search or handling times and increasing capture rate), such that 'inefficient' predators can be sustained at lower prey population densities. Lafferty incorporated a function for energy accrual by the predator to examine the conditions under which a predator should avoid parasitised prey so as to avoid infection. Unless the energetic costs of infection are particularly high, predators are not selected to discriminate against infected prey, especially at high levels of PIVP. In this case, not only does PIVP increase 'catchability' of infected prey, but a predator that ignores infected prey must sample from a reduced population (the uninfected subpopulation), incurring greater search costs. Lafferty (1992) makes the point that in this case, parasites act as a 'delivery service for hard-to-get prey'; by enabling more efficient utilisation of energetically borderline prey species, parasites may thus be important factors determining diet breadth.

The theoretical analyses show that, assuming the parasite can establish, a stable equilibrium may be reached or the populations may exhibit long-term oscillations; a higher, unstable equilibrium is also possible

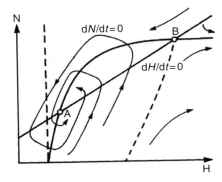

Fig. 3.10. Isocline analysis for a host–macroparasite system with a trophically transmitted parasite. H: population of the intermediate host whose behaviour is affected by parasite manipulation; N: mean adult parasite burden in definitive hosts. Equilibria occur at the intersections of the two zero-growth isoclines ($dN/dt = 0$ for parasites, $dH/dt = 0$ for intermediate hosts); equilibrium A is stable, B is unstable. Lines with arrows show the trajectory of populations near the equilibria. Left-hand dashed line shows threshold intermediate host population for parasite establishment; right-hand dashed line shows boundary between sustained population oscillations (to the left) and exponential growth of the intermediate host (to the right). For isocline analysis see Section 2.3.1. From Dobson (1988).

(Dobson, 1988; Fig. 3.10). Increasing PIVP tends to increase the amplitude of oscillations; in real-world situations this may lead to local extinction of the parasite and/or prey (and predator, if a specialist) when troughs in the cycle are reached (Dobson, 1988; Fenton & Rands, 2006). Parasitic manipulation also tends to reduce equilibrium prey abundance somewhat, whilst greatly increasing parasite prevalence in predators. At the same time, prevalence in prey is reduced, because increased PIVP results in infected prey being removed from the population more rapidly by predation. In these systems, parasite prevalence in prey may be at very low levels, hence a regulatory role for parasites cannot be discounted in these systems just because parasites appear to be rare in the intermediate host (Fenton & Rands, 2006).

The shape of the predator functional response influences the dynamics and coexistence outcomes in these systems. Theoretical analyses show regions with different dynamical behaviours and population outcomes; the boundaries between these regions depend on the type of functional response assumed (Fenton & Rands, 2006). With a saturating (Type II) functional response, there is an interaction between prey handling time and parasite manipulation: increased handling time reduces the range of parameters enabling predator persistence, but increased manipulation ameliorates this effect. Overall a Type II functional response has the effect of increasing the parameter range for coexistence with oscillatory dynamics. A sigmoid (Type III) functional response shifts the boundaries, making parasite persistence more difficult; this is because at low prey densities, the predator ignores the prey, so higher levels of manipulation are required to achieve transmission to the definitive host.

3.6 Applications: predator control and harvesting

An understanding of the potential impact of parasitism on predator–prey interactions is important for our approach to managed populations. One finding from the baseline theory of parasites of prey with density-dependent transmission is that predators can depress prey populations below the threshold size for parasite establishment. This could have two potential consequences for managed populations:

(1) If a managed species is subject to predation, predator removal or reduction (often regarded as a viable management option for protecting game or other managed populations) may lead to an increase in disease incidence.
(2) Harvesting may have a similar effect to predation; a sufficient harvest-induced reduction in host population size could result in extinction of the parasite.

It should also be noted that the baseline predictions above refer to systems with parasites of prey only; if the predator is also a host, increasing predation pressure will generally increase parasite prevalence (at least in the definitive, predator host) (Dobson, 1988; Lafferty, 1992). In this section we discuss theoretical studies that develop these ideas, and also examine the possibility of using parasites of predators as agents of biological control.

3.6.1 Do predators keep the herds healthy?

3.6.1.1 Healthy herds hypothesis

Packer et al. (2003) show that reduced predation can indeed have detrimental effects on prey populations that are also subject to parasitic infection. Assuming a prey population with a generalist predator (with a linear functional response) and parasitism, Packer *et al.* derived equations for the equilibrium abundance of susceptible (S^*) and infected (I^*) hosts. For the microparasite case, the simple relationship $S^* = d/\beta$ holds, where β is the pathogen transmission rate and d the rate at which infected individuals are removed from the population (comprising recovery and mortality). All other things being equal, in this model increased predation will always increase S^* via its effect on mortality, increasing the abundance of healthy individuals and decreasing parasite prevalence. This will occur even if the predator is unselective, although the impact is greater if predators select infected prey. Selectivity is also important in determining total host population size: if predators select infected hosts, they will increase total host population size; if they prefer uninfected hosts, they will reduce population size; and if they are unselective, then total population size remains unaltered. For macroparasites that have an aggregated distribution among hosts, the effect is more extreme. Even a low level of predation could, in theory, result in extinction of the parasite if predators concentrate on prey with a high parasite burden. Selective predation effectively increases parasite-induced mortality, and thus it tends to increase the regulatory impact of parasitism (Packer *et al.*, 2003; Holt & Dobson, 2006). Some support for this 'healthy herds' hypothesis comes from the work of Hudson et al. (1992a) on red grouse and Joly and Messier (2004) on moose; in both cases predators preferentially prey on infected individuals and parasite prevalence is lower at sites estimated to have higher predation levels. There is also evidence from marine systems that a reduction in predators has led to an upsurge in disease incidence for some marine molluscs and echinoderms, possibly because populations have grown above the threshold for establishment of previously unknown diseases (although evidence of selective predation is lacking in these systems) (Lafferty *et al.*, 2004; Ward & Lafferty, 2004).

Ostfeld and Holt (2004) examined the role of predators in the control of rodent populations, many of which harbour zoonotic diseases harmful to humans, including Lyme disease (Box 7.1) and hantavirus (Section 8.6.3). They used similar models to Packer *et al.* (2003) to demonstrate that predator removal is likely to increase both pathogen prevalence in the prey population, and also the absolute number of infected individuals (Fig. 3.11). It is this latter quantity (I^*) that is of interest in multi-host systems in which one or more species act as reservoirs for disease (Section 7.3), as it determines the frequency with which spillover of disease to another species (for instance, a domestic or conserved species, or humans) should occur via cross-species contacts.

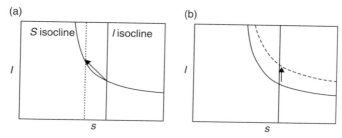

Fig. 3.11. Isocline analysis showing that predator removal increases equilibrium abundance of infected individuals. *S*: susceptible class; *I*: infected class; for a simple SI model of parasitism and predation on the host. Solid vertical line is infected isocline, with predator; dashed vertical line is infected isocline without predator. Solid and dashed curved lines are susceptible isoclines with and without predator, respectively. A stable equilibrium occurs at the intersection of the two isoclines. (a) When predators are removed, this shifts the *I* isocline left, increasing the equilibrium value for *I*. (b) When predators are removed, this shifts the *S* isocline up, also increasing the equilibrium value for *I*. From Ostfeld and Holt (2004).

Whether predators regulate disease in the prey in real-world systems is unclear. For instance, the evidence for predator control of rodents is difficult to interpret as several other factors confound the expected pattern. Human activity has resulted in the loss of many large mammalian top predators, but these species do not generally control rodent populations. Smaller carnivores such as foxes and domestic cats that prey directly on rodents may benefit from the loss of top predators; this 'mesopredator release' may limit rodent populations and their infections more severely (Ostfeld & Holt, 2004). Patterns are expected to differ depending on the type of predator involved. Specialist predators, with their population dynamics coupled to those of the prey, can elicit strong prey population cycles which may increase the amplitude of epidemics during peak prey population growth; generalist predators, which will not invoke cycles, may provide more stable long-term regulation.

3.6.1.2 ... or do predators benefit pathogens?

The healthy herds hypothesis may not, however, be widely applicable. Recent theory shows that when hosts can recover from infection and become immune, predation may have the opposite effect, increasing parasite prevalence (Holt & Roy, 2007). For a broad range of conditions, these models predict a 'hump–shaped' relationship between

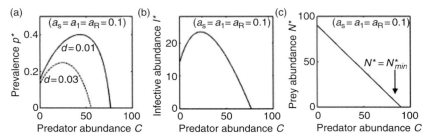

Fig. 3.12. Relationship between predator abundance (C) and parasite prevalence p^* when hosts acquire immunity after infection. Prevalence (a) and abundance of infected individuals (b) is highest at intermediate predator levels; overall host abundance (c) decreases with predator abundance. Text within plots refers to parameter values used in the analyses. From Roy and Holt (2008).

prevalence and predation rate, such that increasing predation leads to an increase in parasite prevalence at low to moderate predation pressure (Fig. 3.12). Two different mechanisms can be responsible for this pattern, depending on the structures modelled. Firstly, since predation occurs on the recovered class (along with predation on susceptibles and infecteds), if the host population is under strong density-dependent regulation, removal of recovered individuals frees up resources for births of susceptibles, indirectly increasing disease prevalence (as susceptibles are themselves the limiting resource for future infection). If, on the other hand, the host is not limited by density dependence (and is instead limited by the pathogen), a cost to immunity can drive a similar hump-shaped relationship. Selective predation alters the pattern: in general, selective predation on infected individuals will drive prevalence down; selective predation on the immune class will drive prevalence up; and indiscriminate predation will result in a hump-shaped relationship with predation pressure. Roy and Holt (2008) demonstrate that this effect can occur for a wide range of models (including predators with a saturating (Type II) functional response, specialist predators and frequency-dependent parasite transmission), and also demonstrate that the equilibrium density of infected individuals (I^*), as well as prevalence (P^*), can exhibit this hump-shaped relationship (Fig. 3.12). Hence, it may not be easy to generalise about the relationship between predation and parasite prevalence without knowledge of parasite, host and predator functional and numerical characteristics.

3.6.2 Biological control

The above discussion focuses on how predator management may indirectly control infection in the prey population. We now turn to alternative applications in biological control, which focus on the direct effects of parasites on predators (and vice versa). Firstly, we examine the use of parasites to directly control predator populations, as a form of classical biological control. In some cases, the opposite application may be feasible; namely, use of predators to directly control parasites. This approach may have potential for vector-borne diseases, where predator control of the vector is possible.

3.6.2.1 Using parasites to control predators

Where other control or management strategies are not cost-effective or practical, parasites may provide an alternative control agent. Biological control using pathogens has been used or proposed for protection of native species in several island ecosystems against feral cat (*Felis catus*) populations. This predator is responsible for some of the most devastating biotically induced losses of diversity ever documented. For instance, from a population size of five, cats introduced in the Kerguelen Islands (southern Indian Ocean) reached several tens of thousands within 50 years (Chapuis *et al.*, 1994); similarly, five cats accidentally introduced to Marion Island (southern Indian Ocean) in 1949 grew to over 2000 individuals over 25 years, causing the local extinction of the common diving petrel (*Pelecanoides urinatrix*). In these and other remote oceanic islands, cats have caused severe depletion of many indigenous vertebrate populations, especially reptiles and birds (Nogales *et al.*, 2004). Because these island cat populations have originated from small founder populations, they tend to be largely free of parasites; this enemy release may have contributed to their successful invasion – see Section 6.3. However, it also means that much of the genetic resistance to parasitism may have been lost, making these populations potentially suitable targets for biological control using pathogens. Cats were eventually eliminated from Marion Island in the 1990s using feline panleukopenia virus (FPLV) in combination with intensive culling (Nogales *et al.*, 2004).

Traits for an effective biological control pathogen include intermediate virulence, so that the pathogen does not 'burn out' before spreading throughout the host's range (see Box 2.1). This may explain why FPLV, a highly virulent pathogen, did not eradicate cats in the absence of other measures; after reducing the population to around 600 individuals, cat population density became too low for effective transmission and FPLV was lost (Courchamp & Sugihara, 1999). Two feline retroviruses (FIV, feline immunodeficiency virus; and FLV, feline leukaemia virus) have been proposed as biological control candidates for remote island systems such as the Kerguelens. Models indicate that these viruses could be more effective as they both persist for a long time before killing the host, allowing multiple rounds of transmission (Courchamp & Sugihara, 1999). However, a recent model for control by FIV predicts that introduction of the pathogen could induce predator–prey population cycles, which could have counter-productive consequences for prey conservation. On the one hand, oscillations bring the predator to low population densities, facilitating final eradication by other means such as culling; however, oscillations induced in the prey increase the extinction risk for these populations though stochastic effects on small populations (Oliveira & Hilker, 2010).

3.6.2.2 Using predators to control parasites

Conversely, biological control of parasites may be possible using predators. Predation may be used to control parasitism through control of vector populations, using introduced arthropod predators or natural enemies. Using predators to control vectors has been proposed for a number of diseases, including malaria, dengue fever and Lyme disease (reviewed in Moore *et al.*, 2010). Like the threshold host population size N_T, a threshold vector population size V_T must be met for establishment and maintenance of a vector-borne parasite. If predation can regulate the vector below V_T, the parasite will be eradicated. This strategy has been investigated theoretically by Moore *et al.* (2010), who show that predation dramatically expands the parameter range for exclusion of the pathogen. This occurs because addition of a predator reverses the relationship between pathogen prevalence and vector fecundity: without predators, increased vector fecundity increases pathogen prevalence in the host, but with predation, higher vector fecundity results in reduced prevalence. This effect reduces R_0 for the pathogen, increasing the threshold V_T. The combined effect of increasing V_T and reducing actual vector population densities results in exclusion of the pathogen (Fig. 3.13). Moore *et al.* point out that even rather limited predation may be beneficial in control: because predation of the vector strongly influences R_0, initial spread of the pathogen (perhaps an emerging disease) will be reduced, providing valuable time for other vector or pathogen control measures to be implemented.

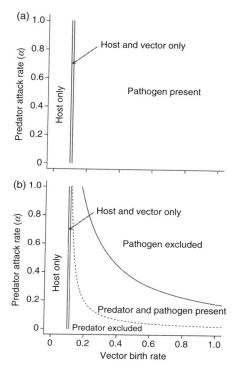

Fig. 3.13. Coexistence outcomes for a vector-borne pathogen system with predation. Outcomes are plotted in relation to vector birth rate and predator attack rate. (a) With no predation, there is a very little likelihood of excluding the pathogen if the vector is able to establish. (b) With predation, the pathogen is excluded for high vector birth rates combined with moderate or high predation rates. From Moore *et al.* (2010).

3.6.3 Harvesting infected populations

As a means of consumption of a 'prey' class, harvesting has similarities to conventional notions of predation, so it is reasonable to consider whether parasitism can interact with harvesting as it can with predation. However, harvesting is not the same as predation for several reasons: there is no numerical response of harvesting to 'prey' population density (unlike a specialist predator); harvesting does not generally preferentially select the infected subclass (indeed, the reverse may be true); harvesting generally concentrates on the adult class, whereas many predators may concentrate on juveniles; harvesting is often heavily constrained seasonally, but predation is usually less so. Despite these differences,

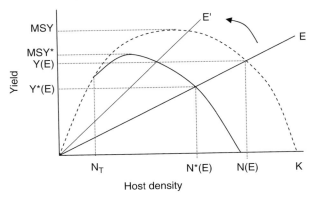

Fig. 3.14. Yield–effort curves for fixed-effort harvesting of an infected fishery. Recruitment curves in the presence (solid) and absence (dashed line) of parasitism cross a fixed-effort harvesting line with effort E, resulting in a stable equilibrium yield – Y*(E) in the presence of parasitism, Y(E) in its absence. Yield under effort E is reduced by parasitism, as is the theoretically maximum sustainable yield (MSY without and MSY* with parasitism). Increasing effort to E′ to achieve the pre-parasitism yield may take the population close to or exceed its MSY* in the presence of parasitism. N_T: threshold for parasite establishment; N*(E): population sizes with harvesting. Modified and redrawn from Dobson and May (1987).

several parallels can be drawn between the theoretical predictions for the effect of parasitism on harvesting and predation, as we examine below.

Dobson and May (1987) adapt fixed-effort harvesting models to consider the effect of parasitism on fisheries, deriving the threshold for disease establishment N_T and comparing this to the equilibrium population size for the exploited host N(E). If N_T exceeds N(E), the parasite cannot persist in the exploited population – it is 'fished out' (Lafferty *et al.*, 2004) as harvesting reduces the host population below the disease establishment threshold. A reduction in normalised reports of disease incidence in several marine fish species appears to support this prediction (Lafferty *et al.*, 2004; Ward & Lafferty, 2004). In the case of an invading pathogen, if N_T is lower than N(E), the pathogen is predicted to establish initially, but if it is sufficiently pathogenic, it will drive host population density down, resulting in a new yield–effort relationship (Fig. 3.14). If harvesting continues at the same rate, a reduced yield will be obtained; for most simple models of recruitment, the maximum sustainable yield (MSY) will also be reduced as a result of parasite-induced mortality. If harvesting effort is raised in order to match the yield in the absence of parasitism, the population may be taken beyond its adjusted MSY in the presence of parasitism; if too great an increase in harvesting effort is applied, it can result in fishery collapse and deterministic extinction of the host. Hence, caution should be exercised with harvesting regimes so that unrecognised (perhaps emerging) disease threats do not have catastrophic effects on the host population in combination with harvesting.

The above models assume no harvest selectivity with respect to infection status or age structure (with which infection status is likely to covary); conditions that may not generally be upheld in a variety of fisheries, particularly where infection affects the quality or marketability of the stock (as is the case for many macroparasites of teleost fish). Kuris and Lafferty (1992), in a model of crustacean fisheries subject to a parasitic castrator (of which there are several economically relevant species), showed that parasite and host life history and management practice can be crucial in determining outcomes. Since castrating parasites of crustaceans only infect females (or feminise males), the standard fishery practice of releasing (gravid) females to boost breeding stocks can exacerbate the effects of parasitism. The best practice to reduce prevalence and increase population growth is, in contrast, to keep and cull infected females, or treat and release them (if possible).

Predicting the outcomes of harvesting in the presence of parasitism is complicated by the scale at which host and parasite recruitment occurs. For enclosed fisheries or reserves free from harvesting, either parasite or host (or both) may be recruited entirely within the enclosure (in which case they have 'closed' dynamics), or they may immigrate into the focal area (in which case, recruitment is 'open' and models must take account of the dispersal process). One consequence of open recruitment is that it may be harder to 'fish out' a parasite as it can immigrate from areas beyond the fishery. For instance, McCallum et al. (2005) show that marine reserves may alter the impact of pathogens on fisheries. These authors show theoretically that reserves can make it harder to 'fish out' a pathogen because the reserve offers a refuge for the parasite in a high-density host population. However, reserves also greatly reduce the sensitivity of host populations to overharvesting, which can be particularly problematic in the presence of a pathogen.

As was the case for the 'healthy herds' hypothesis, recent theory suggests that the 'fishing out' hypothesis (that harvesting will decrease parasite prevalence) does not necessarily hold. In some situations, protected reserves can alter the dynamics of harvesting such that the MSY is higher in the presence of parasitism, due to overcompensation in the density-dependent dynamics of the host (McCallum *et al.*, 2005). Choisy and Rohani (2006) show that harvesting can interact with density dependence in the host to increase disease prevalence P^* and incidence I^*. Key to their assumptions is the inclusion of an immune class, and to some extent the dynamics of this model are driven by similar processes to those underlying the predation model of Holt and Roy (2007): harvesting removes recovered individuals, allowing more births into the susceptible class, which eventually become infected. For seasonal populations, the timing and duration of harvest can have a strong impact on population dynamics without parasitism, in some cases leading to overcompensation of the host population, whereby harvested populations equilibrate at a higher density (Kokko & Lindstrom, 1998). Choisy and Rohani (2006) show that when parasitism is included, the optimal time to harvest needs to be shifted earlier to avoid exacerbating disease epidemics (Fig. 3.15). These models show that the abundance of infected individuals I^*, as well as prevalence P^*, can be increased under harvesting, so careful consideration should be given to harvesting regimes where disease spillover to other species is a concern.

3.7 Conclusions

A general prediction for the outcome of interactions between parasites and predators is hard to make, because many different patterns are predicted (and have been observed). The general models discussed in this chapter predict that parasites may, on the one hand, enhance the population dynamic stability of predator–prey interaction, but in other circumstances decrease it. Equally, predation (or harvesting) may either increase or decrease parasite prevalence and either increase or decrease disease incidence.

An understanding of this important 'interaction between interactions' (Moller, 2008) is clearly relevant for a range of practices in ecology, some of which we have discussed in Section 3.6 (harvesting, predator control of parasites and parasite (bio)control of predators). These processes have further ecological implications which are discussed in more detail later, including ramifications for community structure (Chapter 7) and emerging diseases (Chapter 8). For instance, if predator removal enhances disease incidence in a focal reservoir species, it increases the risk of disease spillover into other species, thus influencing the early stages of disease emergence

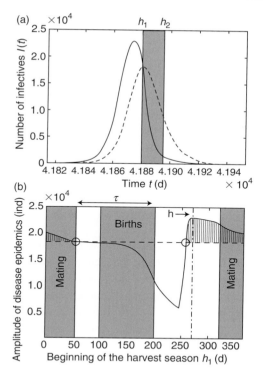

Fig. 3.15. Effect of harvest timing on disease epidemics. (a) Example showing that harvesting can increase the size of an epidemic, and shift its timing; lines plot number of infected individuals with (solid line) and without (dashed line) harvesting. Harvest occurs for a short period each year (shaded, marked h). (b) The effect of harvest timing on the amplitude of disease epidemics. Host mating and birth occurs during the periods marked and shaded; τ represents the gestation period. Lines plot the peak number of infected individuals in the presence (solid line) and absence (dashed horizontal line) of harvesting. Circles show cases of disease full compensation and hatched areas show disease overcompensation (i.e. larger epidemics under harvesting). Harvesting after the birth season and before the next mating season most strongly reduces epidemic size; harvesting just before the mating season actually increases epidemic size. From Choisy and Rohani (2006).

(Holt & Dobson, 2006). Alternatively, the process may be coupled differently, and *increased* harvesting (or predation pressure) may enhance the likelihood of spillover (Choisy & Rohani, 2006).

In order to predict how parasitism will influence a particular predator–prey (or harvest) system, we need to answer the following nine questions:

(1) Is the host population subject to intrinsic density-dependent regulation (and if so, is carrying capacity relatively high or low)?
(2) Is the predator a specialist or generalist?
(3) What shape is the predator's functional response to prey density (or, what is the harvest strategy)?
(4) Does the predator (or hunter) select or avoid infected prey, or are they indifferent to infection status?
(5) Is there seasonal variation in the timing or rate of harvesting (or predation)?
(6) Is parasite transmission density or frequency dependent?
(7) Is the predator also a host to the parasite?
(8) Does the host develop long-lasting immunity to this infection?
(9) Could evolution of parasite or host occur on an ecological timescale?

Data to address, or possibly answer, these questions can be difficult or practically impossible to obtain, so the general utility of parasites in an applied context remains uncertain. Above all, this area would benefit from further robust applied systems (particularly for harvesting and biological control) with which to test general theory or facilitate development and testing of system-specific models. The work on snowshoe hares (Box 3.1), red grouse (Box 3.2) and others in this chapter provide good examples of this approach. The match between a general model prediction (for example, a parasite-induced switch from top-down to bottom-up regulation (Section 3.4.1)) and data (for Isle Royale moose and wolves in Box 3.6) is also very encouraging, but insufficient as a base for a general theory of the interaction between predation and parasitism. Hence a concerted effort to develop the link between theoretical and empirical models is essential to better understand the impact of parasitism on this most fundamental of interactions.

4 · *Parasites and intraguild predation*

4.1 Introduction

Intraguild predation (IGP) is predation among species that are also potential competitors (Polis *et al.*, 1989; Holt & Polis, 1997). In food web terms, IGP is closely related to omnivory, the utilisation of resources from more than one trophic level. However, for an omnivore to be engaged in IGP there must also be an explicit trophic link between the omnivore's prey species (Fig. 4.1). IGP can be unidirectional (i.e. A eats B only), or bidirectional (A eats B, B eats A). With bidirectional IGP, predation rates may be roughly symmetric between the species, or one species may consistently dominate predation of the other (Polis *et al.*, 1989; Dick *et al.*, 1993).

Recent analyses of real food webs indicate that IGP is widespread and important in the structuring of communities (Arim & Marquet, 2004; Bascompte & Melian, 2005; Thompson *et al.*, 2007). It often occurs among closely related species and is frequently associated with cannibalism (Dick *et al.*, 1993; Rudolf, 2008). However, unrelated taxa can also engage in IGP (Hochberg & Lawton, 1990), as its definition extends to similarity of resources used regardless of taxonomy or mode of resource

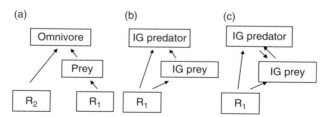

Fig. 4.1. IGP and related modules. Arrows indicate the direction of energy transfer between species. (a) Omnivory without IGP (if the prey shares a resource R_1 or R_2 with the omnivore, this becomes IGP); (b) unidirectional IGP, the classical 'predator–prey' form; (c) bidirectional ('mutual') IGP, with relative predation rates between the more predatory (IG predator) and less predatory (IG prey) species as indicated by size of arrow.

acquisition (Polis *et al.*, 1989). IGP appears to be particularly prevalent where potentially competing species have age or stage structure; smaller stage classes are eaten by the larger classes of species that consume the same general class of resource; in such cases many species may be involved in IGP to varying degrees (Polis & Holt, 1992; Dick *et al.*, 1993).

4.1.1 Parasitism and IGP

Parasites, parasitoids and pathogens are frequently involved in IGP, often with each other. For instance, pathogen–parasitoid interactions in biological control using multiple control agents can frequently be regarded as IGP (Rosenheim *et al.*, 1995). In this class of interactions, the parasite is a participant of IGP, being one of the guild members exploiting the shared resource (Fig. 4.2a); examples are considered in more detail in Section 4.4. Some parasitic plants also engage in IGP with their hosts; hemiparasites (which utilise a host species but also photosynthesise) compete with their hosts for resources (water, light) at the same time as they consume them (Section 5.4). Parasitism can also interact with established IGP relationships, wherein the parasite infects either or both of the consumers (IG predator or prey) of the shared resource (Fig. 4.2b); this relationship is discussed in Section 4.5. Many predation–parasitism examples discussed in Chapter 3 can also be considered in the context of IGP. Predators and parasites sharing a host/prey species are members of the same competitive guild, and predators frequently eat infected prey. This type of 'coincidental IGP' (Polis *et al.*, 1989) may be common and has usually been considered in the context of predation–parasitism interactions (Chapter 3), as the predator was thought unlikely to gain much energetically from consuming the parasite (a key requirement in the original models of IGP (Table 4.1)). However, recent work on the ecosystem roles of parasites suggests that parasites can contribute

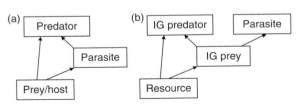

Fig. 4.2. How parasites enter into IGP modules. (a) Parasite intrinsic to IGP (in this case, a parasite and predator share the resource and IGP occurs when infected hosts are consumed; (b) parasites extrinsic to IGP (here, a parasite of an IG prey species shares this resource with the IG predator).

Table 4.1. *Key predictions for IGP from theory (T) and empirical evidence (E)*

Prediction	T: theory E: evidence
1	Coexistence is only possible for a limited range of parameter values and requires that the IG predator is the inferior competitor and benefits substantially from consuming IG prey	T: Holt and Polis (1997) E: Vance-Chalcraft *et al.* (2007; not always supported)
2	The IG prey can invade and persist at a lower productivity than the IG predator	T, E: Dick and Platvoet (1996); T: Holt and Polis (1997) E: Diehl and Feissel (2000)
3	At high productivity, the IG prey is excluded via apparent competition with the IG predator	T: Holt and Polis (1997) E: Diehl and Feissel (2000; not always supported)
4	From 2–3, community structure outcomes vary along the gradient of resource enrichment	T: Holt and Polis (1997) T, E: Diehl and Feissel (2000) E: Borer *et al.* (2003); Vance-Chalcraft *et al.* (2007; not always supported)
5	An IG predator (omnivore) can invade resource-consumer systems at lower productivity than a strict top predator	T: Diehl and Feissel (2000); Mylius *et al.* (2001)
6	IG prey are held to a lower equilibrium density in an IGP (omnivory) module compared to an otherwise identical three-level food chain	T: Diehl and Feissel (2000)

substantially to the diet of predators (Section 7.6), so this assumption needs to be treated with caution. Furthermore, consumption by predators is not always 'coincidental' from the parasite's perspective; trophically transmitted parasites also engage in IGP with the predators of their intermediate hosts (Section 4.1.2). Regardless of the level of direct energy transfer, because the predator is consuming a potential competitor when it eats an infected prey, the dynamics of this interaction differ from those for a linear food chain. In this chapter, we start with a general review of the population outcomes expected in IGP modules, before examining the role of parasitism in IGP in more detail.

4.1.2 Predictions from basic IGP theory

Most mathematical analyses of IGP have considered the dynamics of two competing species engaged in unidirectional IGP (the IG predator and IG prey (Fig. 4.1b)) and their shared resource (Table 4.1). Models predict that IGP can produce complex patterns of population dynamics, including oscillations and unstable equilibria with species exclusion, coexistence or alternative stable states (for a full characterisation, see Ruggieri & Schreiber, 2005). The possibility of alternative stable states (see Table 2.2 for definitions) suggests that the eventual outcome in IGP modules will depend on initial conditions such as the order of arrival of species or their initial propagule size; this might translate into patchwork distributions of particular IGP assemblages within structured habitats (Polis & Holt, 1992).

The outcomes for IGP associations depend on the relative strength of IGP, the relative efficiency of exploitation of the basal resource by the competitors and the relative amount of basal resource available (i.e. degree of enrichment). Most models of IGP in isolation predict IGP persistence over a very restrictive range of parameters; in most instances, exclusion of either the IG predator or IG prey is expected.

One key prediction is that for IGP to persist, the IG prey must be a more efficient utiliser of the shared resource than the IG predator (Table 4.1). Several key points stem from this prediction: firstly, below a critical level of basal resource the IG prey can persist but the IG predator will be extinguished. Secondly, at high resource levels the IG prey will be eliminated via predation by the IG predator; this is effectively apparent competition mediated via the shared resource (Chapter 2). Consequently, the outcome of IGP associations will vary along the gradient of enrichment of the basal resource, running through R (resource) only, to C–R (consumer–resource) communities, to C–P–R (consumer–predator–resource) and eventually P–R communities, as productivity increases (Holt & Polis, 1997; Diehl & Feissel, 2000).

There is some limited and equivocal empirical support for these predictions (Table 4.1). For instance, in laboratory experiments involving two ciliates (*Tetrahymena*, which consumes bacteria, and *Blepharisma*, which consumes bacteria and *Tetrahymena*), *Tetrahymena* persisted alone at the lowest enrichment level, the ciliates coexisted at an intermediate enrichment and *Blepharisma* persisted alone at higher enrichments, in accord with the theory (Fig. 4.3). However, at the highest enrichment level, *Tetrahymena* persisted alone contrary to theoretical predictions (Diehl & Feissel, 2000).

In this study, the two ciliates coexisted in only one treatment, bearing out the basic theory which predicts that IGP will be intrinsically unstable in most cases. However, IGP appears commonly in nature. This has led to the widespread feeling that the basic models of IGP are too simplistic and that

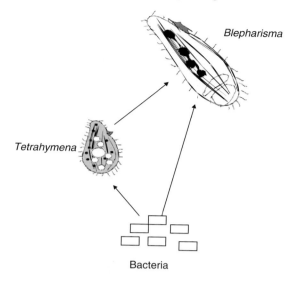

Fig. 4.3. A simple IGP module. The ciliate protozoa *Blepharisma* (the IG predator) also consumes *Tetrahymena* (the IG prey); both consume bacteria (the shared resource).

additional features of population structure or species interactions need to be considered (Rosenheim, 2007). Several recent extensions demonstrate a variety of features that can help stabilise the interaction, increasing the range of parameters that allow long-term persistence of IGP (Box 4.1). Despite these extensions, for much of the biologically feasible parameter space, exclusion of either IG prey or IG predator is still predicted.

4.2 Ecological significance of IGP

A meta-analysis of 20 of the most-studied food webs concluded that, above the levels of basal resource and herbivory, there was substantial omnivory: taxa were no more likely to occupy a 'whole number' (i.e. integer or unitary) trophic position than they were in randomised webs (Thompson *et al.*, 2007). With such a high frequency of omnivory, it is highly likely that IGP is also common, as supported by other food web analyses that take a more explicit approach to the search for IGP (Bascompte & Melian, 2005; van Veen *et al.*, 2006). From this analysis, community structure might be better characterised as a 'trophic tangle' (Thompson *et al.*, 2007) rather than the classical view of a collection of linear food chains.

Box 4.1 Factors influencing species coexistence in IGP

Several extensions to the basic IGP module have been proposed to improve its realism; these also tend to widen the range of conditions leading to the coexistence of IG predator and IG prey (Table 4.2).

Table 4.2. *Extensions to basic IGP models that enhance IGP persistence*

Extension	Prediction	Reference
Invulnerable class of consumers (IG prey)	Qualitative outcomes unchanged; slight increase in range for coexistence	Mylius *et al.* (2001)
Non-predatory class of IG predator	Increased range for coexistence; qualitative outcomes unchanged	Mylius *et al.* (2001)
Stage-structured cannibalism by predator	Increased range for coexistence: IG predator need not be inferior competitor	Rudolf (2007)
Stage-structured cannibalism by prey	Increased range for coexistence: removal of IG predator need not reduce basal resource	Rudolf (2007)
Alternative resource for IG predator	Reduced range for coexistence; IG predator need not benefit from IG prey to persist	Holt and Huxel (2007)
Alternative resource for IG prey	Increased range for coexistence: IG prey need not be superior competitor on shared resource	Holt and Huxel (2007)
Spatial refuge for IG prey	Reduced impact of IG predators can potentially stabilise IGP	Janssen *et al.* (2007)
Shared parasite of IG predator and IG prey	Increased range for coexistence if impact of parasite greater on dominant species (IG predator)	Hatcher *et al.* (2008)

Box 4.1 (*continued*)

These additions help to stabilise IGP because they decouple or weaken links between the focal players (IG predator, IG prey and resource), in some cases by introducing additional trophic interactions or levels (Fig. 4.4). Alternative resources (Fig. 4.4a, b) and invulnerable classes (Fig. 4.4c, d) weaken the feed-back loop between resource, prey and predator, reducing the instability inherent in IGP (Tanabe & Namba, 2005). The inclusion of stage-structured cannibalism in either IG prey or predator (Fig. 4.4e, f) introduces additional trophic structures which can reverse the energetic relationship between the IG predator and basal resource, reversing the indirect impact of top predators on the bottom of the trophic cascade. Species extrinsic to the IGP association, such as parasites of the IG predator and/or prey, can also help to stabilise the association; this type of interaction is considered in more detail in Section 4.5.

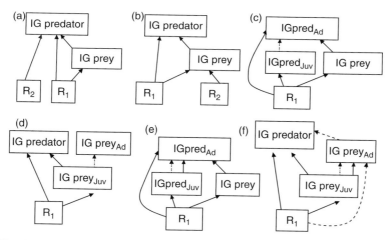

Fig. 4.4. Module extensions that affect persistence of IGP. Alternative resources for (a) IG predator and (b) IG prey; stage structure with (c) non-predatory juveniles (IGpred$_{Juv}$) and (d) an invulnerable adult prey class (IGprey$_{Ad}$); stage-structured cannibalism by (e) intraguild predators or (f) intraguild prey. Solid lines indicate direction of energy flow through consumption; dotted lines indicate developmental transitions; dashed lines in (f) indicate additional likely food links.

Much of trophic theory based on the assumption of linear food chains does not apply if there are significant levels of IGP or omnivory. For instance, IGP can disrupt trophic cascades (Polis & Holt, 1992). When a top predator takes a substantial share of its prey from trophic levels lower than its immediate neighbour, this can alter the sign as well as the strength of the interaction between the resource and intermediate predator. In addition, IGP is predicted to regulate the IG prey to a lower population density than predators would regulate consumers in a linear food chain (Fig. 4.5). Consequently, total community biomass may be less with omnivory than in a food chain (Diehl & Feissel, 2000). Furthermore, IGP models predict competitive displacement of the predator at low productivity and removal of the IG prey at high productivity (Mylius *et al.*, 2001), whereas most simple food chain models predict coexistence (Oksanen *et al.*, 1981) or exclusion of the top predator (Abrams & Roth, 1994) at high productivity.

A consequence of IG predators being inferior competitors is that they are expected to indirectly increase the equilibrium abundance of the shared basal resource; if we remove the IG predator, the IG prey should drive the resource to a lower equilibrium level (Fig. 4.5). There is some

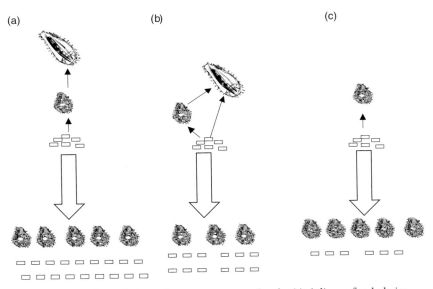

(a) (b) (c)

Fig. 4.5. The effect of IG predators on resource levels. (a) A linear food chain; (b) IGP: the intermediate predator (here, the protozoan *Tetrahymena*) is limited to a lower level via the numerical response of the top predator (*Blepharisma*) to basal resource (bacteria); (c) if the top/IG predator is removed, intermediate predator density can increase, limiting the basal resource to a lower level.

support for this from empirical studies of terrestrial invertebrates. For example, field enclosure experiments on salt marsh aphids *Dactynotus* sp. on marsh elder *Iva frutescens* in the United States showed that parasitoids *Aphidius floridaensis* were more effective at reducing aphid densities than the coccinellid *Cycloneda sanguinea*. Inclusion of *C. sanguinea* reduced the efficacy of aphid suppression by the parasitoid (Ferguson & Stiling, 1996). However, since this experiment was only run for one generation, behavioural interactions may be responsible for the pattern. Polis and Holt (1992) discuss several examples from fisheries management where introduction of a new species as fishing stock or as an additional resource for the stock species failed to have the desired outcome as the influence of the competitive components of IGP was ignored (the IG predators were out-competed as juveniles and thus were eliminated before they undertook substantial predation). In a meta-analysis of empirical studies, Vance-Chalcraft *et al.* (2007) found good evidence for stronger resource suppression by the IG prey in terrestrial invertebrate systems, but the pattern was highly variable in aquatic systems. As these authors point out, most of these empirical systems could not be studied at equilibrium, thus their relevance to equilibrium-based theory is debatable.

The relationship between IG predators, IG prey and resource levels has ramifications for biological control. This is particularly so where multiple control agents are being considered, as these may engage in IGP resulting in less-than-additive control levels or disruption of control (Section 4.4.2). Some authors (e.g. Rosenheim *et al.*, 1995) consider that the combined introduction of several control agents almost inevitably results in some degree of IGP. This problem has also been considered from the perspective of interference competition (Hochberg *et al.*, 1990; Box 2.9). Mechanisms and definitions for IGP and strong interference competition can overlap, especially where parasites are concerned (Holt & Lawton, 1993). For instance, in many pathogen–parasitoid–host systems, the pathogen can infect a parasitised host; subsequently, the parasitoid fails to complete its development and is effectively killed by the pathogen. Although this has traditionally been viewed as interference competition, from a module-level perspective it is identical to IGP (Section 4.4).

4.3 IGP as a unifying framework for competition and predation

There are clearly strong links between IGP and interspecific competition on the one hand and classical predation on the other. One perspective

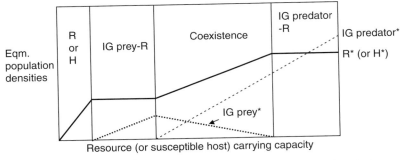

Fig. 4.6. Coexistence outcomes for three generalised IGP systems in relation to resource enrichment. The solid, dashed and dotted lines indicate how equilibrium population densities (as marked, denoted with a *) respond to enrichment. Classical predator–prey IGP, parasitoid–hyperparasitoid–host and pathogen–pathogen–host systems behaved in similar ways along the enrichment gradient. As enrichment is increased, all systems transit through a resource-only equilibrium (R* or H*) to an IG prey/resource joint equilibrium to full coexistence (IGprey/IG predator/resource joint equilibrium) then exclusion of IG prey (IG predator/resource joint equilibrium). Modified and redrawn from Borer *et al.* (2007a).

considers coexistence outcomes along the gradient of resource enrichment. At low productivity, the IG predator cannot invade because it loses out to interspecific competition with the IG prey; at the other extreme, the IG prey is excluded via apparent competition driven by enrichment of the shared resource. At this upper limit, the resource supports predators in sufficient abundance that they exert an overwhelming predation pressure on the prey population. Borer *et al.* (2007a) show theoretically how predator–prey, host–parasitoid and host–pathogen systems can all fall within the remit of IGP and respond in a qualitatively similar way to enrichment (Fig. 4.6). However, caution is needed when interpreting model predictions for outcomes in IGP systems. Although Borer *et al.*'s model shows that disparate IGP systems exhibit similar general behaviour (Fig. 4.6), these systems differ in their dynamical properties. In particular, stability of the three-species coexistence equilibrium depends on the type of association. For the predator–prey system, the joint equilibrium is generally stable (at least, for the parameters investigated by Borer *et al.*); the parasitoid–parasitoid joint equilibrium is always unstable, whereas the pathogen–pathogen joint equilibrium is always stable. Stability differences arise because of inherent differences in the length and strength of the feed-back loops between the three species in the different systems. The parasitoid system contains inherent destabilising lags resulting from the conversion of juvenile stages (i.e. the host–parasitoid complex) to adult parasitoids. Unlike the other systems, pathogen-infected hosts can reproduce and may also recover; this results in weaker feed back between pathogen prevalence and host demographics, reducing the impact of IGP and stabilising the interaction. An unstable equilibrium implies that, if the system is perturbed, it will move away from this equilibrium point so three-species coexistence may be unlikely to persist in real host–parasitoid systems without the stabilising influences of other factors (Box 4.1).

Another way of looking at the outcomes for an IGP module is to consider a gradient of basal resource *utilisation* by the (IG) predator (Mylius *et al.*, 2001). As the IG predator becomes more efficient at exploiting the basal resource, the IGP system becomes dominated by interspecific competition; conversely, less efficient exploitation by the IG predator moves the system towards the status of a linear food chain.

Food web analyses suggest a link between IGP and apparent competition; both appear to be over-represented and tend to co-occur in real food webs (Bascompte & Melian, 2005; van Veen *et al.*, 2006). This link is also clear from theoretical work showing that the inclusion of alternative resources, which can stabilise IGP (Table 4.2), can also lead to degeneration of IGP via apparent competition (Holt & Huxel, 2007). On the one hand, alternative resources for the IG predator can lead to IG prey exclusion via apparent competition (similar to the process discussed above in enriched environments). On the other hand, alternative resources for the IG prey can improve the chances of IG predator–IG prey coexistence, but the shared resource may be excluded. In this case, the module decays into a linear food chain via apparent competition resulting from the numerical response of the IG prey to the alternative resource.

4.4 Parasites intrinsic to IGP

Pathogens, parasitoids and predators can engage in IGP, potentially in any combination. Here we consider the more-common forms where parasites are participants in IGP (Fig. 4.2a). Most recognised examples of IGP involving parasites come from the biological control literature, which we consider in Sections 4.4.2 and 4.4.3. However, we begin by examining an interaction that has been neglected from the perspective of IGP, that of trophic parasite transmission (Section 4.4.1).

4.4.1 IGP in trophic transmission

Parasites with indirect life cycles are often trophically transmitted from the intermediate to the definitive host. These trophically transmitted parasites utilise predation of the intermediate host as a transmission route to the definitive host, where parasite reproduction occurs. If this transmission route is successfully completed, the parasite is effectively an IG predator of the definitive host. In this scenario the definitive host is the IG prey and the intermediate host is the shared resource. On the other hand, intermediate hosts are frequently consumed by non-host species, in which case the parasite is consumed as well. In this case, the parasite is the IG prey. To our knowledge, trophic transmission has not been treated as a canonical IGP module, although the consumption of parasites in general is mentioned as an example of IGP in the original review of this subject (Polis *et al.*, 1989). In many respects, the trophic transmission module in particular provides a good case study for the development of ideas, providing a rich variety of empirical systems and range of relationships between classical 'parasite' and 'predator' roles.

A system studied in our laboratories provides an example of these processes. The acanthocephalan parasite *Echinorhynchus truttae* utilises amphipod species of the genus *Gammarus* as its intermediate host (Box 7.2). Transmission to the definitive host, a salmonid fish, occurs when the fish prey upon an infected gammarid (Fig. 4.7). Parasite cystacanths in *Gammarus* are conspicuously coloured, and this, combined with

Fig. 4.7. The life cycle of the acanthocephalan *Echinorhynchus truttae*. Adult worms are found in the definitive host, a salmonid fish. Parasite eggs are egested in the faeces and are subsequently eaten by the intermediate gammarid host, where they develop into a cystacanth. Transmission to the definitive host occurs when it preys upon an infected gammarid.

parasite-induced changes in activity and photophilic behaviour, enhance the likelihood of predation by the definitive host (MacNeil *et al.*, 2003d).

4.4.1.1 When trophic transmission succeeds

When an infected gammarid is consumed by a host fish species, *E. truttae* becomes the IG predator and the fish is the IG prey (Fig 4.8a). The parasite here is a top predator (or 'top parasite' (Grenfell, 1992)); a position reflected in food webs of this system (e.g. Fig. 1.1). This is an example of IGP with the parasite as IG predator because:

- *E. truttae* accrues a significant portion of its resources from the IG prey (the fish); indeed, the parasite cannot reproduce without infection of its definitive host.
- The IG predator and IG prey potentially compete for resources (gammarids); indeed, sharing a resource (the intermediate host) with a predator is a prerequisite of all trophic transmission strategies.

4.4.1.2 When trophic transmission fails

From the parasite's perspective, inhabiting an intermediate host runs the risk of being consumed by the wrong predator. For instance,

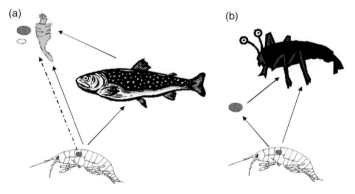

Fig. 4.8. Trophic transmission modules. (a) Successful transmission of the parasite (here, an acanthocephalan) to the definitive host (a salmonid fish) results in the parasite being an IG predator. (b) Failure of trophic transmission: when the infected intermediate host (an amphipod) is consumed by a non-host (crayfish), the result is coincidental IGP for the parasite, which becomes IG prey. Solid lines depict the direction of energy flow; the dot-dashed line depicts developmental transition between parasite life stages.

gammarids are frequently consumed by other gammarids, non-host fish, waterfowl and larger invertebrates such as crayfish. Under these circumstances, the relationship is reversed, with the parasite the IG prey, consumed by the IG predator along with its intermediate prey, the shared resource (Fig. 4.8b). This represents the opposite extreme of IGP from that above, as it is entirely 'coincidental' in nature (Polis *et al.*, 1989). Such coincidental IGP is likely to occur frequently, and can involve parasites with direct or indirect life cycles. For instance, in the amphipod system in Northern Ireland, the native gammarid *G. duebeni celticus* is host to the directly transmitted microsporidian parasite *Pleistophora mulleri*. Infected *G. duebeni celticus* are consumed by the same diverse array of invertebrates, fish and birds, in each case resulting in the probable demise of the parasite, which is specific to *G. duebeni celticus*. *P. mulleri* infects the abdominal musculature, resulting in reducing host motility. This makes infected *G. duebeni celticus* more vulnerable to IGP by other gammarids (Box 4.3), and could also influence risk of predation by other species as it is likely to influence host microdistribution within habitats (MacNeil *et al.*, 2003a). Recent research suggests that such 'accidental' predation might be an underestimated factor in the energy flow of food webs, and in some cases parasite stages form a sizeable part of a predator's diet (reviewed in Johnson *et al.*, 2010).

4.4.1.3 Countermeasures to coincidental transmission

Parasites with complex life cycles frequently manipulate host behaviour to enhance transmission to the definitive host (Section 3.5). It follows that such parasites may also be selected to manipulate behaviour so as to avoid getting eaten by non-hosts, and evidence is accumulating for this process (reviewed in Johnson *et al.*, 2010). For instance, in an amphipod system in France, the acanthocephalan *Polymorphus minutus* has been shown to manipulate the host's antipredator responses and hence reduce the risk of *G. roeseli* becoming the IG prey of the larger, invasive species *Dikerogammarus villosus*. In a laboratory study, both infected and uninfected *G. roeseli* remained in the benthic zone. However, the presence of *D. villosus* induced a change in the behaviour of infected *G. roeseli*, which moved towards the water surface (Medoc *et al.*, 2006) and showed a faster escape response (Medoc & Beisel, 2008). This shift in host behaviour is adaptive to the parasite as it leads to the avoidance of non-host predators and increases the likelihood of transmission to the definitive host. By making *G. roeseli* a much weaker competitor and IG predator, less vulnerable to IG predation by the larger *D. villosus* (Medoc *et al.*, 2006) it also shifts the outcome of IGP.

4.4.2 Parasites and IGP in biological control

Most recognised examples of parasitism within IGP modules come from biological control of arthropod pests (Rosenheim *et al.*, 1995). In many cases IGP arises as a result of employing multiple control agents (predators, parasitoids, pathogens; in this section we use the term 'pathogens' as it is commonly used in the biological control literature to refer to microparasites such as viruses and bacteria). These may engage in IGP via direct predation of one agent on another or via interactions within coinfected hosts. Interactions with native predators and non-target species can also result in IGP. Conservation biological control, where habitats are managed to encourage a diversity of natural enemies, therefore provides many potential opportunities for IGP (see also Section 2.3.2). Examples of biological control tend to involve controlling an introduced or invading pest species (Chapter 6); generally, a control agent from the pest's native range is utilised (classical biological control), although there have also been attempts to use exotic control agents on native pests, pests from different geographical regions (neo-classical control) or native enemies applied at high dose/density or in habitats where the enemy was previously absent (reviewed in New, 2005). These scenarios are instances of man–made or 'deliberate community assembly' (Holt & Hochberg, 2001), and much of the work of screening protocols is to identify (and hence avoid) possible non-target effects of the putative control agent on the other members of the resultant community. There are several avenues for introducing apparent competition and IGP in these man–made assemblages:

- Generalist control agents may attack or infect native species as well as the pest target (resulting in apparent competition mediated by the control agent; the pest may provide a 'reservoir' for the control agent, exacerbating effects on non-target species (Chapter 2)).
- Specialist or generalist control agents may compete with native enemies over the pest or over non-target prey; this can result in IGP (for instance, if the agent is a generalist higher-order predator or if it consumes infected prey). With or without IGP, competitive exclusion of native enemies may be a consequence.
- Higher-order generalist predators may attack the control agent, in addition to native prey. Introduction of the control agent may

therefore result in resource enrichment for native higher predators, with knock-on effects for its other prey. If the control agent also attacks some of these prey species, IGP will occur between the agent and higher-order predator.

Whether or not the control agent is an obligate specialist on the pest species, non-target effects are likely, resulting from the formation of novel competitive, predatory or IGP interactions on release of the control agent (see overview by Cory & Myers, 2000). Even if non-target effects at equilibrium are sufficiently small to enable coexistence, non-target species may be under threat during the establishment phase of the agent, owing to transient fluctuations (under-damping) in the early stages of agent population growth (Holt & Hochberg, 2001). Inevitably, biological control scenarios are embedded within a wider community and additional interactions will also be relevant; we return to these broader implications in the context of parasitism and biodiversity (Section 7.4).

4.4.3 Biological control scenarios with IGP

Outcomes for IGP interactions depend on the players involved, and their propensity to act as an IG predator in the interaction. Here we examine examples involving parasites or parasitoids in biological control of pests (hosts) (Fig. 4.9).

4.4.3.1 Parasitoid–parasitoid

Many parasitoids have parasitoids of their own; these hyperparasitoids can be facultative (which can infect uninfected hosts or those infected by a primary parasitoid) or obligate (which only infect already-parasitised hosts).

To conform to IGP, a facultative parasitoid must be involved (as the IG predator), along with its primary parasitoid (the IG prey); an obligate hyperparasitoid would result in a linear food chain (examined in Holt & Hochberg, 1998). As the primary parasitoid can only produce viable offspring from previously uninfected hosts, it succumbs to competition or direct attack by a developing hyperparasitoid within coinfected hosts (Fig. 4.9a). Population dynamic outcomes are further complicated for some classes of hyperparasitoid in which sex determination depends on the host in which they develop. Heteronomous hyperparasitoids can

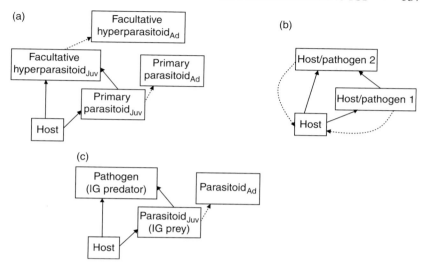

Fig. 4.9. Examples of parasites within biological control IGP modules.
(a) Parasitoid–parasitoid–host; (b) pathogen–pathogen–host; (c) pathogen–
parasitoid–host; dotted lines show developmental transitions from juvenile ($_{Juv}$) to
adult ($_{Ad}$) parasitoids (a) and (c) and recovery/or reproduction by infected hosts (b).

parasitise uninfected hosts or other parasitoids of these hosts. As primary
parasitoids they develop into females; as secondary parasitoids they
become males. Autoparasitoids develop in previously uninfected hosts
as females, or in conspecifics (or other parasitoids), in which case they
become male. Heteronomous and autoparasitic classes of parasitoid
present unique problems for predicting biological control efficacy and
population outcomes because sex ratios and rates of increase for the
secondary parasitoids are coupled to and feed back onto the host and
primary parasitoid frequencies. The disruptive role that facultative
hyperparasitoids can play in biological control (Briggs, 1993; Rosenheim
et al., 1995) has led to the recommendation that all secondary parasitoids
be avoided in multiple-agent control strategies (Kfir *et al.*, 1993).

In the field, parasitoid–parasitoid interactions may enhance or disrupt
biological control. For example, whitefly populations in glasshouses are
managed using the primary parasitoid *Encarsia formosa*. The autoparasitoid *Encarsia pergandiella* uses whitefly larvae that harbour *E. formosa*
parasitoids as hosts for their male offspring (Heinz & Nelson, 1996),
and this IG predation might be expected to disrupt biological control.
However, enclosure experiments revealed a higher level of whitefly

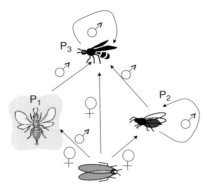

Fig. 4.10. The *Bemisia* parasitoid system. The primary parasitoid (P_1: *Eretmocerus* sp.) produces offspring of both sexes on the whitefly (*B. tabaci*); the obligate autoparasitoid (P_2: *Encarsia* sp.) produces females on the whitefly and males by hyperparasitisation of conspecific females; the facultative autoparasitoid (P_3: another *Encarsia* sp.) develops females on the whitefly and males by hyperparasitisation of conspecific females or males and females of the other two parasitoids.

biological control when both the autoparasitoid *E. pergandiella* and the primary parsitoid *E. formosa* were released, than when either single species was used (Heinz & Nelson, 1996). In contrast, biological control of the silverleaf whitefly, *Bemisia argentifolia*, involves three parasitoid species: a primary parasitoid; an obligate autoparasitoid; and a facultative autoparasitoid (Fig. 4.10). Models developed for this system suggest that the facultative autoparasitoid in particular can be strongly disruptive, because it produces mostly males when parasitisation of the whitefly population is relatively high, reducing the frequencies of female autoparasitoids and primary parasitoids emerging in the following generation (Mills & Gutierrez, 1996).

4.4.3.2 Parasite–parasite

Borer *et al.* (2007a) produced a model in which two directly transmitted microparasites capable of coinfecting a single host effectively engage in IGP (Fig. 4.9b). To conform to (unidirectional) IGP, one pathogen strain must consistently out-compete the other, which consequently fails to produce secondary infections in coinfected hosts. Unlike the other forms of IGP, the dominant pathogen does not necessarily receive an energetic gain from usurping the second pathogen (so, strictly speaking, the 'IG prey' here is not 'consumed'). Nevertheless, Borer *et al.* argue that this interaction represents IGP because of the module's comparable structure and population dynamic outcomes to classical IGP (see below). There are few convincing examples of pathogen–pathogen interactions that directly conform to this model; however, macroparasites frequently engage in this type of interaction within the host (see also Section 2.7). Dobson (1985) reviews cases of interference competition between macroparasites which conform to this structure (i.e. strongly asymmetric, with one parasite consistently eliminating the other from coinfected hosts (Box 2.9)). Some trematodes, which parasitically castrate their snail intermediate host, consume their competitors inside the snail, leading to fewer than expected cases of coinfection (Kuris & Lafferty, 1994).

Other interactions between pathogens that influence the infection process may also impact on population dynamics and community outcomes in similar ways to IGP. For instance, the presence of one parasite may reduce or increase the host's immune response to other parasites, thereby influencing infection probability (Tompkins et al., 2010; Section 2.7.2). In addition, the reduction of host life expectancy by a virulent pathogen may reduce or preclude successful reproduction of other parasites. For example, Bauer et al. (1998) investigated the combination of a microsporidium (*Nosema* sp.) and a nuclear polyhedrosis virus for control of the gypsy moth *Lymantria dispar*. Dual infection was detrimental to the microsporidium as early mortality of the host reduced spore production. The impact on the virus depended on the sequence of exposure; when microsporidian infection preceded viral infection, hosts died sooner and hence viral replication was also reduced. In a laboratory investigation, combined use of a microsporidian (*Paranosema locustae*) and fungus (*Metarhizium anisopliae*) led to higher mortality of the desert locust *Schistocerca gregaria* than did single infections. Whilst spore production by the fungus was unaltered in hosts with dual infections, the fungus out-competed the microsporidian, which produced fewer spores in dual infections (Tounou et al., 2008a; 2008b). As the fungus is saprophytic, producing infective spores upon host death, it could be argued that fungi developing in a dual-infected host are also IG predators of the microsporidian, which can only replicate in living host tissue.

4.4.3.3 Parasite–parasitoid

In many cases microparasites utilise the same arthropod host species as parasitoids. The outcome of this interaction depends on whether one consumer can infect/prey on the other in addition to the shared resource, or on which one can complete its development first. The outcome of this developmental race (Box 4.2) frequently determines the nature of the IGP relationship. In most cases, these races appear to resolve in favour of the pathogen, which is therefore the IG predator (Fig. 4.9c). From the parasitoid's perspective, IGP by another parasite is

---

Box 4.2 Developmental races between pathogens and parasitoids

When pathogens and parasitoids infect the same host individual, the outcome depends on the relative timing of infection, and the duration required for the parasites to complete development in the host. Virulent pathogens are often able to reproduce, killing the host, before parasitoids have completed development; as IG predators, they therefore win the developmental race. However, if pathogen transmission to the host occurs sufficiently late in a parasitoid's development, the parasitoid could successfully emerge, killing its host, before the pathogen can reproduce. Occasionally, pathogens may be able to infect the parasitoid as well as the insect host, in which case they also win the developmental race.

Timing of coinfection: in field and laboratory experiments, parasitoids developing in many aphid hosts fail to complete development before the aphid host dies from other infections (Brodeur & Rosenheim, 2000). For example, although blastophores of the fungus *Verticillium lecanii* infecting the potato aphid do not appear to penetrate and infect the tissues of the parasitoid *Aphidius nigripes*, parasitoid survival depends on the time of exposure to the fungus. In an experiment, 31% and 89% of larvae survived to pupation when potato aphids were exposed to the fungus two and four days post-oviposition, respectively (Askary & Brodeur, 1999). Parasitoid–virus systems can exhibit similar patterns of dominance by the virus; as shown by the moth *Plodia interpunctella*, which is host to a parasitic wasp *Venturia canescens* and a granulosis virus. The parasitoid synchronises its development with the host, taking 21–25 days to complete development, depending on the host instar parasitised. The virus enters the host via feeding contamination, and infection proceeds from the midgut to almost the entire larval body between 7 and 14 days; consequently, any parasitoid larvae also perish within hosts if viral infection occurs sufficiently early in development (Begon *et al.*, 1999).

Box 4.2 (*continued*)

Pathogens infecting parasitoids of their host: pathogens of the host/pest insect may also infect parasitoids of the pest. For example, the microsporidium *Nosema pyrausta* is a principle biological control agent against the European corn borer (*Ostrinia nubilalis*), a major pest of corn in the United States. Studies suggest that IGP between *N. pyrausta* and the braconid parasitoid *Macrocentrus grandii* is determinental to the parasitoids (which become infected within infected hosts) and may drive parasitoid population dynamics, resulting in negative correlation between microsporidium infection prevalence and braconid parasitism over sites and years (Andreadis, 1980; Siegel *et al.*, 1986). Infection with *N. pyrausta* alone produces strong, delayed density-dependent fluctuations, where high infection levels one year reduce corn borer populations the next, so in this case it appears that antagonism is largely asymmetric in favour of the microsporidium, and parasitoid numbers are driven by its prevalence (Siegel *et al.*, 1986). In laboratory studies, entomopathogenic fungi have been demonstrated to act as IG predators by infecting parasitoids in several aphid/fungus/parasitoid systems. Infection results in invasion and consumption of parasitoid tissues by the fungus, killing or delaying development of the parasitoid. However, these studies need to be treated with caution because they may not reflect field conditions (reviewed in Roy & Pell, 2000).

clearly disadvantageous, and some parasitoids are able to detect and avoid infected individuals if infection is sufficiently advanced (Fransen & Vanlenteren, 1993).

Arthropod predators may also interact with fungal pathogens in other ways; for instance, by enhancing fungal pathogen transmission either through incidental contamination of the predator (vectoring) or through predator effects on host behaviour (Roy & Pell, 2000). For example, foraging seven-spot ladybirds (*Coccinella septempunctata*) have been shown to vector fungal (*Erynia neoaphidis*) spores to uninfected aphids in the laboratory (Roy *et al.*, 2001). The risk of predation may cause behavioural responses that, in turn, increase opportunities for fungal transmission. For example, diamond-backed moth larvae (*Plutella xylostella*) displayed increased movement in the presence of foraging parasitoids (*Diadegma semiclausum*), resulting in increased transmission of the fungus *Zoophthora radicans* (Furlong & Pell, 1996). Conversely, fungal infections may cause behavioural changes that increase the host vulner-ability to predation or parasitoids. For example, fungal infection has been shown to reduce aphid responses to alarm pheromones (Roy *et al.*, 1999); this may increase opportunities for transmission of a generalist fungal pathogen to a predator.

4.4.3.4 Predator–parasitoid

In many cases where multiple biological control agents are present, preda-tors consume parasitoid or pathogen-infected individuals. The outcome of this (sometimes 'coincidental') IGP varies; in some cases biological control is disrupted, whilst in others the effect of the two enemies is comple-mentary. For example, the ladybird *Cryptolaemus montrouzieri* is an IG predator of the parasitoid *Leptomastix dactylopii* as a result of predation of infected mealy bugs (Chong & Oetting, 2007). The ladybirds feed on both unparasitised and parasitised mealy bugs. As a result, the level of parasitoid infection decreases in the presence of the IG predator and biological control is disrupted. Similarly, IGP by ladybirds on the parasitoid *Aphidius colemani* decreases the level of parasitism on aphids in short-term experi-ments as a result of mortality and because predator avoidance leads to a decrease in oviposition by the parasitoid. However, longer-term experi-ments revealed that aphid numbers are more strongly affected when both enemies are present (Bilu & Coll, 2007).

Rather than disrupting biological control through predation of para-sitoids, predators may complement their impact as biological control agents. For example, both the predatory harlequin ladybird (*Harmonia axyridis*) and the parasitoid *Aphelinus asychis* are used in aphid (*Macro-siphum euphorbiae*) biological control. Feeding trials revealed that the ladybird is an IG predator of the parasitoid. However, ladybird larvae showed a preference for uninfected aphids and, in caged experiments, aphid densities were lowest when both enemies were present (Snyder *et al.*, 2004). Similarly, the combined effects on the moth *Plutella xylos-tella* of the predatory hemipteran *Podisus maculiventris* and the parasitoid *Cotesia plutellae* were additive, despite preferential feeding by *P. maculi-ventris* on infected moth larvae (Herrick *et al.*, 2008).

4.4.3.5 Predator–pathogen

Fungi used in arthropod pest control may engage in IGP with other natural enemies as IG prey or IG predators (Roy & Pell, 2000). As IG prey, fungi may be consumed by predators of pest species when they consume infected hosts. However, detrimental effects on the fungus may be limited if predators avoid heavily infected or sporulating hosts, or if partial predation of infected hosts permits continued fungal transmission. Studies of aphid biological control in the United Kingdom suggest that the interaction between fungi and ladybird predators can be complex. Although fungal infections can decrease antipredator responses of the aphid (Roy et al., 1999), the (native) predatory ladybird *C. septempunctata* shows a preference for uninfected aphids over those infected with the fungus *Pandora neoaphidis* (Roy & Cottrell, 2008); hence the effects of the two enemies were complementary. In contrast, IGP by the harlequin ladybird (*H. axyridis*, an invasive species initially introduced for bio-logical control) disrupts biological control by the fungus (Fig. 4.11). This is because *H. axyridis* shows little avoidance of infected aphids. Further-more, whilst *C. septempunctata* partially eats infected aphids (Roy et al., 1998), *H. axydris* was more likely to wholly consume infected aphid cadavers, reducing the availability of spores in the environment (Roy & Pell, 2000; Roy & Cottrell, 2008). Hence IGP by *H. axyridis* could disrupt biological control by reducing the impact of the fungus on aphid populations. The situation is further complicated because *H. axyridis* larvae are also predatory on other ladybird larvae (Fig. 4.11).

The above examples highlight that the outcome of the IGP inter-actions between parasites and predators of pests is affected by the preda-tory preferences displayed by the IG predator (Rosenheim, 1998). If the predator preys upon both uninfected and infected prey, then it is likely

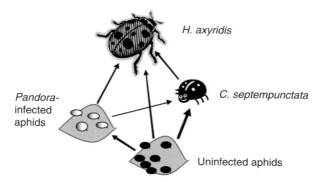

Fig. 4.11. Trophic interactions between aphid enemies. Arrows show direction of energy flow (width depicts relative predatory preference). Although adult ladybirds are depicted, the larval stages are the most voracious predators.

that this IG predation will disrupt the efficacy of biological control by the parasitoid or pathogen. However, if the predator avoids infected prey, it is more likely that the effect of the two enemies will be complementary. Related patterns are found in general models of parasitism and predation; for instance, in relation to parasite-induced vulnerability to predation (PIVP) (Section 3.2). PIVP has been shown to influence host population dynamics, sometimes with counter-intuitive effects on disease prevalence (Section 3.6.1). Straub *et al.* (2008) caution that the practice of conservation biological control cannot ignore the unpredictable influence of IGP arising from the gamut of positive and negative interactions between multiple IG predators. Rosenheim *et al.* (1995) conclude that IGP is generally disruptive for biological control, reducing the combined impact of several agents below that expected additively, or reducing the impact of the key agent deployed. This follows from the theoretical analysis of IGP: as the IG predator is a less efficient utiliser of the shared resource (the target pest species), it indirectly raises the equilibrium abundance of that resource via its effect on the IG prey (Section 4.1.2). Consequently, removal of the IG predator should enable the remaining control agent to regulate the target species to a lower abundance. This reasoning underlies the suggestion that aphid parasitoids will generally be ineffective as biological control tools because they are frequently IG prey to other aphid natural enemies (Brodeur & Rosenheim, 2000). However, IGP can also be a positive factor in conservation, with predation controlling 'mesopredators' that are otherwise detrimental to prey populations (Muller & Brodeur, 2002).

4.5 Parasites extrinsic to IGP

So far we have discussed examples where parasites, pathogens and parasitoids engage in IGP either as IG predator or prey. There is also the potential for parasites to influence IGP associations between other species (Fig. 4.2). This could be important for community structure since all the direct and indirect effects of IGP reviewed above might be changed by the addition of parasitism, acting on the population densities of IGP participants or modifying IGP interaction strength.

Although IGP is widespread and parasites are ubiquitous, there have been few studies that consider the impact of parasitism on hosts engaging in IGP. The potential for parasites to generate both density- and trait-mediated effects on hosts (Box 1.3) means that they could, in principle, alter either the prey or predatory density of behavioural aspects of existing IGP relationships. Parasites are often linked to hosts via trophic relationships and can influence the strength of these relationships. For instance, trophically transmitted parasites frequently manipulate intermediate host behaviour to increase the likelihood of predation by their definitive host (Section 3.5.1). In Section 4.4.1, we considered IGP in trophic transmission. The same features of hosts that make them vulnerable to behavioural manipulation by parasites may render such hosts more (or less) susceptible to IG predators.

4.5.1 Parasite-modified IGP

One system in which IGP and the impact of parasites have been studied is in amphipod Crustacea where native species in Europe are being displaced by invaders (Box 4.3). In this system, competition and IGP are key interactions between the native and invading species, and the frequency of IGP is modified by parasites. Modification of IGP in this system can potentially have consequences for species coexistence patterns (Section 4.6) and for the outcome of invasions (Box 6.3).

In the amphipod system described in Box 4.3, two different parasites influence at least four different aspects of IGP: (1) *Pleistophora mulleri* infection reduces the predatory activity of its host, *Gammarus duebeni celticus*, on its IG prey, and (2) increases the likelihood of being consumed by its IG predators; (3) *E. truttae* reduces the predatory activity of its main

---

Box 4.3 Parasite-modified IGP in native–invasive amphipods

In Ireland, the native amphipod *Gammarus duebeni celticus* is being replaced by the invasive *G. pulex*. A further two invaders occur, *G. tigrinus* and *Crangonyx pseudogracilis*. Observations of uninfected animals reveal a hierarchy of competition and IGP (Fig. 4.12). *G. pulex* is the strongest predator of the four species, and has displaced *G. duebeni celticus* from a series of water courses in Northern Ireland, where the invasion is ongoing. In contrast, *G. duebeni celticus* is larger than the two other invading species, *G. tigrinus* and *C. pseudogracilis*, and is the stronger competitor and IG predator in interactions with these species. Nonetheless, mixed populations of all four species have been reported (Dick, 1996), and the native *G. duebeni celticus* coexists with the strongest predator *G. pulex* in some habitats (Dick, 2008).

Box 4.3 (*continued*)

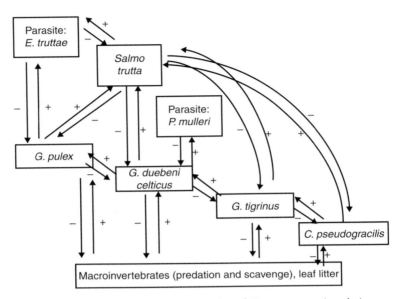

Fig. 4.12. Interaction web for a community of *Gammarus* species, their parasites and predators. Shown are four amphipod species that engage in IGP; their common resource base; two parasite species of gammarids (the microsporidian, *Pleistophora mulleri*, has a direct life cycle, whereas the acanthocephalan, *Echinorhynchus truttae*, uses brown trout (*Salmo trutta*) as its definitive host); and a top predator of gammarids (*S. trutta*). Not shown are other top predators without direct links to these parasites or other guild members that do not participate in IGP with gammarids. Also not shown are links between *E. truttae* and *G. duebeni celticus* and cannibalistic interactions, which occur within each amphipod species and are also modified by parasitism (MacNeil *et al.*, 2003c). Indirect interactions include: *P. mulleri*-mediated IGP between *G. pulex*, *G. duebeni celticus*, and *G. duebeni celticus*, *G. tigrinus*; *E. truttae*-mediated IGP between *G. pulex* and *G. duebeni celticus*; behavioural manipulation of *G. pulex* to enhance transmission of *E. truttae* to *S. trutta* is also documented. From MacNeil *et al.* (2003b; 2003c; 2003d).

Two species of parasite have been shown to influence both the prey and predatory components of IGP in this guild of amphipods (MacNeil *et al.*, 2003c; 2003d): the microsporidium *Pleistophora mulleri* and the acanthocephalan *Echinorhynchus truttae*. *P. mulleri* infects the native *G. duebeni celticus*, infecting the abdominal musculature and reducing host motility (MacNeil *et al.*, 2003c). *P. mulleri*

Box 4.3 (*continued*)

causes little direct mortality or reduction in fecundity. However, this parasite causes increased vulnerability to cannibalism (leading to parasite transmission (MacNeil *et al.*, 2003b)) and reduced predatory ability on sympatric amphipods and isopods (Fielding *et al.*, 2005). In field enclosure experiments, *P. mulleri* had no impact on *G. duebeni celticus* survival, whether in single populations or in mixed–species populations with the smaller *G. tigrinus* (Fig. 4.13). However, parasitism of *G. duebeni celticus* resulted in increased survival of the smaller IG prey

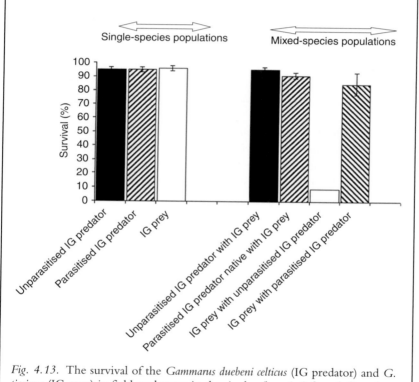

Fig. 4.13. The survival of the *Gammarus duebeni celticus* (IG predator) and *G. tigrinus* (IG prey) in field enclosures in the single- (bars 1–3 from left) and mixed- (bars 4–7 from left) species populations. Mean percentage (+/–SE) survival at two weeks is plotted. Infection with *Pleistophora mulleri* did not affect the survival of *G. duebeni celticus* in either single- or mixed-species populations. However, infected *G. duebeni celticus* were weaker IG predators on the smaller invasive *G. tigrinus*. This led to higher survival of *G. tigrinus* when in sympatry with infected *G. duebeni celticus* than when in sympatry with uninfected *G. duebeni celticus*. Data from MacNeil (2003b).

Box 4.3 (*continued*)

species; whilst *G. tigrinus* kept in mixed populations with *G. duebeni celticus* showed 90% mortality over two weeks, this species was able to coexist with infected *G. duebeni celticus*. Laboratory experiments confirmed that coexistence was facilitated by a reduction in IG predation by *G. duebeni celticus* on *G. tigrinus*. Similarly, infected *G. duebeni celticus* showed less IG predation of *C. pseudogracilis*. In addition to modifying the IG predatory abilities of its host, *P. mulleri* infection increased its vulnerability to becoming the IG prey of the larger species *G. pulex*.

IGP in amphipod communities is also modified by *E. truttae* (MacNeil *et al.*, 2003d). This parasite has an indirect life cycle; fish are the definitive host and the parasite utilises both *G. duebeni celticus* and *G. pulex* as intermediate hosts. This parasite reduced the ability of the larger *G. pulex* to predate *G. duebeni celticus*, promoting their coexistence (Box 7.2).

intermediate host, *G. pulex*, and (4) increases its vulnerability to predation by the definitive host, *Salmo trutta*. These effects only become apparent in mixed-species cultures where there is opportunity for interspecific interactions; such 'cryptic virulence' (MacNeil *et al.*, 2003c) is not apparent from observations of mortality or behaviour of hosts in intraspecific cultures. These effects on IGP are also of interest because they provide further evidence of the trait-mediated indirect effects (TMIEs) of parasitism (Box 1.3). Parasite-induced TMIEs such as these may have broader implications for community processes, as we examine in Section 7.7. Because both TMIEs and IGP operate on short timescales relative to processes such as resource competition, they could potentially have strong impacts on community composition (e.g. Dick *et al.*, 2010).

4.5.2 The potential for parasitism to interact with IGP

The amphipod system in Box 4.3 suggests that the potential for parasites to interact with IGP could be widespread. A promising place to look for such interactions is in systems where parasites are well known, but IGP has not been examined, or where IGP is established, but parasites have not been investigated. Three possible examples are discussed below.

Several papers on apparent competition speculate that parasite-modified IGP might be important, although none have directly investigated this possibility. For example, in a laboratory study, Yan *et al.* (1998) investigated the effect of the tapeworm *Hymenolepis diminuta* on competition between *Tribolium casteneum* and *T. confusum*. *T. casteneum*, the stronger competitor, suffers higher infection and fitness costs from parasitism and thus the parasite should be expected to reverse the outcome of competition. However, the opposite was found, with a faster exclusion of *T. confusum* than occurred in the absence of infection. The authors suggested that parasite-induced changes in surface-seeking behaviour might lead to higher levels of IGP by *T. casteneum*.

Hoogendoorn and Heimpel (2002) found that the North American native ladybird *Coleomegilla maculata* was more susceptible to the parasitoid *Dinocampus coccinellae* than was its invasive competitor, the harlequin ladybird (*H. axyridis*). *H. axyridis* acted as a sink for the parasitoid, and simulations predicted a consequent increase in *C. maculata* density. However, *H. axyridis* larvae are also IG predators of *C. maculata* (Cottrell & Yeargan, 1998), a feature not taken account of in the models. Furthermore, the parasitoid may modify IGP between the native and invasive species, making outcomes for this three-species interaction more complicated than at first realised.

Interactions between mosquito species are often mediated by parasites (reviewed in Juliano & Lounibos, 2005). For example, escape from gregarine parasites facilitates competition by the invading *Aedes albopictus* with *Ochlerotatus triseriatus* (Aliabadi & Juliano, 2002; Chapter 2). Juliano and Lounibos (2005) also propose that IGP may limit the impact of *A. albopictus* as *O. triseriatus* larvae are stronger IG predators. It would therefore be of interest to consider the outcome of IGP and parasitism for this and other larval mosquito communities.

4.6 Models of parasitism extrinsic to IGP

The original analyses of IGP suggest that it is intrinsically unstable in many cases (Section 4.1.2); one way in which it might be made stable is through the effects of parasitism on one or both participants (Hatcher *et al.*, 2008). We investigated this question, stimulated by our empirical studies on amphipods that point to a role for both IGP and parasitism in determining community composition (Box 4.3). General theory for IGP (in the absence of parasitism) would predict that the largest, most voracious predator should replace the smaller native species and that invasions by the other (smaller) species should fail (Section 4.1.2). However, apparently stable mixed communities persist (Dick, 1996;

MacNeil *et al.*, 2003c; 2003d), possibly as a result of the influence of parasitism on IGP modules.

As a first step in understanding how parasitism and IGP interact, we developed a model incorporating a shared parasite of two species engaged in mutual but asymmetric IGP (Hatcher *et al.*, 2008). We took a similar modelling approach to that used for parasite-mediated competition (Section 2.4.2), in which two species sharing a microparasite engage in intraspecific and interspecific competition, but with the additional complexity of predation between host species. This model just looked at the density-mediated effects of parasitism on IGP (Box 1.3); infected hosts had increased mortality, but did not differ from uninfected conspecifics in their predatory or competitive abilities.

Parasitism operating in this way increased the range of conditions leading to host coexistence, provided the stronger IGP participant (determined by the balance between competitive and predatory effects; usually the superior predator) was more adversely affected by the parasite (Fig. 4.14a). Conversely, parasitism enhanced the rate of decline and extinction of the inferior species (usually the less predatory species) if it suffered greater parasite-induced mortality.

The keystone effects of parasitism on IGP shown in this model resemble those described for parasite-mediated competition, where the

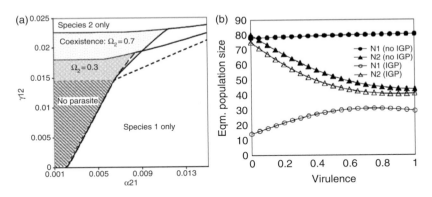

Fig. 4.14. Impact of parasitism on coexistence and population sizes under IGP. (a) A parasite of the IG predator increases the parameter range for IGP persistence (shaded); higher virulence (higher Ω_2) has a greater impact (plotted for IGP by species 2 ($\gamma12$) against interspecific competition by species 1 ($\alpha21$). (b) Effect of parasitism on the competitive and predatory components of IGP. The equilibrium population sizes for the two species (N1 and N2) are plotted against parasite virulence to species 2, for the case where they compete and engage in IGP (IGP) and where they interact only via competition (no IGP) (from Hatcher *et al.*, 2008).

deleterious effects of parasitism on the superior competitor prevent exclusion of the inferior competitor. In competition models, the parasite reduces the population density of the superior species, thereby freeing up resources that can be exploited by the inferior species (Holt & Dobson, 2006). However, parasitism's influence is stronger in the case of IGP because it also reduces predation pressure on the inferior IGP participant. This process is similar to the effect of parasites on top predators in a linear food chain (Chapter 3), facilitating an increase in the equilibrium population density of the prey or inferior predator (Fig. 4.14b). The effects of parasitism on IGP in this model are entirely the result of numerical effects on population density (the parasite only influences population densities via effects on mortality). This may underestimate the impact of parasitism on IGP interactions because it neglects trait-mediated effects of parasitism, which we know occur in the amphipod system (Box 4.3).

4.7 IGP and the evolution of host–parasite relationships

When parasites are involved in IGP interactions, the potential evolutionary outcomes for the parasite range from local extinction to the evolution of new parasite–host relationships. If a parasite is host-specific, then incidental IGP as a result of predation of an infected host will result in the loss of opportunities for transmission. This may lead to selection favouring parasitic manipulation of the host to avoid consumption by non-host predators (Section 4.4.1). However, a more generalist parasite that could adapt to life inside the predator might benefit from being eaten, a scenario recognised in the original treatise on IGP (Polis et al., 1989). Put another way, if parasites can avoid digestion and adapt to a predator's internal environment, enabling their own propagation, they can acquire another host; food webs will bear witness to this 'ghost of predation past' (Johnson et al., 2010). Empirical studies suggest that both patterns (host acquisition and predator avoidance) can arise.

Host acquisition by trophically transmitted parasites

Johnson et al. (2010) discuss examples where predation may have led to host acquisition for trophically transmitted parasites. For instance, one-third of the trophic links involving parasites and predators in a salt marsh food web are between parasites that can exploit the predator as a host (Lafferty et al., 2006a; Box 7.3). However, it is more likely that these links have evolved following exposure of the free-living transmission stages to predation (a frequent event that contributes substantially to energy budgets in these ecosystems (Section 7.6)) than as a result of predation on infected intermediate hosts. Nevertheless, the arrangement of trophic parasites in food webs (for instance, their frequent association with generalist predators (Chen *et al.*, 2008; Section 7.9)) is suggestive of a link between predation and acquisition of new hosts.

Avoidance of predation by trophically transmitted parasites

Parasite stages within the intermediate host can, to some extent, alter the likelihood of their transmission, thereby influencing whether 'accidental' predation by a non-host is likely to occur. Such parasites often manipulate host behaviour so as to enhance the likelihood of predation by the host (reviewed in Thomas *et al.*, 2005a; see also Section 3.5.1). If such behavioural manipulation leads to IG predation by a species to which the parasite is not adapted, then the parasite will not be transmitted. Hence selection on both the parasite and its intermediate host should favour behaviour that reduces the likelihood of the host becoming IG prey, such as the heightened antipredator response observed in *Polymorphus minutus*-infected *Gammarus roeseli* (Medoc *et al.*, 2006; Section 4.4.1).

Host acquisition by directly transmitted parasites

Host acquisition is not feasible for all parasite groups, and is dependent on consumption of appropriate infective stages. For instance, IG predation of parasitoids is unlikely to provide opportunities for transmission, but instead leads to parasitoid mortality (Section 4.4.3); to our knowledge there are no reports of parasitoid transmission to new hosts when the infected host is eaten. IG predation of fungal pathogens, however, can provide opportunities for the predator to become infected, although examples are sparse. For example, parasitoids (*Aphidius nigripes*) that developed on aphids infected by *Verticillium lecanii* had fungal spores in the gut, but there was no evidence for invasion of parasitoid tissues (Askary & Brodeur, 1999). A similar pattern was found for other fungi (Brodeur & Rosenheim, 2000).

IG predation by pathogens, including microsporidia, occurs when parasitoids developing in infected hosts acquire infection or die because they fail to complete their development before the demise of the host (Section 4.4.2). Broad host ranges have been reported for some microsporidia (Futerman *et al.*, 2006), but not all microsporidia can infect parasitoids of their primary host. In the corn borer, for instance, the microsporidian *N. pyrausta* infects *M. grandii* parasitoids developing in the same host (Box 4.2), but does not infect another parasitoid, *Lydella thompsonii*, in similar circumstances (Cossentine & Lewis, 1988). Similarly, although the parasitoid *Glyptapanteles liparidis* oviposited in gypsy moth (*Lymantria dispar*) larvae infected by *Vairimorpha* sp., it did not acquire this infection from its host (Hoch *et al.*, 2000).

Parasites that modify IGP between hosts may benefit from opportunities to infect novel host species if their original host is preyed upon, particularly if IGP involves taxonomically related species. In our native–invasive amphipod system (Box 4.3), the microsporidian *P. mulleri* is directly transmitted between *G. duebeni celticus* amphipod hosts via cannibalism and scavenging. IGP of *G. duebeni celticus* by invading amphipods might therefore be predicted to lead to infection of new host species. However, laboratory transmission experiments and field surveys indicate that this parasite remains host-specific and that IGP has not, as yet, provided a route of transmission to new hosts (MacNeil *et al.*, 2003b).

On balance, the lack of extensive evidence for parasite transmission as a result of IGP suggests that, although IGP may provide opportunities for novel host–parasite relationships to evolve, such events are not common. More work is warranted in this interesting area, particularly for trophically transmitted parasites. Host acquisition might prove to be a rather rare event, because alternative strategies may be under conflicting selective pressures. For instance, parasites in intermediate hosts could be selected to overcome digestion and relocate within the predator, or they might be selected to manipulate intermediate host behaviour to avoid accidental predation by the wrong species. The costs and constraints associated with these alternative strategies will differ.

4.8 Conclusions

IGP represents the first step-up in complexity from modules of competition and predation, as it involves elements of each. Adding parasitism to the mix can further complicate the issue. For this reason we have not introduced equations in this chapter, although they are built and analysed along similar lines to those presented in the previous chapters. An examination of the predictions from mathematical models of IGP (Section 4.2) reveals a complicated array of potential outcomes. This is mirrored in the empirical findings for IGP systems. For instance, from biological control studies (where we have the most information), whilst

IGP of control agents can disrupt biological control strategies in some cases, other studies report complementary effects of (IG) predators and other enemies on target pest species.

In early theoretical treatments of IGP, it was argued that one participant must gain substantially (other than via relaxation of competition) from the consumption of the other guild member (Holt & Polis, 1997), otherwise the interaction can be considered under the remit of interference competition. However, recent theory papers have relaxed this requirement as part of the broader analysis and attempt at synthesis of the disparate fields of host–parasite and predator–prey ecology (Section 4.3). This change of emphasis may provide fresh impetus for the study of systems in terms of IGP, and allow for a more productive cross-fertilisation of ideas. For instance, beyond the biological control literature (Section 4.4.3), there have been relatively few recognised examples of IGP that involve parasites. Feeding on parasites has been recognised as a potential form of IGP, both in the original literature (Polis *et al.*, 1989), and again more recently (Johnson *et al.*, 2010). However, this (arguably) 'coincidental' IGP has received little quantitative empirical treatment, perhaps because of the difficulty in quantifying the energetic gain (if any) of feeding on infected hosts, and the belief that energetic gain to IG predators was crucial to its definition as IGP.

In this regard, a clear example in which parasites are intrinsic to IGP interactions is that of trophic transmission (Section 4.4.1). Here, we argue, parasites can play the role of IG predators (on the definitive host) or IG prey (when infected intermediate hosts are consumed by non-host predators). In the former case, the IG predator benefits distinctly from consuming its IG prey, having zero reproductive fitness without it. It is therefore puzzling that this module has received so little attention from an IGP perspective; research in this area seems likely to yield good rewards.

In this chapter we distinguish two ways in which parasites can enter into IGP interactions: as intrinsic members of an IGP module; or as extrinsic modifiers of IGP between other species (Section 4.1). Most of the research in this chapter has concentrated on parasites as intrinsic participants of IGP. However, the interaction between IGP and extrinsic parasitism may be key to understanding some ecological systems (Section 4.5). The effects of parasitism on IGP systems is a potentially fruitful area for further research, in particular in empirical systems where IGP has been well studied but the impact of parasites has yet to be examined. The few documented cases of the effect of parasitism extrinsic to IGP

(Section 4.5.1) provide evidence that parasitism can alter IGP relationships, and recent theoretical work predicts that parasitism might be an important factor for species coexistence in IGP systems (Section 4.6).

Recent food web analyses find evidence for the importance of IGP in structuring communities (Bascompte & Melian, 2005; van Veen *et al.*, 2006). Contemporary work on parasites also demonstrates that they are involved in a substantial proportion of the interactions in food webs, and that predator–parasite interactions are particularly common (Section 7.5). Examination of the role that parasites play in IGP systems therefore seems to us to be a field ripe for reassessment as the importance of both IGP and parasitism in ecosystems gains recognition. We now need to combine these research agendas to examine the potential for parasites to interact with IGP, which we propose may be of similar importance in community structure as the interactions between parasites and competitors (Chapter 2), or parasites and predators (Chapter 3).

5 · *Plant pathogens and parasitic plants*

5.1 Introduction

In this chapter we consider how parasitism of plants affects ecological interactions directly involving plants, and also some of the population dynamic and community consequences for higher trophic levels. We might expect many of the aspects of plant–parasite interactions to mirror those in animal systems; however, in most ecosystems plants occupy the position of basal resource, so plant disease systems are potentially the best place to look for strong knock-on effects at higher trophic levels. Anticipating the results, we might expect to observe parasite-mediated or apparent competition (Chapter 2) between plant species, knock-on effects of plant disease to herbivores (mirroring knock-on effects to predators as discussed in Chapter 3), intraguild predation-like interactions between parasitic plants and their hosts or possibly between different plant natural enemies (mirroring results in Chapter 4), with effects propagating to higher trophic levels (predators of herbivores) and influencing community structure as a whole. Not all interactions involving parasites and plants are considered here: parasites of herbivores (grazers) can also affect plants via (inverted) trophic cascades (Grenfell, 1992; Dobson & Crawley, 1994); these are considered in Section 7.2.

This chapter bridges the gap between the preceding chapters that examine the mechanistic effects of parasitism within community modules (such as competition and predation) and chapters still to come which examine the broader community-level impacts of parasitism. Plants are singled out for this longer-sighted treatment because the plant literature arguably has the most clear-cut examples of possible community impacts of parasitism, and plant–parasite interactions highlight the broad spectrum of impacts that parasitism can potentially have on communities (Fig. 5.1). Furthermore, most of the theoretical work on community-level processes has focused on animal systems and takes a somewhat different approach to the problems examined empirically here.

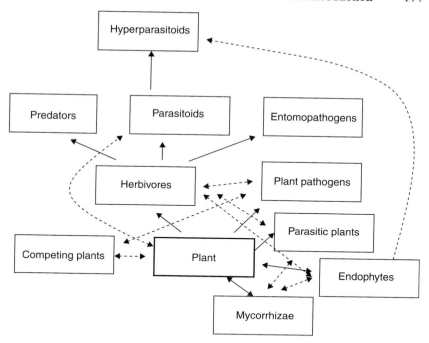

Fig. 5.1. Interaction web illustrating some known effects of plant parasites and endophytes on community members. The focal plant host is depicted in the bold-framed box; other species are connected via direct interactions (solid arrows, showing direction of energy flow) or indirect interactions (which may be positive or negative, symmetrical or asymmetric) mediated via infected plants or their parasites/endophytes. For simplicity, not all conceivable (or demonstrated) interactions are shown.

The reader can thus view this chapter as a largely empirical appetiser for the types of problems examined in more detail in later chapters.

5.1.1 Differences between animal and plant–parasite systems

Whilst above we highlight several similarities between plant and animal disease systems, there are also some important differences. Importantly, the stationary mode of plant life means that disease transmission will rarely involve random mixing (even if the disease is vectored by insects); disease spread will tend to be patchy and localised. Hence the usual assumption of random mixing between susceptible and infectious individuals (density-dependent transmission – Box 1.2) taken in many parasite–host population models will not apply in plant systems. This has consequences for basic predictions about parasite and host population dynamics and co-evolution. For example, coexistence criteria differ as parasite and/or host can become locally extinct, but coexistence across a network of patches can still be maintained; selection on virulence also depends on population structure (Thrall & Burdon, 2003; Section 8.3), and spatial structure can lead to locally coadapted host and parasite populations (Gandon et al., 1996). The consequences of spatial population structure for interactions between multiple host or parasite species has yet to be examined in general (but see Section 5.2 for some specific examples).

In many plant systems, two sources of infection need to be considered (Gilligan & Kleczkowski, 1997): infection from externally introduced sources ('primary inoculum'), perhaps as a result of immigration of an infected vector, dispersal of infected seeds or pathogen spores; and infection from contact between infectious and susceptible individuals ('secondary infection'; this is the sort usually considered in animal models, and usually modelled assuming random encounters – Box 1.2). Furthermore, some parasitic plants can parasitise multiple host individuals (of the same or different species) simultaneously (see Section 5.4.1). This 'mixed diet' parasitic strategy is quite unlike any animal parasitic strategy to our knowledge, and, at the very least, presents considerable problems for mathematical modelling of these systems and for extrapolation of community-level effects.

Plant responses to infection also differ to those of animals. Plants can respond to infection by reducing or increasing the production of susceptible tissue, altering the rate of photosynthesis or the allocation of photosynthate and nitrogen to infected tissue. There is also overlap between plant responses to infection by different parasites: there is a degree of 'cross-talk' between immune responses such that infection by one agent can increase or reduce resistance to infection by a different parasite (or possibly a completely different class of parasite; see Section 5.3). The plant literature divides disease systems into above-ground and below-ground infections; the processes governing infection dynamics, the types of parasites involved and, often, the research laboratories involved in their study can be quite different. This dichotomy has led to the need to study the processes within plants that mediate interactions between above- and below-ground disease systems (Van der Putten et al., 2001; Bardgett et al., 2006).

5.1.2 Parasites of plants

The types of plant parasites are also highly diverse. As with animal systems, we can distinguish between microparasites and macroparasites, but the distinction is not always useful as many other characteristics of life history, infection processes and plant response also differ between types. Microparasites of plants include most of the fungal, bacterial and viral pathogens of plants, of which there are both above- and below-ground specialists (Section 5.2). However, many rust fungi might better be considered as macroparasites as they can form long-lasting chronic infections that repeatedly produce infectious propagules (Dobson & Crawley, 1994). The true macroparasites include below-ground plant parasitic nematodes; arguably, root-consuming insect larvae, aphids and other phloem-sucking insects can also be considered as macroparasites (Section 5.3.1). There are also several thousand species of parasitic plants, species that form intimate parasitic relationships with host plants and are wholly or partly dependent on host metabolism as a source of nutrition (Section 5.4). In addition, there are endophytes: species of fungi (and also some bacteria) that cross the parasitism–mutualism divide, in many environmental conditions conferring advantages to their hosts. These are considered in Section 5.5.

5.1.2.1 Landscape-scale effects

Fungi, bacteria and viruses comprise the majority of plant microparasites or pathogens, and infection or control measures to prevent it are responsible for considerable economic costs in agricultural systems (Pimentel *et al.*, 2000). There are also several well-documented examples of direct effects on natural plant communities as a result of exposure to virulent pathogens, many with dramatic consequences that reach beyond the host species. Classic examples with extreme outcomes frequently involve invasive or emergent pathogen species, often fungal diseases of dominant tree species. These include Dutch elm disease (caused by the fungi *Ceratocystis ulmi* and *C. novo-ulmi*), which radically altered the British landscape in the 1970s through its destruction of mature elm trees (Gibbs, 1978); and chestnut blight (*Cryphonectria parasitica*) in the United States, which essentially removed the American chestnut from eastern deciduous forests (reviewed with others in Loo, 2009). These examples could equally be considered in the context of invasion biology (Chapter 6), ecosystem processes (Section 7.2) or emerging disease (Chapter 8). These virulent plant pathogens have resulted in dramatic and quite likely permanent (in ecological time) changes in communities, apparent even at the landscape scale (for an accessible historical perspective, see Money, 2007); the indirect effects on the constituent species are likely to be far-reaching and diverse.

5.1.2.2 Plant–pathogen population dynamics

General models of plant–pathogen interactions have mostly followed the form developed for animal–parasite systems (Anderson & May, 1981; Box 1.2), with various modifications included in an attempt to deal with some of the differences between plant and animal systems described in Table 5.1. However, many such modifications (for instance, relaxation of the mass action assumption to account for localised transmission) quickly lead to intractable mathematics. Hence general principles gleaned from animal models are often applied (with the addition of caveats that certain assumptions might not be met). Plant models (especially those involving more than one potential host or parasite species) are relatively few and far between, and are more often tailored to specific systems (see examples in Sections 5.3 and 5.4).

Spatially explicit simulation models (e.g. Onstad & Kornkven, 1992; Kleczkowski et al., 1997) demonstrate the importance of spatial structure in limiting the spread of plant diseases. The results of such simulation models are difficult to generalise with rule-of-thumb equations, so alternative approaches have been attempted. Models incorporating non-linear density- or frequency-dependent transmission provide a step towards this goal; for instance, the model of Thrall and Jarosz (1994) provides good fit to experimental data on pathogen spread for one system. Gubbins et al. (2000) model parasite transmission as a functional response to host density in an effort to circumvent the mass action assumption. A decelerating response to host density (probably closest to modelling local spatial effects, but still not explicitly spatial) yields similar results to standard animal systems, with a threshold density of plant hosts and parasite basic reproductive rate $R_0 > 1$ required for pathogen invasion. Gubbins et al. note that in many agricultural situations, crops may be planted in a field already infested with pathogens (this may be the case for many soil-borne generalist pathogens and nematodes). The invasion criteria they derive imply that a threshold density of susceptible hosts is required for pathogen invasion, so if crops are planted at sufficiently low densities (or are inter-cropped with resistant strains or species), an epidemic should be avoided. This type of

Table 5.1. *General models of one-plant–one-pathogen interactions*

System	Outcomes	Study
Simulation to examine threshold iR for foliar pathogen spread (iR = potential reproduction per pathogen; similar to R_0), with terms for frequency of plant leaflet infection (infected leaflets may have latent infection, be infectious or post-infection).	Pathogen spread required $iR > 1$, but this will not guarantee spread. (The threshold for spread depends on loss of propagules during dispersal and infection process.) Cannot predict epidemic on the basis of summed level of disease (latent, infectious and removed lesions).	Onstad (1992)
Spatially explicit simulation to examine endemic persistence of foliar pathogen; host growth response to infection (increased compensatory production of susceptible tissue); localised transmission to neighbouring grid cells.	Requires $iR \gg 1$ (iR = potential reproduction per pathogen; similar to R_0) if there is heterogeneity in host susceptibility. Heterogeneity in iR increases persistence. Increased host response increases persistence.	Onstad and Kornkven (1992)
Non-spatial simulation of equations to examine persistence of sexually (vector) transmitted disease (anther smut fungus). Density- or frequency-dependent transmission; resistant or susceptible populations.	Non-linear (diminishing returns) frequency-dependent transmission provided best fit to data. Knowledge of susceptibility/resistance and between-generational effects (mortality vs sterility) important for predictions.	Thrall and Jarosz (1994)
Spatially explicit stochastic (probabilistic) simulation of soil-borne pathogen (damping-off disease), considering both primary (soil-to-plant) and secondary (plant-to-plant) transmission and non-linearities on secondary transmission	Population-level simulation results (proportion of plants infected) provided good fit to data when transmission rates parameterised from microcosm experiments. Small stochastic differences in primary transmission become	Kleczkowski *et al.* (1997); (A) Gilligan and Kleczkowski (1997)

Table 5.1. (*cont.*)

System	Outcomes	Study
forced by spatial structure. Numerical and analytical methods developed in (A) describe within-season epidemics of this process.	amplified by secondary transmission, reinforcing heterogeneity in disease distribution.	
Transmission as functional response to density; host growth/reproductive allocation response to infection; primary and secondary inoculum.	Non-linearities in transmission and host response can lead to threshold; parasite primary inoculum required for invasion; primary and secondary transmission additive in maintaining infection; in some cases parasites with $R_0 < 1$ can initially invade.	Gubbins *et al.* (2000)

approach has recently been extended to examine invasion of and control strategies for plant pathogens (Gilligan & van den Bosch, 2008).

5.2 Soil-borne pathogens

Here we examine some of the below-ground interactions between plants and their parasites which have measurable population or community effects. We use the term 'pathogen' loosely, following the literature (e.g. Van der Putten & Peters, 1997; Olff *et al.*, 2000) to include soil-borne organisms with deleterious effects on plants that are not necessarily microscopic. These include many nematodes (some of which are endo-parasites, others root-consuming herbivores), fungi, bacteria and oomy-cetes (which are fungus-like organisms, tolerant of many fungicides). To aid comparison, we also discuss the community effects of some beneficial microbes in this section (positive plant–soil interactions (Section 5.2.2)).

There is great scope for interactions between organisms in the rhizosphere, and we have only just begun to elucidate these complex interactions. For most systems, the composition of rhizosphere organisms has not been identified, and evidence for its impact comes from greenhouse culture of plants on 'home' or 'away' soils (soil collected from conspecifics or heterospecifics, respectively), comparing growth or survival relative to sterilised soils. This explains the general use of the term 'pathogen' as a cover-all; in some instances a whole community of deleterious (and some beneficial) species could be involved. For instance, biological control of the leafy spurge *Euphorbia esula/virgata* (an invasive species in North America) is facilitated by synergistic interactions between root-eating larvae of flea beetles (*Aphthona* spp.) and a variety of pathogenic bacteria and fungi. However, in some communities, flea beetle attack is associated with less pathogenic bacteria that inhibit colonisation of the more lethal pathogens (Caesar & Caesar-Ton That, 2008). The complexity of outcomes in such systems is a challenge for future research in this area.

5.2.1 The Janzen–Connell effect

Natural enemies have for some time been implicated in the regulation of plant diversity. Perhaps the best-known formulation of this idea is the Janzen–Connell hypothesis (Box 5.1), which proposed that specialist herbivores are key to maintaining tree diversity in tropical rainforests. In this model, density-dependent species-specific consumption of seedlings close to the parent creates opportunities for colonisation by other species, enforcing a degree of diversity in contrast to the single-species stands that would be expected in the absence of density-dependent mortality effects. Evidence for the Janzen–Connell process as induced by herbivores in tropical rainforests is available (but debatable; see Freckleton & Lewis, 2006); however, the process has been applied with more success to the role of below-ground plant pathogens. In particular, field census and manipulative laboratory studies suggest that soil-borne pathogens can be responsible for 'self-thinning' of seedlings in species of *Prunus* (cherries and plums), greatly reducing the chances of successful establishment of saplings within a considerable distance (15 m or so) of mature conspecifics (Packer & Clay, 2000; Box 5.1). There are a number of reasons why pathogens may be more effective than herbivores at inducing a Janzen–Connell effect. Unlike many herbivores, they do not exhibit a 'satiation response' at high prey/host densities; indeed, dense monocultures are more likely to allow higher rates of pathogen transmission, so exacerbating the density-dependent mortality effect. Many pathogenic soil fungi, including the species involved in *Prunus* thinning, can consume living or dead plant material; there is therefore little selection to reduce virulence in these facultative pathogens, making

Box 5.1 The Janzen–Connell effect

The Janzen–Connell effect was originally proposed as a mechanism to explain high diversity in tropical rainforests, the idea being that natural enemies such as insect seed predators and herbivores cause seedling mortality, which becomes more intense as plant densities increase (Janzen, 1970; Connell, 1971). This mechanism requires two features: (a) mortality is density-dependent (the per capita rate of mortality is an increasing function of density); and (b) the natural enemies involved are specialist (disproportionately attacking a focal species at a locally high density rather than the rare alternatives). If these conditions are met, diversity should be enhanced as the probability of successful establishment is diminished close to conspecific trees. More recently, the focus has turned from herbivores to pathogens as Janzen–Connell agents; in particular, there is evidence that fungal pathogens – particularly Oomycota – which cause damping-off disease in young seedlings may give rise to this effect (reviewed in Freckleton & Lewis, 2006). Some of the first, and best, evidence for this comes from the work of Augspurger (1983; 1984), who showed that recruitment of canopy tree seedlings on Barro Colorado Island, Panama, was almost completely prohibited in the immediate shade of parent trees owing to the high and rapid mortality caused by damping-off diseases.

Packer and Clay (2000) provide evidence for Janzen–Connell in temperate systems. Field censuses of the black cherry (*Prunus serotina*) showed that the mean distance to parents for seedling cohorts shifted away from parent trees over a three-year period. This pattern was produced by strong density-dependent mortality; although the number of seeds germinating was highest close to the parent, the probability of survival to sapling stage was lowest (Fig. 5.2a). Two sets of evidence demonstrated the role of parasitism in observed seedling mortality. Three strains of fungi (*Pythium* spp.) were isolated from the roots of dying seedlings and caused damping-off disease when inoculated into seedlings grown, resulting in a three- to four-fold increase in mortality. In a further experiment, seeds were grown in soil collected close to (0–5 m) or distant from (25–30 m) *Prunus* trees; soil sterilisation improved survival of seedlings grown in close soil but not in distant soil, again suggesting that soil pathogens accumulate close to *Prunus* trees, reducing survival of *Prunus* seedlings grown in that soil.

Box 5.1 (continued)

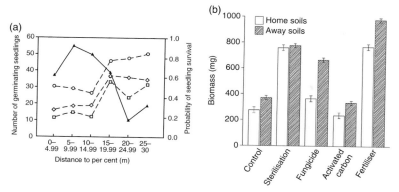

Fig. 5.2. Seedling germination, survival and growth rates in relation to soil. (a) The number of germinating *Prunus* seedlings (black triangles) and probability of seedling survival at different distances from parent trees (circles; 4 months after germination; diamonds: 16 months; squares: 28 months). From Packer and Clay (2000). (b) Biomass of plants raised in 'home' or 'away' soils for various soil treatments. Only sterilisation eliminates the disadvantage of growing on home soils for grass, forb and legume species (combined results). From Petermann *et al.* (2008).

Petermann *et al.* (2008) demonstrated widespread Janzen–Connell-like effects in greenhouse experiments on non-successional temperate grassland species. They grew species from three different functional groups (grasses, forbs and legumes) in soils from field monocultures ('home' soil from conspecifics, or 'away' soil – a mix from the other two functional groups). 'Home' soil reduced growth (measured as plant biomass after eight weeks) for all three functional groups. This effect was stronger when plants were grown with species from the other functional groups, suggesting that the pathogenic effects of soil microbes were exacerbated by competition with healthy (heterospecific) plants. Soil irradiation eliminated the negative effect, demonstrating the involvement of soil microbes. The pathogens involved in this study were unlikely to be damping-off fungi, as similar (in fact, stronger) home–away effects were found in fungicide-treated soils compared to untreated soils (Fig. 5.2b). In this system, it appears that soil fungi may be suppressing other, more serious pests such as bacteria or fungicide-tolerant fungus-like organisms. The exacerbating effect of interspecific competition in this study suggests that pathogens here had trait-mediated indirect effects on plant fitness, affecting plant competitive ability rather than mortality (equivalent to parasite-modified competition in animals (Section 2.5)); nevertheless, if these effects transfer to natural environments, they could be strong enough to structure plant communities.

them effective agents for inducing the strong, compensating density dependence required in the Janzen–Connell model.

5.2.2 Plant–soil feed back

Packer and Clay (2000) and related studies have led to the concept of plant–soil feed back (PSF), in which specialist microbes (having negative or positive effects on the host species) accumulate in the soil over time, altering the soil environment for that plant population. PSFs can be negative or positive; the Janzen–Connell effects described in Box 5.1 are examples of negative PSFs. Negative PSFs, caused by specialist pathogen build-up, may contribute to 'sick soil' syndrome (traditionally assumed to be the result of nutrient depletion, which led to the practice of crop rotation in agriculture). From a PSF perspective, crop rotation avoids negative PSF by planting with species from a different functional group until group-specific pathogen levels have returned to an acceptable level (usually three or four years). Negative PSFs are predicted to maintain diversity by the same mechanism as Janzen–Connell effects (Bever *et al.*, 1997).

5.2.2.1 Negative PSFs

Negative PSFs have also been shown to contribute to plant succession sequences, accelerating deterioration of soil conditions for resident species and preventing their immediate recolonisation. Pathogen-driven succession can operate both in time and space and may be responsible for

---

Box 5.2 Pathogen-driven succession

In coastal foredunes in northwestern Europe, a vegetation succession occurs, driven to some extent by sand stabilisation, mineral/nutrient changes and eventual invasion of woody shrubs. Marram grass (*Ammophila arenaria*) initially colonises sites near the beach which are subject to wind-blown sand and low soil moisture content. After a period of optimal growth, marram grass stands deteriorate and are replaced by fescue (*Festuca rubra arenaria*), then sand sedge (*Carex arenaria*), with sea couch (*Elymus athericus*) moving in from inland of the dune (Fig. 5.3a). Van der Putten *et al.* (1993) demonstrated that

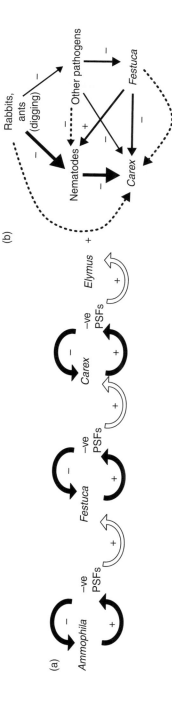

Fig. 5.3. Directional and cyclical succession of grass species. (a) Schematic of directional succession sequence on sand dunes, driven by progressive stabilisation and species-specific negative PSFs (solid arrows indicate direct interactions; clear arrows indicate indirect interactions). (b) Interaction web for cyclical succession of inland pasture, driven by mammalian grazing and negative PSFs (solid arrows indicate direct effects; dashed arrows indirect effects; thicker arrows depict larger effects). From Olff *et al.* (2000).

Box 5.2 (*continued*)

soil-borne pathogens play a part in this succession, accelerating the replacement of the early successional species. They grew seedlings in sand collected from the root zone of natural monocultures of each of the dominant species, comparing biomass from sterilised (gamma irradiated) and non-sterilised treatments. Growth rate for each species was reduced in non-sterile soil from species later in the succession, but was not strongly affected by soil from the species that preceded in the succession. This suggests that specific growth-depressing microbes build up in the soil and contribute to the elimination of each species in turn. A variety of fungal pathogens, endoparasitic nematodes and herbivorous nematodes appear to be involved, in some cases acting synergistically (for instance, root-eating nematodes may increase exposure to fungal infection). For the *A. arenaria/F. rubra* turnover, pathogen-modified competition also occurs, as the pathogens of *A. arenaria* alter the host's root allocation patterns, reducing its ability to compete under nutrient limitation (Van der Putten & Peters, 1997).

Inland, succession to woody species is prevented by the presence of large herbivores, and successional dynamics are locally cyclical between dominant grassland species. Negative PSFs are implicated in driving successions here too, leading to shifting mosaics of patches dominated by different species. Olff *et al.* (2000) examined the dynamics of the two dominant grasses (*C. arenaria* and *F. rubra*) in river margin pasture in the Netherlands over 17 years, finding that the two species locally alternated in abundance, with neighbouring sites out of phase such that *Carex* and *Festuca* patches seemed to move over time as a wave across space. Greenhouse experiments involving culture of seedlings in soil from patches where each species was either declining or increasing were used to determine the role of soil microbes. These experiments demonstrated that both species performed poorly on soil from declining populations of conspecifics, compared to sterilised soil. Further experiments implicated herbivorous nematodes (in the case of *Carex*) and other, different (probably fungal) pathogens (in the case of *Festuca*). There appears to be a complex web of direct and indirect interactions in this grassland system (Fig. 5.3), including apparent competition between the grasses (*Festuca* can support herbivorous nematode populations deleterious to *Carex*) and digging activity by rabbits and ants influencing local densities of soil microbes with indirect effects on the plant species.

succession sequences on sand dunes, and for shifting mosaics of dominant species in grasslands (Box 5.2).

The studies in Box 5.2 illustrate that pathogens causing negative PSFs can have local spatial, as well as temporal, effects on population dynamics. General spatial simulation models of two species competing demonstrate that negative density dependence can generate a range of spatially heterogeneous patterns, including oscillatory or stable clumped distributions and spatially stable barred distributions (Molofsky et al., 2002). However, whether the mechanisms described in this model underlie the patterns observed in Netherlands grasslands is unclear; spatial patterns were sensitive to parameter values and scales of action, and in particular depended on the dispersal distance relative to the scale over which negative effects can act.

5.2.2.2 Positive PSFs

Positive PSFs occur when there is accumulation of microbes beneficial to the host, and can be caused by mycorrhizal fungi and endophytes (see Section 5.5). Positive PSFs can also have consequences for community structure, improving the relative performance of established species and preventing colonisation by alternative species. If positive PSFs occur for several of the dominant species, their action will tend to reinforce abrupt boundary changes between species (Molofsky et al., 2001). Positive PSFs can result in failure of translocated species to establish when introduced without their symbionts. For example, attempts to introduce conifers in the tropics failed until trees were inoculated with appropriate ectomycorrhizal (EM) fungi; now established, these species are potential invasive threats (Rejmanek & Richardson, 1996).

5.2.2.3 Large-scale applications of PSF theory

Reynolds et al. (2003) review positive and negative PSFs as explanations of large-scale diversity patterns in plants. They suggest that extreme environments tend to favour more positive PSFs (for example, more nitrogen fixers in nitrogen-poor environments), whereas in more productive environments, there is less selection for symbiotic associations (indeed, the metabolic cost of supporting such microbes can outweigh the benefit (Section 5.5.2)). Since low-altitude and low-latitude environments tend to be more productive, Reynolds et al. suggest that these will be dominated by negative PSFs. This will result in feed-back mechanisms enhancing succession and diversity in the productive tropics (negative PSFs) and reducing diversity at higher latitudes (positive PSFs). Although an elegant theory, a thorough test with empirical evidence is still required.

In a meta-analysis, Kulmatiski *et al.* (2008) found good evidence that negative PSFs enhance succession and diversity across a wide range of systems, but less evidence for the role of positive PSFs. The strength of PSF was associated with plant type, and patterns of occurrence tended to contradict the pattern predicted by Reynolds *et al.*'s (2003) verbal model. In particular, negative PSFs were strong in grasses (the species for which most data were available), but less apparent in woody species; annual species also exhibited stronger negative PSFs than perennials. Kulmatiski *et al.* interpret their findings in terms of the biology of the host: grasses have relatively high exposure to below-ground enemies – for instance, having high root–shoot ratios compared to woody species; annuals trade off defence for higher growth rates and are therefore more susceptible to soil pathogens. Under this interpretation, early successional species (annuals, grasses) tend to have stronger negative PSFs than late successional species (perennials, woody species). These findings contrast with the Reynolds *et al.* (2003) model, which suggests that late successional communities will be dominated by negative PSFs.

PSFs may also help to explain why some plant species become invasive whilst others remain non-invasive. Invasive species in the novel habitat may have relatively low negative PSFs, having escaped their pathogens (enemy release – Section 6.3.3), whereas non-invasive introduced species will tend to have similar PSFs to native species (Reinhart & Callaway, 2006). From this line of reasoning, we can deduce that early successional species (annuals, grasses) will be successful invasives because they have the most to gain from losing their negative PSFs (Kulmatiski *et al.*, 2008; see also Section 6.4).

5.2.3 Pathogen-modified and apparent competition

The above discussion illustrates how PSF, mediated by specialist pathogens, can alter spatial distribution and successional dynamics in plant communities via the effects of the pathogen on a dominant species. Generalist pathogens have also been implicated as drivers of plant community structure via the processes of apparent competition and pathogen-modified competition (Chapter 2). A well-studied plant–pathogen system, barley

Box 5.3 Barley yellow dwarf virus in grasses

Barley and cereal yellow dwarf viruses (referred to collectively as BYDV) are phloem-limited luteoviruses that infect a broad range of cultivated and wild annual and perennial grasses. BYDV is obligately transmitted by several (at least nine) aphid species and leads to incr- eased mortality and reduced fecundity of host grasses (Malmstrom *et al.*, 2005a). BYDV-induced stunting and yield losses in crops such as wheat, barley and oats make this one of the most economically significant plant pathogens of cereal crops (Pimentel *et al.*, 2000).

Field plot experiments demonstrate how BYDV can mediate apparent competition, altering dominance relationships between annual grasses. Power and Mitchell (2004) grew annual grass species (wild oats (*Avena fatua*; *Setaria lutescens*); brome (*Bromus tectorum*); Italian ryegrass (*Lolium multiflorum*; *Digitaria sanguinalis*; *Echinochloa crus-galli*; *Panicum capillare*)) in plots in Ithaca, New York, introducing a BYDV-carrying or uninfected aphid vector *Rhopalosiphum padi*. Wild oats are strongly susceptible to BYDV and infected plants harbour high concentrations of the virus; experiments revealed that they acted as a reservoir for BYDV infection to other species. In infected plots containing *A. fatua*, virus prevalence in other species was more than five times that in plots without *A. fatua* (Fig. 5.4a).

The consequences for community composition depended on the competitive relationships between the grass species (Power & Mitchell, 2004). In virus-inoculated communities (but not in virus-free plots), adding *A. fatua* decreased the reproductive and vegetative biomass of three other species, demonstrating virus-mediated apparent competi- tion (Fig. 5.4b). In virus-free plots, adding *A. fatua* actually increased the biomass of two of these other species, because competition between *A. fatua* and the otherwise dominant *S. lutescens* released the two subordinate species from interspecific competition. Hence BYDV also modifies direct competition in this system, potentially reversing the sign of the indirect interactions in competitive hierarchies.

---

yellow dwarf virus (BYDV) and its annual and perennial grass host species, shows evidence for BYDV-mediated apparent competition resulting in changes to plant community composition in experimental field communities (Box 5.3, Fig. 5.4).

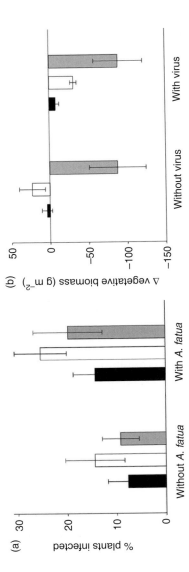

Fig. 5.4. (a) Virus prevalence in three grass species grown in experimental communities with or without the grass *Avena fatua* (black bars: *Digitaria sanguinalis*; white bars: *Lolium multiflorum*; grey bars: *Setaria lutescens*). (b) Above-ground vegetative biomass changes on addition of *A. fatua* with or without the virus (grass species as in (a)). From Power and Mitchell (2004).

In the BYDV system, both apparent competition and pathogen-modified competition are important in determining community outcomes. Apparent competition occurs as some species act as a reservoir for the virus or aphid vector (Box 5.3). In addition, BYDV disproportionately reduces the competitive strength of some species, altering competitive dominance relationships, thereby playing a potential keystone role in grassland diversity. Indeed, such keystone effects of BYDV may underpin the ecological replacement of native perennial bunchgrasses by invasive annual species in the Californian grasslands (Box 6.4).

5.2.3.1 Interactions between pathogens

In the real world, plants are frequently exposed to more than one pathogen at once, and interactions between pathogen species could in principle enhance or interfere with the effects of one pathogen species alone. Regarding the plant host as a shared resource, interactions between pathogens could take the form of intraguild predation (IGP), with all its concomitant complexities (Chapter 4). The relevance of IGP theory to plant–parasite interactions does not seem to have been examined previously, but some empirical studies of multiple plant parasites hint at similar outcomes to those found in animal–parasite IGP systems. A recent meta-analysis revealed no clear pattern of interaction between enemies and effects on plant growth (Morris *et al.*, 2007). However, reminiscent of several IGP systems (Section 4.4.2), individual cases revealed strong synergistic or antagonistic effects on host performance. For instance, Bradley *et al.* (2008) used mesocosm experiments to examine the interaction of a fungal and a bacterial pathogen on a community of four species of *Brassica*. Both pathogens maintained diversity in the short term, reducing the rate at which less competitive species were eliminated. However, subordinate *Brassica* extinction rates were higher in mesocosms inoculated with both pathogen species than in single-pathogen treatments, suggesting some form of interference between the pathogens. This example illustrates another way in which plants can respond to parasitism. The mechanism thought to underlie the patterns observed here involved compensatory reproduction rather than resource competition effects: the seed set of the less abundant species increased significantly in the presence of pathogens. This response may provide an adaptive advantage to the host plant, as it increases the chances of reproduction before succumbing to the infection; such responses have also been reported in plant–herbivore interactions. Fecundity compensation has also been documented in

animals in response to parasitism; for example, snails respond to trematode infections with increased egg production in the weeks preceding
parasite-induced castration (Sorensen & Minchella, 2001).

5.3 Plant defence strategies

Interactions involving parasites and plants are inevitably channelled
through the plant's defence mechanisms. Attack by plant enemies initi-
ates a defence response; a complex series of metabolic steps along one
(or more) of several biochemical pathways. Different classes of enemy
activate different pathways and elicit different responses, but there is also
overlap in the paths activated and the responses these set off (Section
5.3.1). This overlap can lead to 'cross-talk' between defence pathways,
inhibiting or reinforcing plant defence against multiple enemies (Section
5.3.2). Defensive responses include the production of volatile organic
compounds (VOCs), which have been shown to attract natural enemies
of the plant's consumers, but may also be manipulated by some pathogens to enhance their own transmission (Section 5.3.3). Many responses
are systemic (not confined to the point of attack, conferring change
throughout the plant); consequently, below-ground attack has the
potential to influence the success of above-ground attack, and vice versa
(Section 5.3.4). The population dynamic and community consequences
of these defence-related interactions have yet to be examined, and could
prove a fruitful area for future research. Since many of these interactions
involve trait-mediated indirect effects (Box 1.3), we might predict them
to have strong, short-term effects that could have important, potentially
counter-intuitive consequences for population dynamics.

5.3.1 Chemical signalling pathways

Most work has concentrated on two key pathways involved in plant
defence, each triggered by different natural enemies. In general, attack
by wound-inducing herbivores and necrotrophs stimulates the production of jasmonic acid (JA), whereas biotrophic pathogens stimulate salicylic acid (SA) production. JA and SA production result in unique
chemical cascades via changes in gene expression and secondary metab-
olite production. More recently, other pathways have been identified,
stimulated by ethylene (ET) and abscisic acid (ABA); these may work
independently or in concert with JA and SA. Some insects (especially
piercing and sucking species such as aphids and whitefly) induce

pathogen–like responses (stimulating SA and JA), and some pathogens induce wound–like responses (Inbar & Gerling, 2008).

Stimulation of these pathways leads to a complex array of changes in gene expression; for instance, aphid feeding on *Arabidopsis* caused up-regulation of 832 genes and down-regulation of 1349 others (De Vos *et al.*, 2005). Defensive metabolic products include proteinase inhibitors, phenolics and other toxins that alter insect feeding, behaviour, growth and reproduction, VOCs (Section 5.3.3) that repel attackers or attract predators (from the JA pathway), and oxidative and hydrolytic enzymes targeted at denaturing pathogen surface or viral coat proteins (from the SA pathway (Fig. 5.5)). Plant development is also affected; for instance,

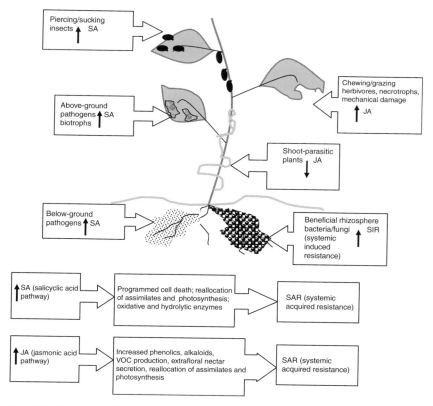

Fig. 5.5. Salicylic acid (SA) and jasmonic acid (JA) defensive pathways stimulated by various plant enemies. There is overlap in the pathways stimulated by different categories of parasite (above-ground vs below-ground, or pathogen vs herbivore) and overlap in the responses elicited by the pathways, leading to the potential for cross-talk and cross-effect in plant defensive responses.

via reallocation of assimilates from roots to leaves, increased photosynthesis in undamaged leaves and reallocation of resources to (or away from) reproduction. Bostock (2005) reviews the genetics and biochemistry of interactions between pathways; for a more ecological perspective see Hatcher *et al.* (2004) and Stout *et al.* (2006); Taylor *et al.* (2004) examine how to integrate between these divergent fields.

5.3.2 Multiple enemies: positive and negative cross-talk

Here we examine how plant defence signalling can alter interactions between multiple enemies. Plants are often under attack from more than one enemy simultaneously. How plants and consumers fare (and, consequently, the fitness outcomes for both) depends, in part, on plant defence signalling and response. We concentrate on phytopathogen interactions with insect herbivores; herbivore–herbivore and phytopathogen–phytopathogen interactions also occur. Such interactions are referred to as *tripartite* (Fig. 5.6a), as distinct from *tritrophic* interactions, which involve plants and other species at different trophic levels (Fig. 5.6b). The population dynamic consequences of plant defence-mediated interactions have not been explicitly studied, but are likely to be complicated, as trait-mediated indirect effects (Box 1.3) are involved: plant traits are modified by attacker 1; this alters the interaction between the plant and attacker 2 (Fig. 5.6). Hence

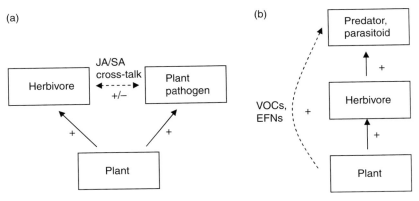

Fig. 5.6. Three-way interactions involving plants: (a) tripartite; (b) tritrophic. Solid lines depict direct interactions, dashed lines indirect interactions resulting from plant chemical defence signalling (JA: jasmonic acid pathway; SA: salicylic acid pathway; VOCs: volatile organic compounds; EFNs: extrafloral nectaries and other structures to encourage enemies of herbivores). Direct negative interactions (not shown) oppose the positive ones.

competition between the two enemies for the plant is mediated by host defences, in a situation that is analogous to host-mediated competition between parasites of animals (Section 2.7.2).

Because plant defence pathways overlap in terms of inducers and products, there is potential for cross-talk, in which defence pathways inhibit or reinforce each other (Fig. 5.4), or cross-effect (in which the defence pathways are independent, but the active products from one pathway are active against alternative attackers). In addition, plant defence systems can be 'primed': the speed and magnitude of response to further attacks is enhanced after exposure to previous attacks (referred to as systemic acquired resistance (Bostock, 2005). Therefore, both sequential and simultaneous attacks by different enemies can be subject to cross-talk or cross-effect.

There are numerous empirical examples of both negative and positive direct and indirect effects of pathogens on phytophagous insects (reviewed in Hatcher, 1995; Stout et al., 2006; Table 5.2). For example, in dock (Rumex spp.) a positive cross-effect reduced the fitness of beetles Gastrophysa viridula on plants infected with the fungal rust Uromyces rumicis. In laboratory manipulations, rust infection of Rumex substantially reduced growth, survivorship and fecundity of the beetle (Hatcher et al., 1994a). In addition, pathogen fitness was affected: field and laboratory studies demonstrated that beetle grazing induces resistance to infection by Uromyces and two other fungal pathogens (Hatcher et al., 1994b).

5.3.2.1 Applications in weed biological control

The existence of bidirectional inhibition of pests in the Rumex and other systems makes it difficult to predict the effect of combined attack by multiple natural enemies on plants, a potentially important strategy in biological control (Hatcher & Paul, 2001; Taylor et al., 2004). Such negative interactions can lead to less-than-additive effects of combined attack on weeds, although the overall effect may still be greater than using a single enemy species (for instance, spatial separation of enemies on individual plants can mitigate against negative effects (Hatcher et al., 1994b)). Pathogen-induced deleterious effects on insect herbivores are, in part, caused by reduced nutritional quality (and, in some cases, quantity) of infected plant parts. Herbivore-induced resistance to pathogen infection would appear to be a case of SAR, mediated by overlap of the herbivore and pathogen defence signals or effects. In other cases, the

Table 5.2. *Examples of inhibitory or synergistic interactions between plant natural enemies (a non-comprehensive list)*

System	Effect	Reference
Dock (*Rumex* spp.), foliar rust (*Uromyces rumicis*), leaf-feeding beetle (*Gastrophysa viridula*)	Bidirectional positive cross-talk or cross-effect: reduced growth, fecundity and survivorship rate of beetles feeding on rust-infected plants; reduced infection probability of rusts on beetle-grazed plants	Hatcher *et al.* (1994b); Hatcher and Paul (2000)
Tomato (*Lycopersicon esculentum*), pathogen *Pseudomonas syringae*, moth (*Helicoverpa zea*)	Positive cross-talk: caterpillar growth rate reduced 50–80% on uninfected leaves of infected plants (indicating systemic acquired resistance)	Stout *et al.* (1999)
Peanut, white mould fungus (*Sclerotium rolfsii*), army worm moth (*Spodoptera exigua*), parasitoid (*Cotesia marginiventris*)	Negative cross-talk: caterpillar feeding rate, survival and weight at pupation 20–25% higher on infected plants; but also increased attraction of parasitoid to dual-infected plants	Cardoza *et al.* (2003)
Rice (*Oryza sativa*), planthopper (*Sogatella furcifera*), rice blast (*Magnaporthe grisea*)	Positive cross-talk or cross-effect: planthopper-infested plants have greatly enhanced resistance to rice blast infection	Kanno and Fujita (2003); Kanno *et al.* (2005)
Alfalfa (*Medicago* spp.), army worm (*Spodoptera ornithogalli*), planthopper (*Spissistilus festinus*), aphid (*Acyrthosiphon pisum*), crown rot (*Fusarium oxysporum*), root rot (*F. roseum*)	Enemy-specific effects: army worm infestation did not affect resistance to crown rot; planthopper infestation increased disease severity of crown rot; aphids increased severity of root and crown rot	Leath and Byers (1977)
Tomato (*Solanum lycopersicum*), dodder (*Cuscuta pentagona*), army worm (*Spodoptera exigua*)	Negative cross-talk, but inhibitory effect: *C. pentagona* reduces JA production and release of VOCs, reducing resistance to army worm; but reduced growth rate of army worm on infected plants indicates reduced nutritional quality	Runyon *et al.* (2008)

opposite pattern is found: negative cross-talk between signalling pathways can lead to inhibition of defence against a second attacker after exposure to a different attacker (Table 5.2). In this scenario, combined attack should result in a better-than-additive effect, as defensive capabilities against one or both enemies in a combined attack are lower than those raised to a single threat. Predicting the outcome of combined control therefore requires careful consideration of the potential for cross-talk and cross-effect: a review of phytopathogen–arthropod laboratory/greenhouse studies found no clear pattern, with no discernable effect in half of the studies, inhibitory effects in one-third of studies and facilitation in the remainder (Stout *et al.*, 2006).

5.3.2.2 Applications in crop protection

Cross-talk or cross-effect may also have applications for the plants we wish to protect. However, the same lessons from weed control apply: there are no clear patterns, and without knowledge of the particular interaction (and the relative costs associated with attacks), outcomes cannot be predicted and strategies cannot be weighed (Table 6.2). Furthermore, plant defensive responses probably come at considerable metabolic cost; otherwise, plants would be selected to mount constitutive defences of this level at all times (Paul *et al.*, 2000). This is one reason why some authors caution against the use of defensively up-regulated engineered plants as a general strategy for crop protection; likely the immediate cost will be paid in terms of reduced yield and product quality (Bostock, 2005; Bruce & Pickett, 2007).

5.3.2.3 Rhizosphere bacteria and fungi

Another strategy for crop protection might be to utilise non-pathogenic bacteria of the rhizosphere (root–soil system); these plant-growth-promoting rhizobacteria can elicit induced systemic resistance (ISR) against viral, fungal and bacterial diseases. Bacterial genera including *Bacillus*, *Agrobacterium* and *Pseudomonas* can cause ISR, and many also have plant-growth-promoting effects (reviewed in van Loon *et al.*, 1998; Bruce & Pickett, 2007). In greenhouse and field trials, a variety of crops, including tomatoes, peppers, water melon and tobacco, inoculated with *Bacillus* spp. became less severely infected or damaged by various pathogens, including viruses, foliar fungi and bacteria, root knot nematodes and damping-off fungi (Kloepper *et al.*, 2004). Treatment of *Arabidopsis* (a model plant system for genetic and biochemical analysis of defence pathways) with *Pseudomonas fluorescens* afforded strong protection against several foliar bacterial pathogens and fungal pathogens. These studies (Pieterse *et al.*, 2002) indicate that rhizobacteria stimulate JA and ET pathways in a different manner to pathogens eliciting systemic acquired resistance (SAR) (Section 5.3.2). Rhizosphere fungi can act in a similar role; species from the genus *Trichoderma* enhance plant growth (by up to 300% in greenhouse and field trials on lettuce, tomato and pepper plants) and provide protection against soil pathogens via ISR and direct attack. *Trichoderma* species are mycoparasites (parasites of other fungi), producing cell wall degrading enzymes and antibiotics before penetrating and killing pathogen cells; their growth-promoting effect and ability to contain pathogen populations has led to research on their use as biopesticides and biofertilisers (Vinale *et al.*, 2008).

5.3.3 Signalling and manipulation

Thirty years ago, Price *et al.* (1980) argued for a three-trophic-level view on plant interactions, noting that plant–herbivore and predator–prey interactions cannot be understood without inclusion of the third trophic level. A tritrophic perspective was required, they suggested, because plants initiate many direct and indirect interactions with the enemies of their enemies; failure to recognise their importance often leads to counter-intuitive or paradoxical observations. Today, the importance of the tritrophic paradigm has not diminished, and the need to include parasites (both as enemies of plants and as enemies of their enemies) is beginning to be recognised (Fig. 5.6b).

Many plants utilise their chemical defence pathways to interact with carnivores as an indirect defence against herbivores (Heil, 2008; Dicke & Baldwin, 2010). Stimulation of the JA and ET pathways by herbivore attack leads to the production of volatile organic compounds (VOCs) some of which repel herbivores, while others act as attractants to natural enemies (most frequently, parasitoid wasps and predatory mites, but flies, nematodes, bugs and beetles have also been implicated (Table 5.3)). Many plant species also have extrafloral nectaries, and stimulation of the JA pathway by herbivory can increase nectar secretion into nectaries, attracting predators (mainly ants). Other plant attributes including food bodies and domatia attract and house predators, but are not thought to respond to defence signals (they are constitutively expressed). These plant-mediated tritrophic interactions are relevant to our overview because they change

Table 5.3. *Effects of volatile organic compound emission on herbivores, predators and plants (not a comprehensive list)*

System	Effect	Reference
Lima bean (*Phaseolus lunatus*), whitefly (*Bemisia tabaci*), herbivorous spider mite (*Tetranychus urticae*), predatory spider mite (*Phytoseiulus persimilis*)	VOCs repel herbivores and attract predators in laboratory manipulations of detached leaves (A); additional whitefly infestation interferes with spider mite induction of JA pathway, reducing attraction of predatory spider mite (B).	(A) Dicke and Dijkman (1992); (B) Zhang *et al.* (2009)
Tobacco (*Nicotaina tabacum*), herbivorous moths (*Heliothis virescens*, *Manduca sexta*, *Helicoverpa zea*)	VOCs released at night in natural populations under moth attack repel ovipositing moths in field conditions.	De Moraes *et al.* (2001)
Wild tobacco (*Nicotaina attenuata*), three herbivores (moth *Manduca quinquemaculata*; leaf bug *Dicyphus minimus*; flea beetle *Epitrix hirtipennis*), predatory bug (*Geocoris pallens*)	VOCs released in natural populations characterised; synthetic versions increased egg predation rates on manipulated plants in a natural population field experiment.	Kessler and Baldwin (2001)
Maize (*Zea mays*), beet army worm (*Spodoptera exigua*), beetle (*Diabrotica virgifera)*, parasitoid (*Cotesia marginiventris*), predatory nematode (*Heterorhabditis megidis*)	Herbivore attack induces large releases of VOCs from plants (C); VOCs attract predators and parasitioids in laboratory manipulations (C); multiple enemies intererfe with VOC signals, confusing predators (D).	(C) Turlings and Tumlinson (1992); (D) Rasmann and Turlings (2007)
Wild tobacco (*Nicotiana attenuata*), sagebrush (*Artemisia tridentata*), generalist herbivore community (six grasshopper species, two moth species)	Sagebrush bushes clipped in the field induce defence in neighbouring tobacco plants. Tobacco plants near clipped sagebrush had reduced leaf damage in the field.	Karban *et al.* (2000)

Table 5.3. (*cont.*)

System	Effect	Reference
Maize (*Zea mays*)	VOCs prime resistance traits in neighbouring plants; plants previously exposed to VOCs from neighbours produce more VOCs when mechanically damaged.	Engelberth *et al.* (2004)

the frequency of encounters of pests and their predators or parasitoids, potentially altering the population dynamic outcomes of these interactions (Fig. 5.6b). Whether plant signals can be used to attract entomopathogens as plant 'bodyguards' is less clear (Cory & Hoover, 2006). There are some examples of attraction of entomopathogenic nematodes to roots damaged by insect larvae, and in some cases the VOCs involved have been identified (Rasmann *et al.*, 2005). Furthermore, plants can influence arthropod–pathogen interactions in a number of ways, altering pathogen persistence on plant tissues, pathogen infectivity and arthropod resistance, potentially providing the raw material for manipulative or other trait-mediated interactions (Cory & Hoover, 2006).

Manipulation (or something similar) has, however, been demonstrated by pathogens utilising negative cross-talk between pathogen-induced and herbivore-induced defence pathways. Belliure *et al.* (2005) demonstrated that tomato spotted wilt virus increases the developmental rate of its vector, the western flower thrip (*Frankliniella occidentalis*), by down-regulating the direct defences of host pepper plants. This reduced the predation rates on infected thrips by two species of predatory mites, because infected thrips reached an invulnerable size class sooner (Belliure *et al.*, 2008). These authors hypothesise that viruses might also utilise cross-talk to reduce JA stimulation, reducing VOC release to limit attraction of predators to their vectors.

Extrapolation of these effects to natural (or field) communities is difficult, however. In the Beillure *et al.* (2008) study, infected thrip predation by a third species was not affected as this species (a bug) was larger and could attack all size classes. If these interactions were translated to the field, potentially complex interactions could result, with the virus facilitating some predators and inhibiting others. Studies of VOC stimulation suggest that plants produce different cocktails of VOCs specific to different herbivore pests (or even different instars of a pest), and that predators and parasitoids can differentiate between these (Heil, 2008; see also Table 5.2). However, VOC signals can be 'confused' (or reduced by negative cross-talk between defence pathways; Table 5.3) under simultaneous attack from different enemies (Rassmann & Turlings, 2007; Zhang *et al.*, 2009; Section 5.3.4). Since in many real-world situations plants are under attack from multiple enemies, the efficacy of VOCs as predator attractants in natural communities requires further work.

5.3.4 Above- and below-ground interactions

Many of the forms of induced defence described above are systemic, operating across the whole plant. They therefore provide a bridge between above- and below-ground communities, two compartments which have more often been studied in isolation by plant biologists. Recently, there has been increasing recognition that these subsystems are interdependent, and of the role of plant defence signals in bridging the gap (Van der Putten *et al.*, 2001; Bezemer & van Dam, 2005). There are several lines of evidence indicating that changes initiated in the rhizosphere affect outcomes above ground (reviewed in Bruce & Pickett, 2007). For instance, on black mustard, *Brassica nigra*, caterpillars of the small white butterfly *Pieris rapae* had lower growth and pupation rates on plants infested with the root-feeding nematode *Pratylenchus penetrans* (van Dam *et al.*, 2005). Non-pathogenic rhizosphere bacteria and fungi (Section 5.3.2) have systemic effects on resistance, enhancing defence against foliar pathogens; similarly, endophytes (Section 5.5) can mediate above-ground–below-ground interactions via enhanced resistance to foliar herbivores and effects on plant growth and competition.

The emission of VOCs attractive to natural enemies can also be altered by above-ground–below-ground interactions. Greenhouse experiments on maize (*Zea mays*) infested with either a root-feeding beetle grub, a foliar-feeding caterpillar, or both, showed that attraction of two specialist enemies (a parasitic wasp and an entomopathogenic nematode) was strongly reduced in doubly infested plants (Rasmann & Turlings, 2007; Fig. 5.7). Emission of the main root VOC was reduced under dual infestation, but there were no measurable changes in foliar VOCs, indicating either predator confusion as a result of mixed signals,

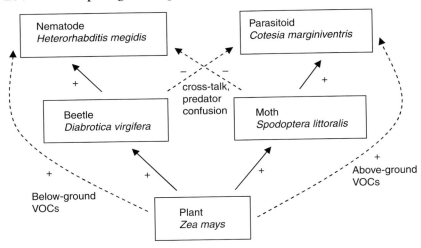

Fig. 5.7. Interaction web for maize herbivores and their enemies, based on the laboratory study of Rasmann and Turlings (2007). Solid arrows: direct interactions (opposing negative interactions not shown); dashed lines: indirect interactions mediated through plant defence signals.

or the importance of other trace VOCs that were not measured. Interestingly, the parasitoid was capable of learning the association between the mixed signal and presence of hosts under dual infection, increasing its response to doubly infested plants after exposure to the mixed odour in a prior encounter with hosts. This study is also important because it provides a clear demonstration that above-ground interactions can have below-ground effects (most other studies concentrate on alteration of above-ground properties by responses induced below ground (Bruce & Pickett, 2007)).

The interactions discussed above need to be placed in context with the overall response of the plant to attack (Dicke & Baldwin, 2010). One strategy available to plants is not to respond at all (plants can afford to lose some leaves to herbivory, for instance); another is to mount permanent constitutive responses (such as the production of tannins) to deter attackers. Karban and Baldwin (1997) distinguish between 'defence' and 'civilian' (or 'housekeeping') responses of plants to attack, with immune responses and VOC production being defence responses. Civilian responses include remobilisation of resources from storage tissues or shifting resources to or from the roots (depending on the point of attack), and up-regulation of photosynthesis by undamaged leaves.

The consequences of these civilian responses for interactions between plant enemies has received relatively little attention in comparison to the defence responses discussed above (reviewed in Paul *et al.*, 2000).

5.4 Parasitic plants

There are around 4400 species of parasitic plants (Nickrent *et al.*, 2010), approximately 1% of all angiosperms. Parasitic plants are defined as plants that make haustorial connections to other plants from which they obtain some or all of their water, carbon or nutrients. Parasitic plants occur in at least 20 families and take a broad range of life-forms from annual and perennial herbs to vines, shrubs and trees; the parasitic habit appears to have evolved independently many times (Nickrent *et al.*, 2010). They are defined as either root or shoot parasites, depending on where the parasite haustorium (a modified root) attaches to and penetrates the host. These parasites are further classified as either hemiparasitic or holoparasitic, depending on the presence or absence of functional chloroplasts; hence hemiparasites obtain some and holoparasites all of their photosynthates from the host. As with animal parasitic systems, there are also some hyperparasitic plants, which make specialised connections to other parasitic plant species. For an introduction to parasitic plants, see Musselman and Press (1995); for many useful current resources see http://www.parasiticplants.siu.edu.

Recent research reveals that parasitic plants can have similar population and community effects to those of animal parasites, via apparent competition (Chapter 2) and keystone effects on community diversity (Press & Phoenix, 2005). In addition, parasitic plants can influence the chemical and physical structure of communities (they can act as ecosystem engineers (Section 7.7) through their effects on nutrient cycling, water relations and as a result of effects on the host plant's growth form. For instance, in laboratory culture, *Striga hermonthica* (a major parasite of cereal crops (Section 5.4.3)), increases host plant transpiration rates to such an extent that it can cause a 7°C reduction in leaf temperature (Press *et al.*, 1989). High transpiration rates allow the parasite to divert water and photosynthates from the host plant to the parasite; potentially, this could affect water and nutrient relations in the surrounding soil (Phoenix & Press, 2005). Because parasitic plants like *Striga* frequently have a strong negative impact on host growth and reproductive output, their impact on populations has been compared to that of herbivores, although there are some important differences (Pennings & Callaway, 2002).

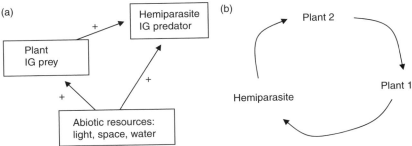

Fig. 5.8. Plant–hemiparasite interactions. (a) Module of the interaction, showing similarity to IGP (direct interactions depicted by solid arrows). (b) Rock–scissors–paper oscillations in dominance of each species in theoretical communities of two host plant species and a hemiparasite (Smith, 2000). Owing to asymmetries in interspecific competition and parasitism, each species is dominant over another and is dominated by another, leading to out-of-phase cyclical oscillations in the density of each.

A further feature often overlooked is that parasitic plants are often in direct competition with the plants they parasitise, competing for nutrients, water, space (or light, in the case of hemiparasites). This means that in some cases their interaction with the host will resemble that of intraguild predation (IGP) (Fig. 5.8a). The interaction can be further complicated by the habit of some parasitic plants to form multiple connections with different host individuals or species. Smith (2000) analysed the population dynamics of hemiparasitic plants with one or two host species (although joint parasitism of different host species by individual parasites was not assumed to occur simultaneously). The model demonstrated that parasitic plants can play keystone roles in maintaining or reducing diversity in a similar way to models of apparent and parasite-mediated competition. However, because hemiparasitic plants can potentially survive without a host, the dynamics are more complicated and include the possibility of non-transitive competitive relationships, leading to rock–scissors–paper-type oscillations (Fig. 5.8b). To our knowledge, there have been no empirical studies of hemiparasitic plant systems demonstrating this type of oscillation, although cyclical dynamics occur in a system where vegetation cycling is driven by holoparasitic plants (Box 5.4).

Population and community outcomes in this model share several similarities with models of IGP (Chapter 4). Although Smith (2000) notes similarities between parasitic plant habits and omnivory, he does

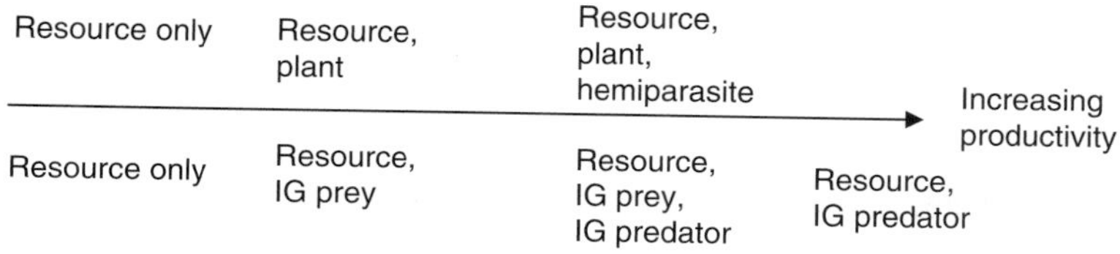

Resource only	Resource, plant	Resource, plant, hemiparasite	
			→ Increasing productivity
Resource only	Resource, IG prey	Resource, IG prey, IG predator	Resource, IG predator

Fig. 5.9. Theoretical changes in plant–parasitic plant community composition along a resource gradient (from the model of Smith, 2000). Below the line of increasing productivity shows the predicted community structure for IGP systems; above the line depicts composition for plant/parasitic plant communities. Like IG predators, parasitic plants are predicted to become more common in high-productivity environments; in natural parasitic plant systems, the converse seems to occur. Smith (2000) did not report on the predicted occurrence of hemiparasite-only communities under very high enrichment.

not make explicit comparisons to IGP, but because the comparison is instructive for community problems, it is worth examining here. In particular, community composition changed along productivity gradients (Fig. 5.9). Hemiparasitic plants compete with hosts for resources, in addition to feeding from host plants; they can thus be seen as the IG predator, while the host plant is the IG prey. Mirroring IGP theory, low resource levels supported the host (IG prey) but not the parasite (the less efficient competitor and IG predator). As resource levels increase, IG predators (parasites) can be supported, leading to the prediction that hemiparasitic plants will be more frequent in productive environments. However, empirical observations suggest that the converse is true: parasitic plants tend to be more common in low-productivity environments (Musselman & Press, 1995). There may be several explanations for this discrepancy; for instance, the assumptions of the model may neglect fundamental aspects of parasitic plant ecology and evolution; or real-world environments may not span the range examined in the model. In particular, the selective pressures leading to a parasitic habit may need to be considered; in relatively unproductive environments, the benefits of parasitism compared to the costs of competition for scarce resources may be relatively high. It may still be the case that in extremely unproductive environments, putative hosts cannot support a parasite and susceptible host plants are eliminated. The theoretical underpinnings of hemiparasitism and IGP seem remarkably well matched, so why natural systems fail to show the predicted pattern would seem a worthy area for theoretical and empirical research.

In the following sections we review recent work on three genera of parasitic plants with very different habits, illustrating the range of community effects that can occur. There are several other systems in the parasitic plant literature worthy of examination in this context; for these we direct the interested reader to the reviews of Pennings and Callaway (2002), Press and Phoenix (2005) and Phoenix and Press (2005).

5.4.1 Dodder (*Cuscuta*)

The genus *Cuscuta* comprises about 170 species of mostly annual, twining holoparasitic herbs. Upon germination, seedlings develop a rudimentary root, and a stem which twines on contact with a host plant. Once haustoria have become established above ground, *Cuscuta*'s below-ground tissue degenerates, resulting in an entirely aerial plant sustained by haustorial uptake from the host. Dodders occur throughout temperate and subtropical regions and include species having significant economic impacts in agriculture. Mature plants consist of a mass of orange or yellow stems which can cover substantial areas (tens of square metres) and make hundreds of connections with many different host individuals, often of different species. *Cuscuta* masses can eventually smother host plants, depriving them of light and thereby also leading to the eventual demise of the parasite. This degree of pathology may drive vegetation cycles and enhance plant community diversity in salt marshes (Box 5.4). A further interesting feature of dodder foraging behaviour is host choice; emerging shoots have the capacity to choose between hosts species, growing towards or away from potential host plants (Pennings & Callaway, 1996). Laboratory experiments have shown that *Cuscuta* seedlings use volatile organic compounds (VOCs) (Section 5.3.3) to locate and choose among hosts (Runyon *et al.*, 2008); this ability to seek new hosts sets these parasites apart from many other parasites and pathogens, and is likely to influence population and community dynamics, such as by reinforcing local spatial patterns in plant distribution (Box 5.4).

Box 5.4 Parasite-mediated vegetation cycles and zonation in salt marshes

Cuscuta salina (marsh dodder) is a widespread parasitic plant in salt marshes on the west coast of North America. Marshes exhibit a strong pattern of zonation across a gradual elevation gradient, their distribution determined largely by physicochemical factors (salinity,

Box 5.4 (*continued*)

relative time submerged). Moving along this gradient, there are abrupt changes in the dominant plant species, with almost mono-specific stands of some species (Fig. 5.10). In Californian salt marshes, *C. salina* may reinforce this pattern as a result of its preference for the pickleweed *Salicornia virginica*, which dominates throughout most of the lower saline zone. Correlative field censuses demonstrate that *Cuscuta* reduces *Salicornia* biomass and range expansion, thereby facilitating expansion of another pickleweed species, *Arthrocnemum subterminale*, which dominates the higher elevation saline zone (Callaway & Pennings, 1998). Locally, parasitism of *Salicornia* may reinforce the boundary between these two pickleweeds, enabling *Arthrocnemum* expansion to lower elevations in this zone.

Host preference for *Salicornia* may also drive vegetation cycles, as the parasite reaches high densities and smothers its host. Field studies indicate that as *Cuscuta* depresses *Salicornia*, two rarer species, *Limonium californicum* and *Frankenia salina*, increase in abundance (Pennings & Callaway, 1996). Areas with little *Salicornia* then lose *Cuscuta*, allowing eventual reinvasion by *Salicornia*. This process results in a shifting mosaic of plant species in the lower saline zones, with enhanced overall diversity.

Parasite removal experiments by Grewell (2008) provide add-itional evidence for these processes and demonstrate a further possible keystone effect of *Cuscuta* in another Californian salt marsh. Here, *C. salina* had a strong negative impact on growth of

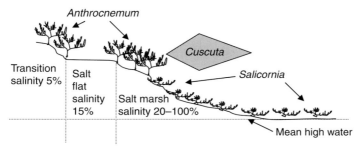

Fig. 5.10. Salt marsh zonation and *Cuscuta* frequency. The grey kite depicts relative frequency and occurrence of *Cuscuta*, towards the upper elevations of the *Salicornia* zone. Within this zone, the frequency of subordinate species *Limonium* and *Frankenia* is positively associated with *Cuscuta* cover (not drawn to scale). Redrawn and modified from Pennings and Callaway (1996).

Box 5.4 (*continued*)

the dominant species, *Plantago maritima*, facilitating recruitment of a rare hemiparasitic species, *Cordylanthus maritimus*. After two years, the effect of *Cuscuta* removal on most other plant species was neutral, with no significant differences in their abundance between *Cuscuta*-free and control plots. However, this may mask short-term dynamic effects, in which *Cuscuta* alters species composition in local patches which then change again as the vegetation cycle continues.

5.4.2 Mistletoe (Santalales)

Mistletoes are shoot parasites attaching to the stem of other plants, usually a tree or shrub. With over 1000 parasitic species spread across several families within the sandlewood order (Santalales), there are a variety of morphological forms and life habits, and parasitism appears to have evolved independently several times (reviewed in Mathiasen *et al.*, 2008). Despite quite severe direct effects on host fitness in many cases, they also appear play keystone roles in communities; and perhaps more than any other parasitic group, their positive effects in ecosystems have been recognised (Watson, 2001). Many of the community-level effects of mistletoes result from direct interactions between mistletoes and other species, including pollinators and seed dispersers, as well as the host species; these direct interactions can have knock-on effects for other community members (reviewed in Press & Phoenix, 2005). The keystone roles of mistletoes can be broadly categorised as effects-mediated via provision of additional food sources, changes in habitat structure and feed-back interactions.

5.4.2.1 *Additional food sources*

Unlike most other parasites, the mistletoes are food sources in their own right for many insects, birds and mammals, via their requirements for pollination and seed dispersal. As such, they may attract and maintain certain species in a locality, and forests with larger mistletoe populations may support greater numbers or a richer variety of wildlife (Watson, 2001). Many tropical species of showy mistletoes (*Loranthaceae*) have evolved close associations with particular bird species involving elaborate flower opening and pollination mechanisms. These species have bright flowers offering large quantities of sugar-rich nectar; some require birds to open the flower buds, eliciting an explosive release of pollen in exchange for access to further nectar supplies. The association with seed dispersers is equally strong, with most mistletoes relying on birds for dispersal. Fruits are often adapted for bird dispersal, being large, brightly coloured and nutritious, and many bird species are specialist consumers of mistletoe berries; a prolonged, asynchronous fruiting period may further strengthen these associations as mistletoe berries can be available throughout the year (Watson, 2001; Mathiasen *et al.*, 2008). The need to manage mistletoe populations in Australia and New Zealand for conservation and preservation of pollinators and seed dispersers (as well as the mistletoes themselves, many of which are threatened species) has been identified (Norton & Reid, 1997).

5.4.2.2 *Changes in habitat structure*

Many mistletoes produce dense clumps of leafy shoots, or affect host morphology such that the physical structure of the environment is altered by their presence; mistletoes can thus be considered as ecosystem engineers (Thomas *et al.*, 1999; Press & Phoenix, 2005). Dwarf mistletoes (*Arceuthobium* spp.) affect host phytohormone balance, inducing the growth of dense, twiggy masses ('witches' brooms'). Witches' brooms and mistletoe clumps are used by many bird and mammal species for concealment, shelter or nesting (Watson, 2001). The beneficial effect of mistletoes in providing shelter and nesting habitat is now recognised in forest wildlife conservation policy. For instance, in northeastern Oregon, northern flying squirrel populations declined in stands where dwarf mistletoes were thinned, leading to the recommendation that patches of infected trees be retained (Bull *et al.*, 2004). In northwestern Oregon, 90% of observed nests of the northern spotted owl are in witches' brooms; the owl is a federally protected species and conservation recommendations include maintaining dwarf mistletoe populations (Marshall *et al.*, 2003). In Colorado, Ponderosa pine (*Pinus ponderosa*) forests infected with the dwarf mistletoe *A. vaginatum* tend to support more diverse bird communities and greater populations of elk and deer than uninfected stands (Bennetts *et al.*, 1996). In conifer forests in the northern United States, dwarf mistletoes have been shown to have far-reaching effects on ecosystem function through their effects on host morphology and fitness (Mathiasen *et al.*, 2008). Severe mistletoe infestation can have strong detrimental effects on the host, including branch die-back terminal to the site of infection, treetop die-back and even death. This creates gaps in the canopy and also results in a shift of plant matter so that more foliage is concentrated closer to the ground (Godfree *et al.*, 2002). This vertical reorganisation of canopy structure, and the increased accumulation of dead wood from host die-back and death, can have important consequences for forest response to wildfires. For instance, low-intensity forest floor fires (which tend to be localised with short-term effects, and are often used as a conservation or management tool) are more likely to transition into damaging and difficult-to-control crown fires in forests with altered canopy structure and a predominance of dead wood and witches' brooms at intermediate heights (Mathiasen *et al.*, 2008).

Mistletoes may also be responsible for physicochemical changes in habitats, leading to altered community composition or productivity via their effects on nutrient cycling. In eucalypt forests in Australia, trees infected with box mistletoe (*Amyema miquelii*) contributed up to 189% more litter biomass than uninfected trees. The increased litter resulted from the long duration and high rate of mistletoe leaf turnover compared to that of host trees, litterfall from which was not affected by infection. Mistletoes hence altered the spatial and temporal distribution of litter, distributing litter in patches of varying size and depth across the landscape, and extending the season of litterfall; consequently, under-storey plant biomass was positively associated with mistletoe density (March & Watson, 2007). Strong impacts of mistletoes on host plant water usage are also well documented, and may be associated with changes in under-storey community composition or competitive relationships between the host and other tree species (Pennings & Callaway, 2002; Press & Phoenix, 2005).

5.4.2.3 *Feed-back interactions*

As a result of limited or directed dispersal (see below), mistletoes tend to enhance heterogeneity in forests, with aggregated patches of infection (and concomitant habitat changes) interspersed between relatively unin-fected patches. Mistletoes of the Viscaceae family (including the common European mistletoe, *Viscum album*) have seeds with a sticky coat of viscin, which allows the seed to stick to host branches following regurgi-tation or defecation by seed-dispersing birds. The viscin coat induces preening behaviours to dislodge seeds from bills or feathers, which may further enhance transmission to appropriate hosts (because viscin elicits a modification in the vectors' behaviour, mistletoe dispersal can be regarded as 'directed' (Press & Phoenix, 2005)). As birds specialising in mistletoe seeds spend more time on mistletoe-infected trees, they are more likely to dislodge seeds there, and infection becomes strongly aggregated. This can lead to a positive feed back between infection intensity and the number of dispersers attracted, further enhancing the patchy distribution of infection. For instance, Townsend's solitaires (*Myadestes townsendi*) disperse seed of the mistletoe *Phoradendron juniperinum* and its juniper host *Juniperus monosperma*. More birds are attracted to mistletoe-infected juniper stands, improving dispersal of juniper seeds and resulting in higher juniper seedling density in infected stands. However, positive feed back between mistletoe and solitaire density leads to heavily infected stands; heavily infected individuals suffer greater physiological effects, which eventually outweigh the positive effects of mistletoes on seed dispersal (van Ommeren & Whitham, 2002). Further feed-back interactions may occur between dwarf mistletoes and the below-ground fungal community associated with conifers. In pinyon pine (*Pinus edulis*) stands in Arizona, ectomycorrhizal (EM) colonisation was positively associated with prevalence of dwarf mistletoe (*Arceuthobium divaricatum*). This relationship could result because infected trees have increased nutrient demands, selecting for greater EM colonisation, or because high-level EM trees are nutritionally superior and can support greater mistletoe colonisation (Mueller & Gehring, 2006). Soil beneath infected trees supported 33% more pinyon seedlings and a different ascomycete fungal community than uninfected trees. Both of these effects may result from increased EM density; feed backs between EM and mistletoe density seem likely to increase heterogeneity between infected and uninfected stands.

5.4.3 Broomrape (Orobanchaceae)

The broomrape family (Orobanchaceae) is by far the largest family of parasitic plants, comprising nearly half of all recorded parasitic plant species. These root hemiparasites are generally small, often annual herbs with a broad host range, although some host species (in particular, legumes and grasses) are preferred. Many have a marked deleterious effect on host plant productivity, and some species are major parasites of agricultural crops. Chief among these is witchweed (*Striga*), with three species (*S. hermonthica* in particular) now affecting up to 50 million hectares of cereal (maize, pearl millet and sorghum) and legume crops in Africa, causing an estimated $7 billion in crop losses (reviewed in Parker, 2009). Several species of *Orobanche* may cause comparable economic losses through parasitisation of a wide range of crops in southern

Europe, north Africa and west Asia (for instance: *O. crenata* on legumes and umbellifers (carrots, parsley); *O. cumana* on sunflowers; *O. cernua* and *O. ramosa* on Solanaceae (tomato, potato, tobacco) (Parker, 2009)). *Striga* and *Orobanche* problems appear to have increased in developing countries over the past 20 years, presumably in association with expanding crop monocultures. Little is known about their interactions in the broader context of community ecology, although plant community composition is relevant for their control. For instance, one strategy for their control involves intercropping with 'trap' species that stimulate germination of parasite seedlings but do not allow attachment of the developing haustorium (reviewed in Hearne, 2009).

Another genus, the Rattles (*Rhinanthus* spp.), has a more direct relevance in community ecology. Rattles are widespread and occasionally common throughout temperate and sub-Arctic meadowlands in North America and Europe. Rattles reduce host productivity and generally appear to be associated with relatively nutrient-poor, species-rich grasslands; they may be restricted to these habitats via competition with host plants for light (Matthies, 1995). Because of their capacity to affect competitive relationships between plant species, in particular reducing the vigour of dominant grasses, recent research has focused on their use as a tool in grassland restoration (Box 5.5).

Box 5.5 Grassland restoration using *Rhinanthus*

Early studies provided evidence that *Rhinanthus* substantially reduces host growth in the laboratory (Matthies, 1995), and because of this alters the competitive balance between host species, influencing plant community structure in meadowlands (Gibson & Watkinson, 1992; Davies *et al.*, 1997). As *Rhinanthus* appears to 'prefer' dominant host grasses (perhaps by choice, perhaps because grasses are more prevalent and have extensive root systems, so establishment of haustorial connections are more likely), their use in grassland restoration to enhance wildflower diversity was recommended by Davies *et al.* (1997). Recent manipulative experiments (Joshi *et al.*, 2000; Pywell *et al.*, 2004; Bardgett *et al.*, 2006; Westbury & Dunnett, 2008) provide support for this idea, and there is now a burgeoning literature on the application of *Rhinanthus* in grassland restoration (Bullock & Pywell, 2005; see also other papers in the same journal issue).

Box 5.5 (*continued*)

The rattle *R. alectorolophus* reduced host plant biomass in plots sown with rattle seed in Swiss upland meadows, particularly when sown into species-poor communities (Joshi *et al.*, 2000). Outcomes were dependent on the functional diversity of experimental plots (the richness of different plant functional groups – grasses, legumes and herbs): in more diverse plots, there was no significant effect on overall community biomass in the summer following rattle seed sowing. Dominant grasses were most strongly negatively affected, and in diverse plots the percentage cover of herbs increased, presumably due to competitive release (Fig. 5.11a). Establishment of rattle plants was less successful in highly diverse plots (probably due to competition for light and space at the seedling stage), although once established, parasite biomass and seed production per individual were higher in functionally diverse plots (perhaps due to the 'mixed diet' possible when simultaneously parasitising different host species).

Pywell *et al.* (2004) performed manipulative experiments in a productive, species-poor meadowland in Oxfordshire, seeding plots with different densities of *R. minor* and, two years later, oversowing with a native wildflower seed mixture. Parasite establishment was slower at low sowing densities, but due to self-seeding there was little eventual difference in *R. minor* density between plots (including control plots into which *R. minor* also self-seeded). In the following two summers after wildflower seed was sown, there was a strong

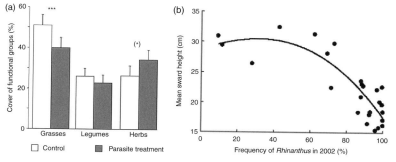

Fig. 5.11. Effects of *Rhinanthus* on plant communities. (a) Percentage cover of three functional groups in plots formerly parasitised with *R. alectorolophus* and control plots (from Joshi *et al.*, 2000). (b) Relationship between *R. minor* frequency and grass growth rate (sward height). From Pywell *et al.* (2004).

Box 5.5 (*continued*)

negative relationship between *R. minor* density and sward height (Fig. 5.11b), indicating parasite-induced reduction in host growth rate. Pywell *et al.* also detected a significant positive relationship between plant species richness and *R. minor* density the previous year, suggesting that diversity is enhanced by reduced competition faced by herb species during establishment as seedlings.

Westbury and Dunnett (2008) compared use of *R. minor* versus a selective graminicide in promoting grassland diversity, using experimental plots on recently established grassland in Yorkshire. *R. minor* performed better than the graminicide, producing a greater increase in species richness over the two years plots were monitored (indeed, half-strength application of the graminicide to control but not eliminate grasses had no effect on species richness compared to controls).

The studies in Box 5.5 suggest that *Rhinanthus* impacts diversity via two distinct mechanisms: firstly, it suppresses growth of dominant grass species; and secondly, on its annual senescence in late summer, it creates gaps in the sward for colonisation by other species (Pywell *et al.*, 2004). The hemiparasite's eventual impact on community structure depends on whether it mainly parasitises dominant or subordinate grassland species; in an earlier correlative study *R. minor* presence was associated with reduced diversity (Gibson & Watkinson, 1992). Another important consideration in its use as a tool for restoration is the management regime: in semi-natural (mown) grasslands, hay-making must be delayed until late summer to allow *Rhinanthus* to set seed (although this management strategy is also required for enhancement of other flowering herb populations); livestock must also be excluded for this period as *Rhinanthus* is toxic to grazers.

An additional factor is the effect of *Rhinanthus* on nutrient cycling: field mesocosm experiments demonstrate that *R. minor* increases the availability of inorganic nitrogen, possibly via effects on soil microbe community composition (Bardgett *et al.*, 2006). However, despite this apparent increase in soil productivity (generally a negative factor in diversity restoration), mesocosms with *R. minor* attained higher plant species diversity and reduced biomass. Ameloot *et al.* (2008) have also shown that *Rhinanthus* increases the rate of nitrogen cycling in semi-natural (mown) grasslands, partly because nitrogen-rich resources are released into the soil on senescence of the rattle, which occurs before nutrients are removed by mowing. Other members of the Orobanchaceae family have been shown to influence nutrient cycling, demonstrating their role – like mistletoes – as ecosystem engineers releasing nutrients that may sustain populations of other plant species in low-productivity environments (reviewed in Quested, 2008).

5.5 Endophtyes

Endophytes are organisms that live inside plants, and nearly all plants that have been examined appear to have them. Many fungi are endophytic, infecting a wide range of plant species without external symptoms or negative effects on the host. Many further species are largely asymptomatic, but occasionally outbreak into evident disease (Arnold et al., 2000). Here, we consider foliar fungal endophytes of grasses, which generally confer positive benefits on their hosts. These endophytes are vertically transmitted via seeds from generation to generation of hosts. Up to 30% of grasses may harbour endophytes, which form a systemic infection throughout the above-ground tissue of the plant, growing interstitially between the host plant cells. These endophytes gain nutrition and protection from their host in exchange for a variety of services, including defence against herbivores, pathogens, drought and competition. Endophytes of the genus *Neotyphodium* have radiated widely with their hosts, cool-season grasses of the subfamily Pooideae; they confer resistance to herbivores via production of high levels of bioactive alkaloids (reviewed in Clay, 1996; Clay & Schardl, 2002). Although endophytes typically benefit the host, the distinction between mutualist and parasite in this group is somewhat blurred, with some interactions more parasitic in nature under some circumstances (Saikkonen et al., 1998). For this reason, and because the contrast in effects of endophytes versus those of parasites may be informative for community ecology, we feel justified in considering endophytes (as 'honorary parasites') in the following sections.

5.5.1 Endophyte effects on communities

5.5.1.1 *Effects on non-host plants*

Because endophytes can alter growth rates and stress tolerance, they can potentially influence intra- and interspecific competition between host and non-host plants. Greenhouse mesocosm experiments demonstrate that tall fescue grass (*Lolium arundinaceum*) infected with the endophyte

Neotyphodium coenophialum has a higher growth rate and fares better in interspecific competition than uninfected fescue (Clay *et al.*, 1993; Rudgers *et al.*, 2004; 2005). These effects, together with differential grazing pressure (see Section 5.5.1.2), can translate into effects on plant community composition in the field. In a long-term field experiment in old field meadowland in Indiana, plots were ploughed and sown with either endophyte-infected (E+) or uninfected (E−) tall fescue. Community composition was monitored over 12 years, during which time other plant species were allowed to colonise naturally. After 4 and 12 years, E+ plots had significantly lower plant species diversity than E− plots, and after 12 years a clear reduction in the rate of succession was apparent in E+ plots, which had lower tall forb cover and reduced tree-species richness and cover (Clay & Holah, 1999; Rudgers *et al.*, 2007). These effects probably resulted from a combination of mechanisms, including modified competitive ability resulting from endophyte infection, inhibition of tree seedling establishment (Rudgers & Orr, 2009; see below) and, most importantly, endophyte-modified herbivore pressure (Clay *et al.*, 2005; see Section 5.5.1.2). Tall fescue, a native of the Mediterranean, is invasive across much of the eastern United States, and endophyte infection may facilitate this process (see Section 6.4). Indeed, because E+ strains are at a competitive advantage over E− strains and natives, they weaken the link between community diversity and invasibility (E+ strains can establish with relative ease, irrespective of diversity (Rudgers *et al.*, 2005)). Thus endophytes are implicated in altering basic aspects of ecosystem function (Section 7.8).

5.5.1.2 *Effects on other trophic levels*

Although not typical of all endophytes, the foliar fungal endophytes of grasses enhance the levels of alkaloids and other toxins in leaves, reducing host plant nutritional quality and making them less palatable to insect herbivores (Clay, 1996; Clay & Schardl, 2002). Greenhouse mesocosm experiments demonstrate that *Neotyphodium* sp.-infected (E+) populations of the Italian ryegrass *Lolium multiflorum* support a lower abundance of aphids than E− populations, with aphid species differentially affected (Omacini *et al.*, 2001). This change in the aphid guild has knock-on effects at higher trophic levels, reducing parasitoid and hyperparasitoid abundance and altering guild composition (Fig. 5.12). Abundance of aphids and hyperparasitoids was higher on E− plants, whereas the numbers of intermediate parasitoids were hardly

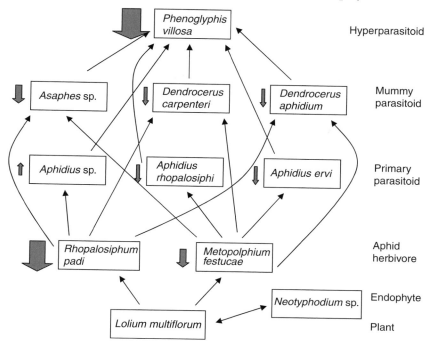

Fig. 5.12. Interaction web for grass/aphid/parasitoid mesocosms, from the work of Omacini *et al.* (2001). Solid line arrows indicate the direction of positive direct interactions (energy flow). Grey block arrows depict the effect of adding the endophyte to communities (size reflects relative effect on population densities).

affected, producing an alternating pattern of abundance reminiscent of that predicted by food chain theory for bottom-up trophic cascades (Oksanen *et al.*, 1981). In this example, endophyte presence was not associated with changes in plant biomass, so the cascading trophic effects were the result of changes in plant nutritional quality alone.

Field experiments in Indiana on the *L. arundinaceum/N. coenophialum* system (see Section 5.5.1.1) suggest that the interaction between endophytes and herbivory is sufficiently strong to drive selection on symbiosis and successional dynamics of plant communities (Clay *et al.*, 2005). Control field plots exhibited a 2.5-times greater increase in E+ plant frequency compared to plots where mammalian and invertebrate herbivores were excluded over four years, with a concomitant reduction in forb and non-fescue grass biomass. Comparison of control plots to those in which voles (*Microtus pennsylvanicus* and *M. ochrogaster*) were excluded indicated that voles, via their avoidance of host plants in favour of E− fescues or other plant species, were responsible for a 48% increase in biomass of E+ plants.

Endophytes have also been shown to affect the composition of the detritovore and generalist predator assemblages, probably via the direct or indirect effects of increased alkaloid and mycotoxin content of the host (Lemons *et al.*, 2005; Finkes *et al.*, 2006). Manipulative field experiments using the *L. arundinaceam*/*N. coenophialum* system revealed reduced species richness of spiders and reduced abundance of two out of three dominant spider families in E+ compared to E− plots. Spiders could be affected for a number of reasons, including reduced prey availability owing to endophyte-mediated effects on insect herbivores, or altered plant architecture for web-building resulting from endophyte-mediated effects on plant diversity (Finkes *et al.*, 2006). Endophyte-infected litter has also been shown to decompose more slowly in this and another endophyte–grass system, with related changes in the composition and activity of the detritovore community (Omacini *et al.*, 2004; Lemons *et al.*, 2005).

These studies indicate that endophytes can have influences ramifying throughout the food web. It is tempting to conclude that the influence of endophytes on community structure in particular might be a template for understanding the possible community consequences of pathogens and parasites in general, but some important distinctions must be remembered:

(1) Endophytes generally have a positive effect on their host's fitness, so community consequences are expected to be the reverse of those predicted for parasites. For instance, by promoting host dominance, endophytes reduce plant diversity − the opposite occurs for those parasites playing a keystone role in maintaining diversity because they reduce fitness of a dominant host.

(2) Endophytes, as organisms associated directly with plants, have direct effects at the base of the food web. Because they affect producers, their indirect effects on dependent trophic levels may be more severe (or at least different) to those of symbionts associated with consumers.

(3) Endophytes may interact with other symbionts and parasites of plants, which may result in further complications when predicting the sign and strength of indirect interactions and their community consequences (see Section 5.5.2).

5.5.2 Endophyte interactions with plant parasites and mutualists

The mutualistic relationship between plants and endophytes can be reversed whenever the costs of nutrient provision to the endophyte exceed the gains in terms of enhanced competitive ability or protection from enemies (Saikonnen *et al.*, 1998). For instance, when the grass *Lolium pratense* is infected with the endophyte *Neotyphodium uncinatum* and parasited by the hemiparasite *Rhinanthus serotinus*, growth rate of the host plant is significantly reduced compared to that of E− grasses. Furthermore, *R. serotinus* acquires defensive mycotoxins produced by the endophyte via its host, protecting it from herbivory by generalist aphids. Thus the joint demands of endophyte and hemiparasite on host photosynthate exceed the benefits of endophyte infection in this case, and endophyte infection enhances hemiparasite growth at the expense of the host, effectively becoming parasitic in the presence of the hemiparasite (Lehtonen *et al.*, 2005). The community consequences of this interaction will depend on the type and level of herbivore pressure experienced (as different herbivores are differentially affected by the mycotoxins and/or have differential preferences for host plant and hemiparasite), and on plant nutrition. For instance, foliar endophytes and mycorrhizae both negatively affect larval growth and survival of the moth *Phlogophora meticulosa* feeding on perennial ryegrass *Lolium perenne*, but effects are generally less-than-additive and also depend on the supply of phosphorus to the plant (Vicari *et al.*, 2002).

Endophytes also interact with arbuscular mycorrhizal fungi (AMF), another class of plant symbionts that generally form mutualistic relationships with their hosts, but under some conditions can also be parasitic. Two recent studies suggest that endophytes of grass are effectively in competition with AMF for access to host plant resources, and that basic asymmetries in this relationship favour the endophyte (Omacini *et al.*, 2006; Mack & Rudgers, 2008). In both cases, E+ plants had a lower density of AMF colonisation than E− plants, possibly because endophytes have priority in accessing host photosynthates due to their location in the leaves as opposed to the roots. In mixed E+/E− mesocosms, colonisation of AMF was inhibited in E+ plants but stimulated in their E− neighbours (Omacini *et al.*, 2006). In another system, increasing soil fertility had an effect reminiscent of a bottom–up trophic cascade: endophyte density was increased, and AMF colonisation reduced; there was only a slight increase in host plant biomass, suggesting that extra resources were shunted to the endophyte in favour of the host

(Mack & Rudgers, 2008). The community consequences of these inter-actions have yet to be established: the results suggest that outcomes will depend on productivity gradients, and that endophytes may reinforce spatial heterogeneity of AMF colonisation (or vice versa). It has been estimated that AMF associates with the roots of approximately 80% of angiosperms (Johnson *et al.*, 1997). Given that grass–endophyte associ-ations are also common (Clay & Schardl, 2002), there is clearly great scope for the occurrence of endophyte–AMF interactions, the commu-nity consequences of which we have barely begun to elucidate.

5.6 Conclusions

Pathogens and parasites of plants are a diverse group, and their effects on communities are equally so. Perhaps more than any other area, plant–parasite systems provide evidence of the impacts of parasitism on com-munity structure and ecosystem function, yet this research is generally rather poorly rooted in theory. The lack of organising principles from theory make it difficult to establish the full range of community pro-cesses in which plant parasites are involved; we are left with having to catalogue examples, and perhaps for this reason plant parasitic systems are given less weight than they deserve in mainstream reviews of parasite community ecology. Theory is lacking in a number of areas, including the role of plant pathogens and endophytes in diversity/invasibility relationships, population-level effects of multiple host use by plant parasites, similarity to IGP for plant parasites and the population conse-quences of cross-talk and cross-effect in plant defence systems.

In this chapter we have seen that many systems mirror examples from animal literature. For instance, there are examples of parasite-mediated apparent competition (Pennings & Callaway, 1996; Section 5.2.3) and IGP (Smith, 2000; Section 5.4). However, systems differ in several key respects. Firstly, the stationary mode of plant life introduces strong spatial heterogeneity; consequently, the influence of parasitism may be reflected in spatial as well as temporal dynamics, with effects of disease on succession, thinning and vegetation cycles (Section 5.2). Secondly, the nature of plant biochemical defence and signalling pathways introduces a further dimension to interactions between natural enemies: in addition to the indirect interactions resulting from competition for shared host resources, there may be plant-mediated resistance/susceptibility, signal-ling interactions and manipulations (Section 5.3). Thirdly, the great range of fitness effects on the host, from the strong negative effects of some parasitic plants (Section 5.4) through to the generally positive effects of endophytes (Section 5.5). This range provides scope for comparison of community impacts across the mutualism spectrum, where a general pattern of parasite-related increases in diversity to endophyte-related decreases in diversity is apparent. However, we must caution against drawing many general conclusions until mathematical and empirical models have been developed and tested in a number of areas.

6 · Parasites and invasions

6.1 Introduction

Biological invasions are widespread in natural and managed habitats, where they may drive changes in biodiversity and community structure, and are frequently of great economic importance (Mack *et al.*, 2000; Pimentel *et al.*, 2000; Lockwood *et al.*, 2007; McGeoch *et al.*, 2010). Understanding the mechanisms that underpin invasions is essential in predicting and managing invasion outcomes. There is growing evidence that parasites can modify the success of an invasion and its impact on the community. Furthermore, the study of biological invasions provides examples of ongoing, natural experiments in which to observe the effect of parasites on novel hosts and the impact of parasites on the wider community (Prenter *et al.*, 2004; Dunn, 2009).

A biological invasion can be defined as the spread of a non–indigenous species from the point of introduction to some substantial level of abundance (Elton, 1958). When the invasive species is a parasite, or host to a parasite introduced with it, the novel host–pathogen interactions may sometimes also result in emerging infectious diseases (EIDs) (Chapter 8). These can include the spatial range expansion of a parasite within existing host species (e.g. the spread of malaria linked to climate change – discussed in Chapter 8), or a species range expansion to a new host species (such as spread of squirrel pox virus from grey squirrels to red squirrels – discussed in Section 2.6.1).

We can distinguish several stages in a potential invasion (Fig. 6.1). Of the many species that are introduced to a new area, relatively few may actually become established. Of these, fewer still will become invasive species, spreading out from the initial population and colonising new territories, sometimes far from the point of introduction, often at high densities seldom seen in the natural range. Each of these different processes (translocation, introduction, establishment, invasive spread) is governed by different factors, and require different management approaches (Kolar & Lodge, 2001; Fig. 6.1). Parasites are just one of

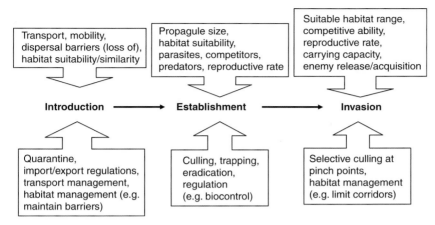

Fig. 6.1. Schematic of the phases of a biological invasion. There is a lack of consensus over the terminology to describe the various processes in an invasion. Throughout this text, terminology used is from Kolar and Lodge (2001); *Introduction*: a species is introduced into a new habitat where it may or may not become established and spread (become invasive). *Establishment*: a species with a self-sustaining population outside of its native range. *Invasion*: an introduced species that then spreads in the new habitat. Many of the species reaching an earlier phase will not proceed to later phases. Some of the factors influencing the likelihood of reaching each phase are shown in boxes above that stage; potential management and control options are outlined in the boxes below. At successive stages, the number of factors influencing outcomes increases, but the general options for control decrease.

the factors that can influence, or be influenced by, invasion processes, as we discuss in this chapter.

The ecological impact and management of biological invasions is a large and growing research area and the subject of complete texts. The ecology of introduced plants is considered by another book in this series (Myers & Bazely, 2003), whilst the ecology of animal and plant invasions is considered by Lockwood *et al.* (2007). In this chapter we focus on the role of parasitism in invasion processes, beginning by considering changes in parasite diversity and distribution that result from a biological invasion. An invader may introduce novel parasites to the recipient community; it may acquire parasites from its new environment or it may lose its parasites during the process of invasion (Fig. 6.2). The second part of this chapter considers the role of parasites in determining the success of a biological invasion and

(a)

(b)

(c)

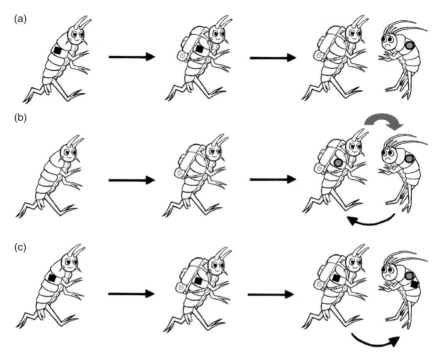

Fig. 6.2. Changes in parasitism following the invasion of a new habitat. (a) Enemy release; an invader may lose parasites through stochastic effects or selective pressures during the invasion. (b) Parasite acquisition; parasites may be acquired in the new habitat. The invader may then dilute the infection (black arrow) or act as a reservoir for the infection (spillback, grey arrow). (c) Parasite introduction; introduced parasites may go on to infect native host species in the new habitat (spillover). Thanks to Emily Imhoff for the drawing.

its impact on the community. Here we consider how parasites can alter the outcome of an invasion both through direct effects on the host and by modifying host–host interactions such as predation or competition.

6.2 Parasite introduction and acquisition

A biological invasion often provides opportunities for novel host–parasite associations. Invasive species may be exposed to novel parasites in their new habitat, or they may introduce novel parasites to the invaded community. An invader can:

- *introduce new parasites to indigenous hosts*. If an invader introduces a new parasite to an indigenous host, the parasite may itself influence the outcome of invasion either through direct effects on the novel host, or through apparent or parasite-modified competition (Chapter 2). Introduction by an invader of a novel parasite provides a novel host–parasite relationship and subsequent evolution of the parasite could lead to emerging disease (EIDs are considered in more detail in Chapter 8).
- *act as an intermediate host, allowing parasites to establish in a new range*. Many parasites have indirect life cycles that involve two or more host species. The introduction of a new species to a community may provide an intermediate host, allowing completion of a parasite's life cycle, enabling their establishment in new range.
- *acquire parasites from the invaded range*, possibly altering the existing population dynamic relationship between indigenous hosts and parasites.

An invasive species may bring with it parasites from the native range. In many cases, these may remain specific to the host. However, introduction of a new parasite by an invader can cause the decline and even extinction of indigenous species. For example, over the last 40 years, amphibian populations have undergone declines and extinctions on a global scale. A key factor in amphibian declines is Chytridiomycosis (Box 8.6), caused by the fungal pathogen *Batrachochytrium dendrobatidis*, which has been introduced as a result of global trade in the American bullfrog *Rana catesbeiana*. The fungus is avirulent in the bullfrog but is highly virulent in many species of frog and is highly transmissible between hosts. The bullfrog has a worldwide distribution as a result of farming and accidental release, and the parasite has been detected in numerous populations, suggesting that the bullfrog is the source of the fungal infection in native amphibians (Garner *et al.*, 2006). In accordance with this idea, a recent study of the genetic diversity of *B. dendrobatidis* from diverse geographical areas indicate that it has experienced a genetic bottleneck and has undergone rapid and recent range expansion (James *et al.*, 2009).

The introduction of a novel parasite to one species in a community will have knock-on effects on its competitors and predators. In a study of amphibian communities in Spain, Bosch and Rincon (2008) found that the common toad (*Bufo bufo*) avoided laying its eggs in ponds occupied by the midwife toad (*Alytes obstetricans*), avoiding larval competition and predation. However, when an outbreak of *B. dendrobatidis* led to massive declines in midwife toad numbers, the numbers and range of the common toad expanded.

A well-known example of the impact of an introduced parasite is that of rinderpest, which is a virus of ungulates that causes high fever, mouth sores, diarrhoea and dehydration, leading to high levels of mortality. At the end of the nineteenth century, rinderpest was introduced to wild ungulates in sub-Saharan Africa through accidental introduction of infected cattle (Section 7.2.1). Empirical and mathematical studies of population dynamics before and after vaccination control of rinderpest reveal that the virus is not able to persist in wild ungulate populations as the hosts are at low densities. However, domestic cattle act as a reservoir for the virus, which has a strong regulatory effect on wild ungulate populations. This regulatory effect of the introduced virus was revealed when the virus was controlled in the reservoir host via vaccination. A vaccination programme for domestic cattle was introduced in the 1950s in the Serengeti national park. As a result, the virus disappeared from domestic cattle. It was also eradicated from wild ungulate populations, despite the fact that only domestic cattle were vaccinated; this demonstrates the role of domestic cattle in maintaining the virus. Eradication of the virus led to dramatic increases in wildebeest (which saw a 15-fold increase in numbers) and buffalo, as well as increased abundance of lions and hyenas, which prey upon these ungulates.

There are many instances where spillover of a novel parasite to indigenous hosts facilitates the invasion process and, in some cases, the resulting extinction of native species. In the United Kingdom, the native red squirrel (*Sciurus vulgaris*) is largely restricted to northern Scotland and has been replaced by the American grey squirrel (*S. carolinensis*) over the last 70 years. A poxvirus introduced by the grey squirrel causes high mortality in the novel host (the red squirrel) and has hastened its decline and replacement. The dynamics of this interaction are considered in detail in Boxes 2.5 and 6.3. Similarly, replacement of the imperilled native UK crayfish *Austropotamobius pallipes* by the American signal crayfish *Pacifastacus leniusculus* is mediated by crayfish plague (*Aphanomyces astaci*), which was introduced along with the invader in the 1970s (Box 6.2). There may of course be many instances where introduced parasites go on to infect native species, but have little impact on the fitness of their new hosts. However, such an outcome is less likely to be detected during studies of biological invasion.

An invader may also introduce a parasite to an indigenous host by acting as an intermediate host for the parasite. For example, the tapeworm

Echinococcus multilocularis is widespread in populations of the Arctic fox (*Vulpes lagopus*). However, it is absent from Finland, Norway and Sweden as there are no suitable intermediate rodent hosts in these regions. In Spitzbergen, the recent introduction of the sibling vole (*Microtus rossiaemeridionalis*) has provided an intermediate host for this parasite, which has since been detected in foxes in this region (Henttonen *et al.*, 2001).

An invasive species may also acquire parasites from the invaded range, possibly altering the existing population dynamic relationship with the indigenous hosts. The outcome depends upon the impact of the parasite on native versus invading hosts. If the invader is a less competent host, then parasite dilution may occur, benefiting the native host (Norman *et al.*, 1999; Ostfeld & Keesing, 2000). Alternatively, the new host may act as a reservoir, increasing the impact on the parasite via spillback to the native host population (Daszak *et al.*, 2000; Kelly *et al.*, 2009). For example, *Thelohania contejeani* is a second parasite that affects the imperilled white-clawed crayfish. Signal crayfish appear to have acquired this parasite in their new range. However, rather than slowing the invasion, the parasite may in fact facilitate it. This is because the parasite is less pathogenic to the invader, which may act as a reservoir for infection (Dunn *et al.*, 2009). This example is considered in more detail in Box 6.2. Another example of acquired parasites is discussed in Box 6.4. Barley yellow dwarf virus (BYDV) modifies competition between grasses in California, leading to the replacement of native perennial grasses by invasive annuals. BYDV was acquired by the invasive annuals. As the invasive species support greater numbers of the aphid vectors, this increases the likelihood of BYDV infection in the natives.

6.3 Loss of parasites by invaders: enemy release

Invading species are often present at higher densities than are similar species in the invaded habitat (Torchin *et al.*, 2003). The enemy release hypothesis (ERH) proposes that the success of an invader is related to a reduction in natural enemies such as herbivores, predators and parasites (Keane & Crawley, 2002; Torchin *et al.*, 2002; 2003). In addition, the host may benefit if resources that would be used for defence are reallocated (evolution of increased competitive ability (EICA) (Section 6.4.2)). Although there is evidence for enemy release in a number of animal and plant systems, some may have alternative interpretations and few studies have tested the impact of parasite loss on host fitness and invasion success (Colautti *et al.*, 2004).

There are several reasons, however, why parasites in particular may be easier to lose during the initial stages of invasion than other natural enemies. Loss of parasites may result from sampling effects and from selective pressures encountered during introduction and establishment (Torchin *et al.*, 2002; Colautti *et al.*, 2004). An introduced population represents a subsample of individuals (and genotypes) from the source population. This subsampling may lead to a genetic bottleneck in the host and to loss of parasites, in particular for parasites that are present at low prevalence in the source population (Drake, 2003; Colautti *et al.*, 2004). During translocation and colonisation of the invader, selective pressures may lead to a greater loss of parasitised (less fit) than unparasitised hosts, favouring resistant host genotypes (Colautti *et al.*, 2004). In addition, if parasite transmission is density dependent, a threshold host population size is required for parasite maintenance (Anderson & May, 1981; Box 1.2). Low host density, particularly during the initial stages of colonisation, is therefore predicted to result in parasite loss even if the parasite was translocated with the initial colonisers. Different types of parasite may be more or less prone to loss in the exotic range. Parasites with life cycles involving vectors or multiple host species may also be introduced but then lost if they are unable to be transmitted to a vector or the next host in their life cycle. In Section 6.3.3 we examine evidence that vertically transmitted parasites are harder to lose than horizontally transmitted ones during biotic invasions.

Drake (2003) used a general probability model to explore the theoretical effect of enemy release and population size on the likelihood of initial establishment of an introduced species. On the one hand, the likelihood of an introduced species becoming established should increase with propagule size (the number of individuals introduced). However, the larger the initial propagule, the more likely it will be that parasites are also introduced with it. Hence, release from enemies becomes less likely with increasing propagule size, and the two effects are traded off against each other, potentially cancelling out. In fact, Drake found that highly virulent parasites lead to a curved relationship, with an optimum propagule size for successful establishment (Fig. 6.3). This effect applies only to the initial establishment phase of an invasion; once established, enemy release (if it has occurred) may continue to enhance the spread of an introduced species, possibly leading to invasive status.

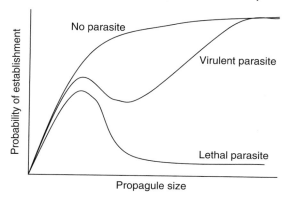

Fig. 6.3. Predicted relationship between propagule size and establishment of an introduced species with parasitism. Larger propagules are less likely to escape from parasitism, but may nevertheless establish as uninfected or immune genotypes are selected. The humps at low propagule size indicate the effect of enemy release on the probability of establishment. Modified from Drake (2003).

Nonetheless, surveys and meta-analyses of parasitism in native and invasive species do suggest that enemy release has occurred and may influence the invasiveness of many animal and plant species (Mitchell & Power, 2003; Torchin *et al.*, 2003). Studies of enemy release use two approaches: comparison of parasitism in invading species and their indigenous counterparts (community studies – Section 6.3.1); and comparison of parasite diversity and prevalence in its native and invasive range (biogeographical studies – Section 6.3.2). Both approaches shed light on enemy release and invasion success. The community approach may also develop our understanding of the role of parasites in native–invader interactions.

6.3.1 Community studies of parasitism in invasive versus indigenous species

Parasites introduced with an invader may remain specific to this host, or may be introduced to native species. Similarly, parasites that are indigenous to the invaded habitat may be acquired by the invading species. A number of community studies report higher parasite diversity in native species in comparison with invasive species (Table 6.1). However, patterns of parasite prevalence vary; some studies report higher prevalence in the native species, whilst others (fewer) report higher prevalence in invaders.

Table 6.1. *Examples of community studies of enemy release*

Species compared	Parasites studied	Evidence	Reference
Native amphipod *Gammarus duebeni celticus* and three species of invasive amphipods in the UK	Various endo and ecto parasites	*Enemy release.* The native species suffered a higher parasite diversity. Prevalence of two parasite species was higher in the native, but prevalence of a third was higher in an invader.	Dunn and Dick (1998); MacNeil *et al.* (2003a); MacNeil *et al.* (2003b); MacNeil *et al.* (2003c)
Invasive brine shrimp *Artemia francisca* and natives *A. parthenogenetica* and *A. salina* in the Mediterranean	Cestodes	*Enemy release.* Native species of brine shrimp suffer a higher diversity and prevalence of cestodes than does the invader.	Georgiev *et al.* (2007)
Invasive and native morphs of the snail *Melanoides tuberculata* in Africa	Trematode *Haplorchis* sp.	*Enemy release.* Parasites absent from invasive morph. Invasive morph resistant to native parasites.	Genner *et al.* (2008)
Native gecko *Lepidodactylus lugubris* and invader *Hemidactylus frenatus* on Pacific islands	Cestode *Cylindro-taenia* sp.	*Mixed evidence.* In sympatric populations in the field, parasite prevalence was higher in the native species.	Hanley *et al.* (1995); Hanley *et al.* (1998)

These studies compared parasite diversity and/or prevalence in native versus invasive hosts within an invaded community.

In natural systems, invasive (and native) species have a variety of natural enemies, and invasion success is likely to be determined by their combined effect. For instance, Agrawal and Kotanen (2003) and Agrawal *et al.* (2005) examined the impact of natural enemies for 30 taxonomically paired native and introduced old–field plant species raised in experimental field plots in Ontario. In the 2003 study they reported higher levels of herbivore damage for introduced species than for native congeners, in contrast to the predictions of the ERH. In the extended study of 2005, incidence of foliar pathogen attack was also assessed, and soil microbial activity was examined by measuring plant soil feed back (see Sections 5.2.2 and 6.4.3). In these new experiments, introduced plants averaged lower levels of herbivory, half the negative soil microbial feed back, and tended to have lower incidences of infection by foliar fungal and viral pathogens compared to native congeners. Taken separately, individual taxonomic pairings often showed strong differences in enemy attack levels, but these were rarely consistent across all enemy guilds. Hence, escape from one guild of enemies does not necessarily imply escape from other guilds, and invasion success will be determined by the net effect of all enemies. Because this may vary over time and space (as illustrated by the differing impact of herbivores in the two studies), there may be limited 'invasion opportunity windows' during which introduced species can spread in recipient communities.

In a study of invaded freshwater communities in Ireland, we compared parasite diversity and prevalence in native and invasive amphipods. Parasite diversity was higher in the native *Gammarus duebeni celticus* than in three species of invading amphipods. However, patterns of parasite prevalence varied. Of the three parasite species that affected both native and invasive hosts, two were more prevalent in the native species, whilst a third reached higher frequencies in the invasive *G. pulex* (Box 6.1). The differential impact of these parasites on native and invasive hosts is discussed further in Box 6.5.

Box 6.1 Enemy release in amphipod invaders in the United Kingdom

In rivers and streams in Ireland the native amphipod *Gammarus duebeni celticus* is being displaced by the European *G. pulex*. A further two species of invader also co-occur; *G. tigrinus* and *Crangonyx pseudogracilis*. Studies of several invaded amphipod communities revealed that parasite diversity is higher in the native *G. duebeni celticus* than in the invading species. Five species of parasite were detected, of which three were shared but two – the microsporidian *Pleistophora mulleri* and the acanthocephalan *Polymorphus minutus* – were restricted to the native host. For two species (*Embata parasitica* and *Epistylis* sp.) that infected both native and

Box 6.1 (continued)

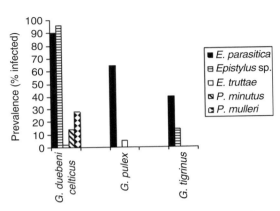

Fig. 6.4. Parasite diversity and prevalence in three species of amphipod in Northern Ireland: the native *Gammarus duebeni celticus* and the invaders *G. pulex* and *G. tigrinus*. Where prevalence was measured in several field sites, the average prevalence is plotted. Data taken from Dunn and Dick (1998); MacNeil *et al.* (2003a; 2003b; 2003c).

invading species, parasite prevalence and burden was higher in native hosts (Fig. 6.4). In contrast, prevalence of the acanthocephalan *Echinorhyncus truttae* was higher in the invasive *G. pulex*. The role of these parasites in mediating interactions between native and invasive gammarids is discussed later in Section 6.5.4 and in Chapter 4.

Similarly, studies of native and invasive mud snails in North America revealed that the native snail (*Cerithidea californica*) was host to a greater diversity (ten species) of trematode parasites than was the invasive snail *Batillaria cumingi* (Torchin *et al.*, 2005). Although overall trematode prevalence was similar in native and invasive hosts, the invasive snail harboured only one trematode species, which is thought to have been introduced with the host, using native birds and fish as definitive hosts.

6.3.2 Biogeographical studies of parasitism in the native versus invasive range

Biogeographical studies of enemy release compare patterns of parasitism in hosts from their native habitat versus their novel habitat (Colautti *et al.*, 2004). There are numerous examples of decreased parasite diversity in the invaded habitat (Table 6.2). For example, the rabbit fish *Siganus rivulatus* harboured only seven species of parasites in its invasive Mediterranean range, compared with 24 species in native Red Sea populations (Pasternak *et al.*, 2007). Although parasite diversity was lower in the invasive populations, the prevalence and abundance of a single species of flatworm, *Polylabris cf. mamaevi*, were three times higher (Pasternak *et al.*, 2007). Similar patterns are found for plant systems – for example, in the wild campion *Silene latifolia*, which is native throughout much of Europe but invasive with pest-level status in parts of the United States. Wolfe (2002) compared 80 native and invasive populations and found that native populations suffered substantially higher attack rates by generalist herbivores and by two specialist enemies (a seed predator and the anther smut fungus).

In a meta-analysis, Torchin *et al.* (2003) compared parasite diversity and prevalence in the native and invasive ranges of 26 invasive species (including molluscs, crustaceans, fish, birds, amphibians, reptiles and mammals). For most species, parasite diversity was reduced in the invasive range, with an average of 16 parasite species found in native populations, versus seven species in the invasive range. Of the species in the invasive range, on average three had been introduced with the invader and four acquired in the new habitat. Although invasive species experienced reduced parasite diversity, for those parasites that were introduced with the host, prevalence was similar in the native and invasive ranges.

Mitchell and Power (2003) took a similar approach to examining evidence for ERH in plants. They surveyed 473 plant species naturalised from Europe to the United States, finding that introduced populations supported on average 84% fewer foliar fungi and 24% fewer virus species. The degree of enemy release depended on two processes: the loss of enemies from the native range and the acquisition of novel enemies from the invaded range. Species that were more completely released from pathogens were also more widely listed across the United States as pest species in agricultural and natural ecosystems over a larger geographic range. Furthermore, the composition of pathogen assemblages was altered in the novel habitat as a result of differing patterns of enemy loss and acquisition (Fig. 6.5).

Mitchell and Power's (2003) and Torchin *et al.*'s (2003) analyses suggest that invasiveness is determined both by release from enemies and by acquisition of enemies in the novel habitat. This latter process can

Table 6.2. *Examples of biogeographical studies comparing parasite diversity in the native versus the invasive range of the host*

Biogeographical comparison	Parasites studied	Evidence	Reference
Native versus invasive populations of marine animals	Various	*Enemy release.* Meta-analysis of the literature revealed a higher average parasite diversity in native populations	Torchin *et al.* (2002)
Native versus invasive populations across a range of species	Various	*Enemy release.* Meta-analysis of the literature revealed a higher average parasite diversity in native populations	Torchin *et al.* (2003)
Native versus invasive populations of the rabbit fish *Siganus rivulatus*	Monogenean	*No evidence for enemy release.* Prevalence and intensity of infection higher in invasive hosts in the new range	Pasternak *et al.* (2007)
Native and invasive populations of the tree frog *Eleutherodactylus coqui*	Various	*Enemy release.* Parasite diversity higher in the native range.	Marr *et al.* (2008)
Native and introduced populations of the common myna *Acridotheres tristis*	Malaria *Plasmodium* and *Haemoproteus*	*Evidence for enemy release inconsistent.* Lower total parasite prevalence in invasive populations.	Ishtiaq *et al.* (2006)

(cont.)

Table 6.2. (*cont.*)

Biogeographical comparison	Parasites studied	Evidence	Reference
Native and invasive populations of the starling *Sturnus vulgaris*	Helminths	However, the pattern held for only two–sixths of invasive populations *No evidence for enemy release.* Average parasite diversity did not differ between native and invasive populations when the source of the invasion was taken into account	Colautti *et al.* (2005)
Native and invasive populations of the amphipod *Crangonyx pseudogracilis*	Microsporidia	*No evidence for enemy release.* No difference in parasite diversity in the native versus invasive populations when the source of the invasion was taken into account	Slothouber-Galbreath *et al.* (2010)
Native and invasive populations of the amphipod *Dikerogammarus villosus*	Microsporidia	*No evidence for enemy release.* Parasite diversity higher in the invaded range. Parasite prevalence of the dominant species of microsporidian did not differ in the native and invasive ranges	Wattier *et al.* (2007)

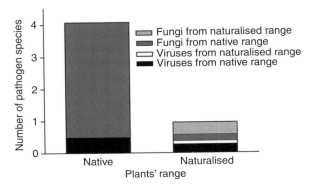

Fig. 6.5. Composition of pathogen assemblages in the plants' native and naturalised ranges, averaged across 473 species surveyed. Enemy release depends on escape from native enemies and avoidance of novel enemies. From Mitchell and Power (2003).

be thought of as an example of biotic resistance, another process thought to determine invasion outcomes. The biotic resistance hypothesis proposes that invasion outcomes are determined by interactions between the invader and members of the invaded community (for instance, competitors and enemies) – hence ecosystem-level characteristics such as diversity may influence invasibility; we return to this subject in Chapter 7.

Although decreased parasite diversity in the invasive range has been reported for many species, the prevalence of those parasites occurring in the invasive range is often similar to or higher than that in the native hosts, leading Torchin *et al.* (2002) to suggest that introduced populations do not differ in their susceptibility to infection, but that parasites are lost as a result of the invasion process. In addition, there have been few studies which consider the mechanisms that may underpin enemy release. In a recent study of the potential mechanisms for enemy release, MacLeod *et al.* (2010) looked at chewing lice on 36 species of bird introduced to New Zealand. They found that few parasites were lost as a result of sampling effects, but that most parasites were lost because their hosts or the parasites themselves failed to establish. For hosts that became established, parasite establishment was positively associated with host body mass and the number of hosts introduced, both factors linked to transmission.

Although there are many examples of lower parasite diversity in invasive populations, Colautti *et al.* (2004) note two potential sources of bias in biogeographical studies of enemy release. Firstly, native populations may be more extensively studied than invading populations, and so sampling effort should be taken into account – as, for example, in Torchin *et al.* (2003). Secondly, comparing parasites in the native versus the invasive range may be biologically unrealistic, as invading propagules may originate from a subset or even a single native population. Tests for enemy release should therefore compare parasitism in the invasive range with that in the likely source population. Empirical studies that apply this more conservative test suggest that the role of enemy release may be less important than previously thought. For example, Colautti *et al.* (2005) investigated enemy release in European starlings, which were introduced to North America during the nineteenth and twentieth centuries. A comparison of the parasite fauna in the invasive and native ranges found apparent enemy release from parasitic helminths. However, historical records suggest that starlings were introduced from Britain, and the similarity of helminth fauna between British and American populations also support this. When parasites were compared for the invasive and likely source populations in Britain, there was no difference in mean diversity. Although the birds appeared to have lost parasites from the native range (13/20 parasite species found in Britain were present in North America), 17 parasite species had been acquired from the new habitat.

We recently tested for enemy release in the invasive amphipod *Crangonyx pseudogracilis* (Slothouber-Galbreath *et al.*, 2010), which was introduced from its native range (the east coast of North America) to the United Kingdom in the 1930s and has since spread to France and the Netherlands. Phylogenetic analysis revealed that this amphipod had suffered a post-invasion genetic bottleneck since its introduction to Europe and suggested a single area of origin for the invasion in Lake Charles, Louisiana, an area of heavy shipping traffic. In the native region, eight species of microsporidian parasites were detected, whilst four were detected in the invasive range. However, when we compared parasite diversity between invasive populations and native source populations, we found no evidence for enemy release. Two parasite species were detected in the source population and these also occurred in the invasive range. The invader had also acquired two species of parasite post-invasion.

Similarly, there was no evidence that the Ponto–Caspian *Dikerogammarus villosus* had benefited from enemy release during its invasion of Europe. Wattier *et al.* (2007) found no evidence for a genetic bottleneck in the invasive population, a finding in accord with a pattern of recurrent introductions via rivers and canals. A single parasite species was detected in both the native and invasive range, whilst a further four species may have been acquired from native species in the new habitat.

6.3.3 Enemy release from vertically transmitted parasites

Horizontal transmission is the more common mechanism of parasite transmission and is the mechanism used by the majority of parasites discussed in the previous section. However, a number of parasites are vertically transmitted from generation to generation of hosts, usually via the host gametes (Fig. 6.6). This route of transmission may provide an effective mechanism of dispersal over long distances or in patchy habitats (Dunn *et al.*, 2001). Vertical transmission imposes different selection pressures on the host. As these parasites depend on host reproduction for their transmission to the next host generation, they typically cause low virulence (Bandi *et al.*, 2001). Furthermore, opportunities for vertical transmission do not depend on host density. Therefore we predict that, although vertically transmitted parasites may be lost from invasive populations due to sampling effects, they are unlikely to be lost through low host density or through differential mortality of infected hosts in founder populations. Similarly, Mitchell and Power (2003) propose that invasive plants are less likely to escape from viruses than fungal parasites as the former are often seed transmitted and have lower virulence.

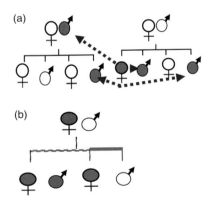

Fig. 6.6. (a) Horizontal parasite transmission. Transmission can occur between individuals in the same or different generations of hosts. (b) Vertical parasite transmission. Parasites are passed from generation to generation of hosts, often via the host gametes. Owing to the differences in gamete size, such parasites are usually maternally inherited.

A small number of studies consider enemy release for vertically trans- mitted parasites, and the observations vary (Table 6.3). The strongest evidence comes from Mitchell and Power's (2003) study of enemy release in US invasive plants (see Section 6.3.2). Their results supported their prediction that viruses would be less easy to escape than fungi as they are often seed transmitted and have lower virulence (Table 6.3). Enemy release has also been explored for intracellular *Wolbachia* and microsporidia of arthropods. Release from *Wolbachia* was reported for three invasive ant species. In contrast, in the European paper wasp, *Wolbachia* seems to have invaded along with its host, reaching similar prevalences to those observed in native populations (Table 6.3). A study of enemy release from microsporidia (Section 6.3.2) also found evidence of invasion of the parasite along with its host. The microsporidium *Fibrillanosema crangonycis* appears to have invaded Europe along with its crustacean host *Crangonyx pseudogracilis*, reaching higher prevalence in invasive populations.

The invasion of *Wolbachia* and microsporida along with their hosts is of particular interest, as these parasites are widespread in arthropod hosts, where they cause reproductive manipulation (Terry *et al.*, 2004; Duron *et al.*, 2008). Owing to the differences in gamete size, only female hosts transmit the infection and these parasites often manipulate host reproduction to enhance their own transmission – for example, by inducing cytoplasmic incompatibility or by feminising their host (Bandi *et al.*, 2001). These reproductive parasites can have dramatic effects on their host ecology and evolution and have been implicated in speciation and the evolution of novel mechanisms of sex determin- ation (Bandi *et al.*, 2001). We propose that reproductive manipulation may also play a role in biological invasions, enhancing the likelihood of invasion by the host.

Invasion along with the host has been reported for *F. crangonycis*, which feminises its host, whilst there is evidence that *Wolbachia* causes cytoplasmic incompatibility (and hence female-biased sex ratios) in the paper wasp. If hosts that harbour such reproductive parasites are included in an introduced propagule, then reproductive manipulation should drive the parasite (and the infected host genotype) through the introduced population. Furthermore, both of these parasites lead to female-biased sex ratios in the host. Theoretical models predict that this overproduction of females under parasitic sex ratio distortion should cause a higher rate of population increase, at least initially (Werren & Beukeboom, 1993; Hatcher *et al.*, 1999). Hence infection

Table 6.3. *Studies testing for enemy release from vertically transmitted parasites*

Invader	Vertically transmitted parasite	Test for enemy release	Reference
Argentine ant *Linepithema humile*	*Wolbachia*; effect on host unknown	Populations from invasion source infected. However, *Wolbachia* invasive populations less frequently infected.	Tsutsui *et al.* (2003)
Fire ant *Solenopsis invicta*	*Wolbachia*; effect on host unknown	*Wolbachia* common in native populations but found only in one-fifth of invasive populations. Prevalence lower in invasive populations.	Shoemaker *et al.* (2000); Bouwma *et al.* (2006)
Fire ant *Solenopsis richteri*	*Wolbachia*; effect on host unknown	*Wolbachia* found in native populations but absent from introduced populations.	Shoemaker *et al.* (2000)
Garden ants *Lasius neglectus* (invasive species) and *L. turcicus* (not invasive)	*Wolbachia*; effect on host unknown	*Wolbachia* measured for 30 populations. Prevalence lower in invasive than native species.	Cremer *et al.* (2008)
European paper wasp *Polistes dominulus*	*Wolbachia*; effect on host unknown; male killing, feminisation and parthenogenesis induction ruled out; however, infection	Prevalence of infection varies across invasive populations, but is similar to that in populations in native range. The same strain detected in	Stahlhut *et al.* (2006)

(cont.)

Table 6.3. (*cont.*)

Invader	Vertically transmitted parasite	Test for enemy release	Reference
	persistence and its occurrence in several host haplotypes suggestive of cytoplasmic incompatibility	native and invading populations.	
Crangonyx pseudogracilis	Microsporidium *Fibrillanosema crangonycis*; evidence for feminisation of the host	Parasite detected in invasive and likely source population. Parasite detected at high prevalence in all invasive populations screened.	Slothouber Galbreath *et al.* (2010); Galbreath *et al.* (2004)
Review of 473 invasive plants	Comparison of viruses (lower virulence, often vertically (seed) transmitted) versus fungi (higher virulence, more horizontal transmission)	A greater release from fungi (84% fewer fungi) than viruses (24%).	Mitchell and Power (2003)

by such parasitic sex ratio distorters may, in fact, enhance the rate of initial invasion. However, the long-term outcome of the invasion will depend, among other factors, on the invasive's mating strategy; if the parasites attain a high prevalence, and male re-mating potential is low, males may become the limiting sex, potentially reducing the per capita reproductive rate (Hatcher *et al.*, 1999).

Feminising parasites could also be acquired in a new habitat (there is evidence for horizontal transmission events in reproductive parasites, e.g. Terry *et al.*, 2004; Werren *et al.*, 2008). Two species of microsporidium have been acquired by the invasive *Dickerogammarus villosus* in its new range in France (Section 6.3.2). The close relationship of these parasites to the feminiser *Nosema granulosis* suggests that they too may feminise the host. It is therefore interesting to speculate that acquisition of feminising parasites could facilitate the ongoing invasion of *D. villosus*.

Vertically transmitted beneficial microbes that travel with the host to a novel habitat may also be implicated in the invasion process. Many plants harbour fungal endophytes, which usually confer a variety of benefits to their hosts, including reduced palatability to herbivores and increased drought tolerance (Section 5.5). Hence, invasive plants arriving with their endophytes will be at an advantage as natural enemies in the new location are poorly adapted to novel plant–endophyte complexes; there are several examples of invasive plants where the possession of endophytes seems linked to their success in the novel habitat (Evans, 2008; see also Section 6.4.3).

6.4 Invasions and host–parasite co-evolution

Biological invasions may lead to novel host–parasite associations, whether through the introduction of new parasites by an invader or the acquisition by an invader of parasites in the invaded range. Thus invasions provide an opportunity to compare parasite fitness in hosts with which they have a co-evolutionary history with fitness in novel host species. An evolutionary perspective on adaptation of hosts and parasites to each other in their native or invasive range also provides alternative explanations to enemy release for the success of some invaders in terms of their interactions with parasites.

6.4.1 Local adaptation

Local adaptation theory predicts that selection should favour a parasite that specialises on common host genotypes, leading to local adaptation; parasites should have higher fitness in local hosts with which they have a co-evolutionary history than in hosts from other populations (Dybdahl & Lively, 1995; Kaltz & Shykoff, 1998; Gandon, 2002). In the context of an invasion, a native parasite will have co-evolved with native hosts and thus should have higher fitness in native rather than invasive hosts (Gandon *et al.*, 1996; Moret *et al.*, 2007). Alternatively, Colautti *et al.* (2004) predicted that invaders should be more susceptible to native enemies, either because they are naïve to the parasites, or because genetic bottlenecks experienced during invasion might reduce the ability to evolve disease resistance.

Studies of the effect of native parasites on invasive species tend to support the hypothesis that the parasite is locally adapted to the native host (Firlej *et al.*, 2005; Tain *et al.*, 2007; Genner *et al.*, 2008). For example, the native acanthocephalan parasite *Pomphorhynchus laevis* infects both the native *Gammarus pulex* and the invader *G. roeseli* in rivers in eastern France. The parasite alters the phototactic response of the native host, increasing its vulnerability to predation by the definitive fish host, but does not manipulate the invasive host's behaviour (Tain *et al.*, 2007). Similarly, the parasitoid *Dinocampus coccinellae* was found to be locally adapted to the native ladybird *Coleomegilla maculata*, which suffered higher parasite attack and a higher rate of successful parasite development than did the introduced species *Harmonia axyridis* (Firlej *et al.*, 2005).

6.4.1.1 *Novel weapons hypothesis*

Another possible explanation for the invasiveness of some species is provided by the novel weapons hypothesis (Callaway & Ridenour, 2004). As originally conceived, this supposes that some plant species become invasive because they possess novel biochemicals (usually root exudates) not previously encountered by potential native competitors in the introduced range. Callaway and Ridenour (2004) suggest that many root exudates that are relatively ineffective against competitors in the plant's natural range due to adaptation may be highly effective in the newly invaded community. This mechanism, which can be seen as another example of local adaptation (of native competitors to each other), can be extended to include pathogens harboured by the invasive species. For instance, SQPV can be seen as a 'novel weapon' employed by grey squirrels against red squirrels in the invaded range (Box 6.3). The mechanism can also be applied to soil microbes in the native and invaded range, affected differentially by root exudates of the invading species. For instance, garlic mustard *Alliaria petiolata* is native to Europe and invasive across North America. It produces allelopathic root exudates which are harmful to AMFs in soils from the invaded range. *Alliaria* itself does not form mycorrhizal associations and may obtain a competitive advantage by damaging the beneficial symbionts of its competitors in the invaded range (Callaway & Ridenour, 2004).

6.4.2 Evolution of increased competitive ability

The ability of a potential host to defend itself against parasitism should be selected when parasites are common and costly. However, such defences can be costly, and invasive species may benefit from enemy release. Blossey and Notzold (1995), in a study of plant invasions, predicted that enemy release may enhance invasive potential if resources that would be used for defence are reallocated to growth and reproduction. In this scenario it is assumed that plants trade off one function (enemy defence) against others (e.g. growth). Selection for reduced enemy defence would thus lead to EICA. This argument has been extended to animals by Lee and Klasing (2004), who proposed that successful (vertebrate) invaders should reduce costly inflammatory responses to parasites and rely on antibody-mediated immunity. This is because invaders might lose parasites and might also acquire novel parasites that could induce costly immune responses. Similarly, they predicted that successful invertebrate invaders may have lower (non-specific) immune responses, thereby freeing up resources for growth and reproduction.

Evidence for EICA in animal systems is mixed. For example, Lee *et al.* (2005) measured the responses to inflammatory challenges of the house sparrow (*Passer domesticus*) and the tree sparrow (*P. montanus*). Whilst the invasive house sparrow showed no changes in metabolism, reproduction or behaviour, the challenge was more costly to its less invasive congener, the tree sparrow. The lower immune response of the house sparrow may enhance its invasion success. However, a comparison of phylogentically related pairs of bird populations in Hawaii (Matson, 2006) found no evidence for an attenuated immune response in island birds (subject to lower parasite diversity) than continental birds, but found an increase in innate inducible defences. Rigaud and Moret (2003) found that the native acanthocephalan *Pomphorynchus laevis* was found to induce a greater immune response in the invasive *G. roeseli* host than the native *G. pulex*.

6.4.3 Plant–soil feed back

For introduced plants, feed back between plant biochemistry and soil microbes encountered in the new environment or transported along with the plant can occur in ecological or evolutionary time, influencing invasion outcomes. Such plant–soil feed back (PSF – Section 5.2.2) can influence the dynamics of succession and vegetative cycles in plant communities.

PSFs may help to explain why some plant species become invasive whilst others remain non-invasive. Some species may make successful invaders if they lose the negative PSFs on introduction to the novel habitat (through enemy release, for example), or if they gain positive PSFs through novel associations with beneficial symbionts (Reinhart & Callaway, 2006). Early successional species with high growth rates and low defensive allocations (annuals), or those with high root–shoot ratios (grasses) are more likely to be successful invasives because they have the most to gain from losing their negative PSFs (Section 5.2.2). There is some support for this pattern from a recent meta-analysis (Kulmatiski *et al.*, 2008), but in earlier reviews this pattern of enemy release is less clear-cut (Kolar & Lodge, 2001; Keane & Crawley, 2002). Some well-known instances of plants with negative PSFs in their native range also show evidence for enemy release leading to reduced or neutral PSFs in the invaded range (reviewed in Reinhart & Callaway, 2006). For instance, the black cherry *Prunus serotina* is native to the United States, but invasive in parts of northern Europe. Strong negative PSFs regulate population density in the native range (Box 5.1), but PSFs in the invasive range appear to be neutral (Reinhart *et al.*, 2003), suggesting enemy release. Similarly, marram grass *Ammophila arenaria* is a native of northern Europe, where its replacement is driven by negative PSFs (Box 5.2), but invasive in coastal dunes in the United States, South Africa and New Zealand. It appears to have escaped some of its negative PSF in South Africa (Knevel *et al.*, 2004), but not in California (Beckstead & Parker, 2003).

On an evolutionary timescale, PSFs provide evidence for local adaptation. In some cases, invasive species may gain an advantage if they bring with them soil microbes that differentially impact on competitors. For instance, the tall fescue (*Lolium arundinaceum*), an invasive grass across much of North America, harbours an endophyte (Section 5.5) which reduces the growth and survival of several species of native tree saplings (Rudgers & Orr, 2009). Thus the endophyte (which can be seen as a 'novel weapon' to which invaded communities are poorly adapted (Section 6.4.1)) facilitates the invasion and inhibits succession of grass-land to scrub and woodland (Rudgers *et al.*, 2007). The probable mechanism here is endophyte inhibition of the beneficial arbuscular mycorrhizal (AMF) associated with the native trees (Mack & Rudgers, 2008). For other species, AMFs may be important drivers of positive PSFs acquired in the novel habitat by introduced species. For instance, the knapweed *Centaurea maculosa*, a native of Europe, is an aggressive invader across western North America. It cultivates a strong negative PSF in European soil, which may be important in regulating population densities in the native range. However, when grown in US soil, it has a positive PSF, possibly contributing to its occurrence at high densities in invaded habitats (Callaway *et al.*, 2004). This result suggests both escape from enemies and acquisition of mutualists; many plant soil pathogens are specialist and when lost on translocation are not quickly replaced in the new range, whereas many AMF are generalists, so novel beneficial associations are readily established.

From an evolutionary perspective, then, the outcome of an invasion will depend on: the evolutionary responses of the (introduced) host to the novel parasites encountered; the evolutionary response of these parasites to the novel host; or the evolutionary response of native host species to novel parasites (and vice versa). Such novel host–parasite encounters can potentially lead to the evolution of emerging diseases (Woolhouse *et al.*, 2005), a topic examined in Chapter 8.

6.5 The impact of parasitism on biological invasions

Parasites may be specific to one species of a native–invader pair, or may affect both species. The impact of a shared parasite can be complex. For example, an invader may benefit the native host if it creates a dilution effect whereby transmission to the resistant, invasive host lowers the infection prevalence in the main, native host (Section 7.3.2). Conversely, there can be a reservoir effect whereby introduction of an alternative (invading) host, rather than diluting the effects of the parasite, may act as a reservoir for infection, a factor that may be exacerbated by high densities of the invading hosts. The arrival of an alternative host could thus result in an increase in abundance of the native parasite, resulting in reduced population growth of susceptible hosts. Such 'apparent competition' (Chapter 2) between host species can result in the extinction of the host suffering the greatest negative impact of the shared parasite (Holt & Lawton, 1993).

6.5.1 Parasite dilution by invading hosts

In a situation where a parasite is locally adapted to the native host, transmission to a less suitable, invasive host may create a dilution effect, thereby lowering infection prevalence in the main, native host (Section 7.3.2). For example, the presence of an alternative, invasive host species was found to reduce the impact of bacterial (*Bartonella birtlesii* and *Bartonella taylorii*) infection on wood mice (*Apodemus sylvaticus*).

The parasites are vectored by fleas. Telfer *et al.* (2005) observed a decline in parasite prevalence in wood mouse populations as densities of the invading bank vole, *Clethrionomys glareolus*, increased. The infection was detected in wood mice, but not in bank voles, suggesting a dilution effect in which transmission to the less competent invader has led to a decline in parasite prevalence in the native wood mouse.

A dilution effect may contribute to the coexistence of native and invading snails in New Zealand (Kopp & Jokela, 2007). A study of parasite transmission revealed that, whilst the native *Potamopyrgus antipodarum* developed the infection following ingestion of *Microphallus* eggs, an invading European species, *Lymnea stagnalis*, did not develop the infection. However, the invader acted as a sink for the parasite; prevalence was twice as high in single-species *P. antipodarum* populations as in populations where *P. antipodarum* was kept in sympatry with the invader.

Sink effects have also been documented for deliberate introductions. The ladybird *Harmonia axyridis* was introduced to North America for biological control and has since become invasive. This ladybird benefits the native ladybird as it acts as a sink for the parasitoid *Dinocampus coccinellae*. The parasitoid oviposits in both native and invasive ladybirds, but successful maturation is more common in the native host. Hence the presence of the invader leads to a decrease in parasitoids and is predicted to lead to an increase in the density of the native species (Hoogendoorn & Heimpel, 2002). This scenario may, however, be further modified by IGP between the hosts and between hosts and parasitoid (Chapter 4).

Invasive species often do well in a novel habitat because they are resistant to the parasites in that habitat. However, if they act as less competent targets for native parasites, then this dilution effect can in fact benefit native species by disrupting host–parasite interactions. Whilst dilution effects appear to be less frequently documented in the context of invasions than are reservoir effects, this may reflect study effort, and we suggest that such effects warrant further empirical and theoretical study.

6.5.2 Invading hosts as infection reservoirs

Novel hosts can, alternatively, act as reservoirs for infection, resulting in apparent competition with the native host (Section 7.3.2). If the invader introduces a novel parasite, then this parasite may spillover to infect native hosts. In addition, an invader may provide an additional host from which native parasites can spillback to infect native species (Daszak *et al.*, 2000; Kelly *et al.*, 2009). The first observation of apparent competition in an invasion scenario was Settle and Wilson's (1990) study of the interaction between the native leafhopper *Erythroneura elegantula* and the invasive variegated leafhopper *E. variabilis* in North America. Field experiments revealed that direct competition between the species was not important for the outcome of the invasion. However, the parasitoid *Anagrus epos* affected both species. Establishment of the invasive species provided a reservoir host for the parasitoid and led to an increased parasitoid population. As the native leafhopper suffered higher parasitoid attack rates, this shared parasite shifted the competitive balance towards the invader.

In the United Kingdom, the decline of the native white-clawed crayfish (*Austropotamobius pallipes*) and its replacement by the North American signal crayfish (*Pacifastacus leniusculus*) is mediated by parasitism (Box 6.2). As well as competing with the native species for resources and refuges, the invader acts as a reservoir for two species of parasite. Crayfish plague was introduced to the United Kingdom by the signal crayfish. In contrast, porcelain disease is endemic in the United Kingdom, but recent work suggests that the signal crayfish has acquired the parasite and may act as a reservoir.

Box 6.2 The effects of parasites on crayfish extinction and replacement

The white-clawed crayfish (*Austropotamobius pallipes*) is listed as endangered by the IUCN. It is the only native crayfish in the United Kingdom, where it is threatened with extinction by habitat destruction and by the invasion of the North American signal crayfish (*Pacifastacus leniusculus*). The species compete directly for resources (food and refuges). In addition, the signal crayfish acts as a reservoir for two species of parasite that are more virulent in the native host.

Crayfish plague (caused by the fungus *Aphanomyces astaci*) is fatal to white-clawed crayfish, and plague outbreaks have led to extinction of local populations (Holdich, 2002; Holdich & Poeckl, 2007). The signal crayfish is resistant to crayfish plague, but can transmit the infection to the native species in areas of co-occurrence. Populations of the invader occur at higher densities than those of the native, which will also increase opportunities for parasite transmission. Crayfish plague was introduced to Europe in 1865 along with the first introduction of an *Orconectes* sp. Both signal crayfish and plague were introduced to Britain in the 1970s.

Box 6.2 (*continued*)

Porcelain disease is caused by the microsporidian parasite *Thelohania contejeani* and causes chronic disease and eventual death in white-clawed crayfish (Oidtmann *et al.*, 1997). Infected muscle becomes opaque, leading to the porcelain appearance and death after several years. Signal crayfish have not been found to be symptomatic for porcelain disease. However, we have recently detected the infection in the invasive crayfish in the United Kingdom using PCR, which allows detection of parasites even at low burden (Dunn *et al.*, 2009). Its impact on the native–invader interaction is currently under investigation. On the one hand, acquisition of this parasite from the native range might limit the spread of the invader. However, preliminary data suggest that the infection causes lower mortality in the invader than in the native. In addition, *T. contejeani* prevalence was high (26–75%) in the *P. leniunsculus* populations studied (Dunn *et al.*, 2009), and the signal crayfish occurs at much higher densities than its rival (Peay, personal communication). These factors suggest that *P. leniusculus* may also act as a reservoir for porcelain disease to spillback to the native crayfish.

Understanding the spread of the signal crayfish and its role in transmitting plague has informed strategies for conserving the native crayfish. In the United Kingdom, recent conservation efforts have concentrated on protecting the species from invasive crayfish and the risk of disease by creating Ark sites (Whitehouse *et al.*, 2009). A number of Ark sites have been set up in the United Kingdom by translocating animals from existing sites (often sites where the threat from invasion or parasitism is high). The establishment of Ark sites in isolated water bodies – for example, disused aggregate or minerals workings – provides a refuge from invasion and parasitism.

A further probable example of an invader acting as a reservoir host is the grey squirrel, a native of the United States that has invaded in the British Isles, replacing the native red squirrel throughout much of its UK range. English populations of the grey squirrel harbour a virus (squirrel pox virus: SQPV (Thomas *et al.*, 2003), previously considered to be a parapox), thought to have been introduced along with the squirrel. The infection causes no discernable pathology in greys, but is highly virulent in reds; the squirrels compete directly for food resources and nesting sites, but SQPV enhances the rate of replacement of red squirrels via apparent competition (Tompkins *et al.*, 2003; Box 2.5). Recent empirical data are strongly suggestive of a role for SQPV in red squirrel decline in England. Rushton *et al.* (2006) compared the rate of spread of greys in Cumbria and Norfolk (where greys are SQPV-positive) to those in Scotland and Italy (where greys do not carry the virus) and found invasion rates up to 25 times faster in localities with infected greys. Spatially explicit individual-based models (IBMs – Box 2.5) are being applied to predict the course of the invasion through the remaining red strongholds in the British Isles (Box 6.3). This modelling approach lacks generality but has advantages in an applied context when considering strategies for the control and management of endangered species in particular habitats. The models demonstrate that grey squirrel populations provide a reservoir for successive infection of newly encountered red squirrel populations, and suggest that SQPV may 'burn out' in habitats unfavourable to greys, providing a potential management approach for red squirrel conservation (Box 6.3).

Box 6.3 Squirrel pox virus containment and red squirrel conservation

Over the last century, grey squirrels harbouring SQPV have spread across much of England and Wales (Tompkins *et al.*, 2003; Box 2.5), and have now established expanding colonies in previous red squirrel strongholds across northern England, including the lowlands of the Lake District. From the Lakeland valleys, SQPV-infected greys have spread along wooded river corridors to north Cumbria and have now entered Scotland. The first reports of seropositive greys in Scotland came from the Lockerbie area in 2005, with the first recorded outbreak of SQPV disease in Scottish red squirrels nearby in 2007. Current research focuses on monitoring and containment of SQPV-carrying greys through the remaining red strongholds of northern England and Scotland (for updates see http://www.snh.org.uk/ukredsquirrelgroup).

Rushton *et al.* (2006) applied a spatially explicit model (Box 2.5) to examine management of the ongoing expansion of grey squirrels in Cumbria. The model predicted that both species would overlap in suitable habitat for only 3–4 years before the red squirrel was eradicated from these zones (consistent with rates observed in the field), despite low between-species contact and disease transmission rates. Rushton *et al.* examined possible strategies for red squirrel conservation in south and central Cumbria through culling of grey

Box 6.3 (*continued*)

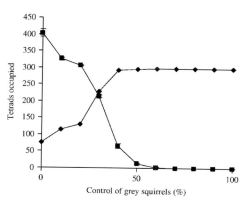

Fig. 6.7. Projected effect of grey squirrel culling on red squirrel populations in northern Cumbria. Graph shows the number of population cells occupied by red squirrels (circles) and grey squirrels (squares) given the percentage of adult grey squirrels culled. From Rushton *et al.* (2006).

squirrels. Model projections suggested that more than 60% of all grey squirrels in all habitat patches would have to be culled to prevent red squirrel decline (Fig. 6.7). This strategy would require continual implementation as SQPV-infected greys continue to enter the region via habitat corridors from Lancashire; therefore, salvation of the red squirrel in this region now appears to be an unrealistic goal, although the species may persist at low levels in habitat less favourable to greys (Gurnell *et al.*, 2006; Rushton *et al.*, 2006).

At the English–Scottish border along Northumberland and County Durham, Kielder National Forest is likely to be the last refuge of red squirrels in England. This area of more than 50 000 ha has mostly coniferous forest, which can support low-density red squirrel populations but is largely unsuitable for greys. Gurnell *et al.* (2006) used spatial models at different scales to identify the main habitat corridors through which grey squirrels will likely enter this region and examine the spread of SQPV into red squirrel populations. They identified four possible corridors through which infected greys can reach Kielder; the first within the next few years, with larger numbers arriving within ten years (Fig. 6.8). Assuming the central Kielder habitat remains unsuitable for greys, the model predicts that SQPV will infect red squirrel populations

Box 6.3 (*continued*)

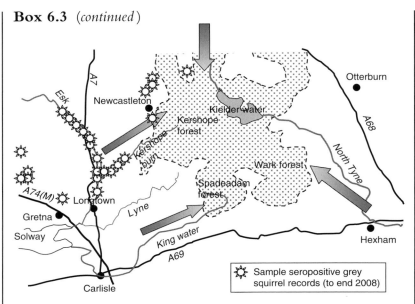

Fig. 6.8. Schematic of Kielder poxvirus invasion scenario. Block arrows show the four invasion corridors along river valleys into Kershope, Wark and Spadeadam forests. Central Kielder (stippled) may retain uninfected red squirrel populations if grey squirrel control at corridor pinch points is successful and habitat remains unsuitable for greys. Based on Gurnell *et al.* (2006) and data from the UK Red Squirrel Group (http://www.snh.org.uk/ukredsquirrelgroup).

at the forest margin, causing substantial mortality. However, infection would not become endemic deep within the forest because disease progression in reds is too rapid to sustain its spread without the presence of greys. Effectively, the disease is predicted to 'burn out' at the forest margins, but there will be repeated outbreaks as naïve red squirrels from the central forest recolonise the margins where they come into contact with infected greys.

Although SQPV enhances the rate at which red squirrels are replaced in the United Kingdom, grey squirrels are also a threat throughout mainland Europe, where the disease is currently absent. Greys were introduced in northern Italy in the early twentieth century, and there are now three established populations, all sero-negative for SQPV. Projections from models suggest that greys will spread rapidly through favourable habitat along river corridors and continuous decidous forest, establishing populations in Switzerland

Box 6.3 (*continued*)

within 15–40 years (Tattoni *et al.*, 2006) and France within 30–75 years (Bertolino *et al.*, 2008). Colonisation of the rest of Eurasia appears to be only a matter of time, although eradication of the two smaller Italian populations would buy valuable time (*c.*100 years) to formulate an international control strategy (Bertolino *et al.*, 2008). These projections assume that no further introductions occur, and that SQPV remains absent from the continent. Since the species is still sold in the pet market in continental Europe, further escapes are likely and it is perhaps only a matter of time before infected grey squirrels are introduced in Continental Europe.

The models discussed in Box 6.3 highlight a number of key points for conservation:

- Disease limitation by culling of grey squirrels from already established populations (e.g. Cumbria) is impractical given the high-percentage cull required.
- Control into refuges may be possible by selectively trapping or culling at identified 'invasion pinch points' (e.g. the four corridors identified for Kielder).
- 'Burn out' may limit the impact of disease in red-only colonies. Conservation measures should thus target retaining habitat suitable only for reds (e.g. limit mixed planting in the heart of Kielder, retain pure stands of small-seeded spruce species unsuitable for greys).
- The high virulence of disease in reds may be key to sustaining red squirrel refuges; if the effects of SQPV become attenuated and disease progression to death slows, per capita within-species transmission will be increased potentially to the point where SQPV can be sustained within pure red populations.

Furthermore, conservation of the red squirrel by reintroduction of wild-caught or captive-bred animals to boost local breeding populations is not likely to be effective in the presence of SQPV (worse still, it can risk introducing new outbreaks of disease). Carroll *et al.* (2009) document several unsuccessful introductions to the East Anglian refuge population of Thetford Chase; in each case, animals designated for release contracted the disease through contact with grey squirrels.

6.5.3 Native hosts as infection reservoirs

It is also possible for native hosts to act as reservoirs for pathogens that also infect introduced species. For instance, BYDV may be involved in the replacement of native perennial grasses by invasive annuals in Californian grasslands (Box 6.4). In this example, invasion is driven by

---

Box 6.4 Barley yellow dwarf virus in grasses

Barley and cereal yellow dwarf viruses (referred to collectively as BYDV) infect a broad range of cultivated and wild annual and perennial grasses. Field plot experiments demonstrate how this aphid-vector-borne disease can mediate apparent competition, altering dominance relationships between annual grasses (Power & Mitchell, 2004; Box 5.3). BYDV-induced stunting and yield losses in crops such as wheat, barley and oats make this one of the most economically significant plant pathogens of cereal crops (Pimentel *et al.*, 2000), and recent studies suggest it may be important in driving the invasion of annual grass and replacement of native perennial bunchgrasses (*Nassella pulchra* and *Elymus multisetus*) across the Californian grasslands (Malmstrom *et al.*, 2005a; 2005b; 2006).

European settlers introduced oats (*Avena* spp.) and bromes (*Bromus* spp.) several hundred years ago, and these exotics have now almost completely displaced native bunchgrasses in the Californian Great Valley grasslands. BYDV-infected bunchgrass from natural populations exhibits stunting, reduced fecundity and increased mortality (Malmstrom *et al.*, 2005a), and the presence of *Avena* and *Bromus* enhances the likelihood of BYDV infection because these annuals support greater numbers of the aphid vector (Malmstrom *et al.*, 2005b). In field plot experiments, BYDV infection did not affect bunchgrass survivorship when grown alone, but halved survivorship of bunchgrasses grown with exotics (Malmstrom *et al.*, 2006). This suggests that BYDV alters the competitive dominance relationship between the native perennials and invasive annuals (Fig. 6.9).

Box 6.4 (*continued*)

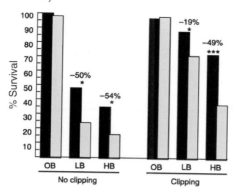

Fig. 6.9. Survivorship at the end of the growing season for native bunchgrass seedlings inoculated (grey bars) or mock-inoculated (black bars) with BYDV. Survivorship of bunchgrass growing in the absence of competitors (OB) did not depend on infection status, but was reduced substantially for infected plants in the presence of annual *Bromus* spp. sown at low (LB) or high (HB) densities. Clipping (to simulate grazing) had a significant impact on survivorship, improving the relative performance of infected bunchgrass (cf. right panel, clipped, to left panel, not clipped). From Malmstrom *et al.* (2006).

A recent mathematical model quantified with life history and infection data suggests that BYDV is key in determining the invasion outcome (Borer *et al.*, 2007b). In the absence of BYDV, the long-lived perennial grasses (*N. pulchra* can live for several hundred years) are predicted to resist invasion by annual species as gaps for establishment of new seedlings are seldom available and uninfected bunchgrass out-competes annuals. BYDV infection reverses competitive dominance (Box 5.3) and also enhances gap creation through infected plant mortality. These processes combined enable the spread of annual species and both perennial and annual species are predicted to coexist, the perennials at low frequency with 100% infection prevalence (Fig. 6.10).

This model demonstrates that it is actually the bunchgrass that acts as a long-term reservoir for infection, because the perennial population allows BYDV persistence through California's seasonal drought when annual species are not available. The annual species nevertheless provide an important function as amplifiers of the aphid vector, enhancing transmission efficiency to perennials.

Box 6.4 (*continued*)

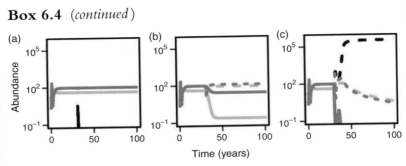

Fig. 6.10. Invasion outcomes for BYDV and annual grasses into native perennial populations. The figures plot the predicted trajectories for (a) uninfected annuals, (b) infected perennials and (c) infected annuals into uninfected perennial populations. Healthy annuals cannot invade healthy bunchgrass (a), but the infection can be maintained in bunchgrass-only populations (b). Infected annuals can invade, resulting in long-term coexistence of annuals at high density and bunchgrass at low density with all plants infected. Solid lines: uninfected plants; dashed lines: infected. Annual population in black, perennials in dark grey (adults) and light grey (seedlings). From Borer *et al.* (2007b).

Borer *et al.*'s (2007b) analysis suggests that annual grasses could not have invaded without the effect of BYDV on competitive relationships; similarly, BYDV could not persist in annual populations alone. Hence, the virus must either have been present at low frequency in native bunchgrass populations prior to the introduction of annual species (which then altered pathogen population dynamics); or, but less likely, BYDV arrived around the same time as the annuals, perhaps accidentally imported in infected plant material or with infected aphid vectors.

parasite-modified competition (Box 5.3), with the native species (which are perennial) acting as a temporal reservoir for infection through California's drought period, when annual grasses are not available. The deleterious effects of BYDV on the native species then enable establishment of the invasive (annual) species.

BYDV disproportionately reduces the competitive strength of infected natives against invaders, stunting growth (thereby reducing effectiveness in competition for light) and reducing seed production (Box 5.3). The pathogen thereby reduces the likelihood that the native perennial species, as opposed to the invasive annuals, will succeed in exploiting gaps in the turf. Indeed, gap creation through mortality of infected plants may be of primary importance in driving this invasion, given the long-lived nature of the native grasses.

6.5.4 Native–invader interactions mediated by parasites

In most of the previous examples, parasites influence invasion outcomes through differential effects on the fitness, and hence population density, of native and invading hosts. Parasites can also act directly on the life history traits or behaviour of the host, as seen in the case of BYDV (Box 6.4), modifying native–invader interactions with important potential effects on invasion. In the previous chapters, we discussed examples in which parasitism may reduce the host's competitive ability (Chapter 2), increase its vulnerability to predation (Chapter 3) or alter the strength of its role as an intraguild predator or prey (Chapter 4). These trait-mediated indirect effects (TMIEs) (Box 1.3) of parasitism can potentially affect the outcome of an invasion.

For example, the outcome of competition between the invading mosquito *Aedes albopictus* and the native North American mosquito *Ochlerotatus triseriatus* is modified by the gregarine parasite *Ascogregarina taiwanensis*. *A. albopictus* is the superior competitor. However, *Ascogregarina taiwanensis*-infected *A. albopictus* have a lower effect on the survivorship of *O. triseriatus* than do uninfected individuals (Aliabadi & Juliano, 2002). In the years following introduction of *A. albopictus* to North America, infection was low. This enemy release may make *A. albopictus* more competitive during the initial phase of invasion (Juliano & Lounibos, 2005).

Parasites with indirect life cycles are often trophically transmitted from the intermediate to the definitive host (Fig. 6.11). Many such parasites have been found to manipulate the host's antipredator responses so as to increase the opportunities for transmission to the definitive host. This can mediate invasion success if native and invading hosts are differentially affected. For example, rivers in France are being invaded by the amphipod *Gammarus roeseli*. A native acanthoceplanan, *Pomphorhynchus laevis*, infects both the native *G. pulex* and the invader *G. roeseli*. The parasite increases the vulnerability of *G. pulex* to predation by fish, the definitive host of the parasite, by inducing positive phototaxis. However, the parasite does not manipulate the behaviour of the invasive host (Bauer *et al.*, 2000; Tain *et al.*, 2007). Hence, predation is likely to be higher on the native than on the invasive amphipod. Similarly, the invasive American brine shrimp *Artemia franciscana* has acquired cestode parasites from natives in its new

Mediterranean range. These parasites cause reversed phototaxis and colour-change in the native species, but do not alter the antipredator behaviour of the invading species (Georgiev *et al.*, 2007).

Parasites may also modify intraguild predation (IGP; Chapter 4); that is, predation amongst species that are also potential competitors. We have shown theoretically (Section 4.6) that shared parasitism can increase the conditions under which IG predators may coexist (Hatcher *et al.*, 2008); coexistence is made more likely when the parasite exerts a greater deleterious effect on the 'stronger' species in terms of the combined effects of competition and predation. Such a keystone effect of parasitism may be important in determining the outcome of native–invader interactions, in particular between taxonomically related native and invading species that interact through both competition and IGP (Hatcher *et al.*, 2008).

We have carried out empirical studies of the impact of parasitism on IGP and the outcome of invasions, focusing on amphipod invasions in the United Kingdom. In Ireland the native *Gammarus duebeni celticus* is being replaced by the invasive *G. pulex*. *G. pulex* is the larger amphipod and is a stronger competitor and IG predator than the native. The native *G. duebeni celticus* is larger than two other invaders, *G. tigrinus* and *Crangonyx pseudogracilis*, and is a stronger competitor and predator upon these invaders. However, IGP is strongly influenced in these communities by the acanthocephalan parasite *Echinorhynchus truttae*, which reduces the strength of IGP by the invasive *G. pulex*, thereby promoting coexistence with the native *G. duebeni celticus* (Box 6.5).

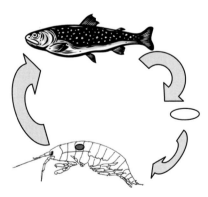

Fig. 6.11. The life cycle of *Echinorhynchus truttae*. The definitive host is a salmonid fish. Parasite eggs are egested in the faeces and are subsequently eaten by the intermediate gammarid host. Transmission to the definitive host occurs when it preys upon an infected gammarid.

Box 6.5 The role of parasites in amphipod invasions

In freshwater communities in Ireland, the invasive amphipod *Gammarus pulex* is replacing the native species, *G. duebeni celticus* through IGP and competition (Dick, 2008). IGP between native and invader is asymmetrical; the invader is a stronger IG predator. However, this interaction is mediated by parasitism (Fig. 6.12). Both the native and the invasive species serve as intermediate hosts for the acanthocephalan parasite *Echinorhynchus truttae*. Transmission to the definitive fish host occurs when the fish preys upon an infected gammarid. Parasite-induced changes in activity and photophilic behaviour enhance the likelihood of predation by the definitive host (MacNeil *et al.*, 2003c). Although both species are used as intermediate hosts, *E. truttae* is more frequently found in *G. pulex* populations and, in parasitised populations, prevalence is higher in the invasive species.

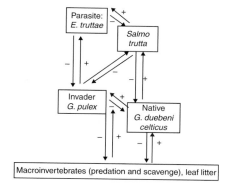

Fig. 6.12. Interaction between the native *Gammarus duebeni celticus* and the invasive *G. pulex* and their parasites and predators. Shown are the two amphipod species which interact via competition and IGP, their resource base, the shared parasite *Echinorynchus truttae* and the brown trout predator, which is also the definitive host for this parasite. Not shown are other top predators without direct links to these parasites or other guild members that do not participate in IGP with gammarids. Also not shown are probable links between *E. truttae* and *G. duebeni celticus* and cannibalistic interactions, which occur within each amphipod species. *E. truttae* mediates IGP between *G. pulex and G. duebeni celticus*; behavioural manipulation of *G. pulex* to enhance transmission of *E. truttae* to *S. trutta* is also documented (MacNeil *et al.*, 2003a; 2003d). This is a simplified web for this community of native and invading amphipods; a more complete picture including two additional invaders and a further species of parasite is found in Box 7.2).

Box 6.5 (*continued*)

Infection by *E. truttae* increases the predatory abilities of *G. pulex* on the endemic prey, *Asellus aquaticus* (Dick *et al.*, 2010) but decreases intraguild predation on the native *G. duebeni celticus* (MacNeil *et al.*, 2003c). Field enclosure experiments were used to assess the impact of parasitism on the native and invading hosts in single- and mixed-species populations. In single-species populations, *E. truttae* did not directly affect the survival of the native or the invader. In mixed-species populations, the survival of *G. pulex* was again unaffected. However, parasitised *G. pulex* were less likely to attack the native species, leading to higher survival of the native species (MacNeil *et al.*, 2003c; Fig. 6.13); this reduction in IGP might facilitate species coexistence (Hatcher *et al.*, 2008).

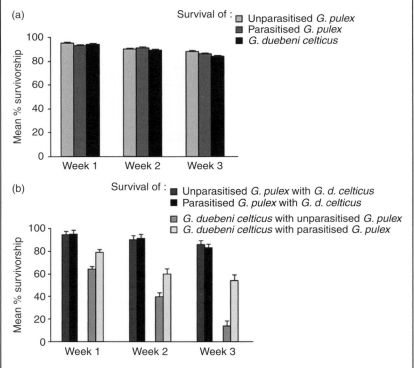

Fig. 6.13. Survival of the native *Gammarus duebeni celticus* and of uninfected and *Echinorhynchus truttae*-infected *G. pulex* in the field in (a) single- and (b) mixed-species enclosures (tubes). Parasitism did not directly affect the survival of the invasive *G. pulex*, but parasitism of the invader enhanced the survival of the native species. From MacNeil *et al.* (2003c).

The microsporidian *Pleistophora mulleri* was shown to mediate IGP between the native *G. duebeni celticus* and another species of invader, *G. tigrinus*. Although the parasite caused no direct mortality or reduction in fecundity in its *G. duebeni celticus* host, it reduced the ability of *G. duebeni celticus* to predate *G. tigrinus*. As a result, survival of *G. tigrinus* over ten weeks was less than 10% in sympatry with uninfected natives, whereas the invader coexisted with infected natives (MacNeil *et al.*, 2003b; Box 4.3).

6.6 Conclusions

Invasive species are one of the key drivers of biodiversity loss and changes in ecosystem services, as recognised by the Millennium Ecosystem Assessment and the IUCN (http://www.millenniumassessment.org/en/index.aspx; http://www.iucn.org/what/tpas/biodiversity). Human activities such as agriculture and transportation mean that the rate of invasions has increased markedly over the last 100 years, with this increase likely to continue. A recent analysis of data from 57 countries found an average of 50 (ranging from 9 in Equatorial Guinea to 222 in New Zealand) non-indigenous species per country that have a negative impact on biodiversity (McGeoch *et al.*, 2010). Invasive species have major economic costs through their impact on ecosystems, agriculture and human health. For example, Pimentel *et al.* (2000) estimated that 50 000 species have been introduced to the United States, causing environmental damage and economic losses of over $130 billion per year. In the United Kingdom, the Minister for Biodiversity stated in 2007 that invasive non-native species cost the British economy approximately £2 billion per year (http://www.parliament.uk/documents/post/postpn303.pdf).

This chapter illustrates the key roles that parasites can play in an invasion. Release from the regulatory pressure of parasites may explain the success and impact of invasive species (Section 6.3). However, invaders may also introduce or acquire parasites in the new range, leading to novel parasite–host associations. The spillover of introduced parasites and their impact on native species is well known (Box 6.3). However, a recent review (Kelly *et al.*, 2009) found that 67% of parasites infecting invasive species were acquired in the novel range, indicating that parasite spillback to the native hosts may be an important and understudied phenomenon. In addition to their impact on host density, study of parasite-mediated native–invader interactions is an exciting area that is likely to shed light on the mechanisms driving invasions (Hatcher *et al.*, 2006; Dunn, 2009).

From a theoretical perspective, there appears to be quite a gulf between empirical examples and general theoretical underpinnings. For instance, much of the enemy release 'model' is based on elegant reasoning and lacks theoretical verification. Drake (2003), in a very general and simplified model, showed that enemy release may occur only for a very limited range of established populations, owing to interactions with propagule size (Section 6.3). We currently do not know whether a relaxation of some of Drake's simplifying assumptions will produce similar outcomes. Whilst we have some quite detailed models for specific systems (e.g. Box 6.3), further general models of all stages of the invasion process are clearly required. The trait-mediated indirect effects (TMIEs) of parasites are also key factors mediating the outcome and impact of invasions. TMIEs act on a shorter timescale than do density-mediated effects, and there is a need for theoretical exploration of these effects.

The theme of parasites and invasions recurs throughout this book. The changes in the parasite fauna as a result of an invasion may have far-reaching effects on a community, a topic that we return to in Chapter 7. In this chapter, we have looked mostly at examples where parasites mediate interactions between invasive host species and the native biota. However, parasites can themselves be invaders. The introduction of novel parasites to a habitat can lead to emerging infectious diseases (EIDs). For example, malaria is undergoing range expansion in Africa and India as a result of climate change and socioeconomic factors (Box 8.7), whilst the recent H1N1 influenza pandemic resulted from introduction of a novel strain of the virus into human populations (Box 8.3). The ecology and impact of EIDs are explored in Chapter 8.

7 · *Ecosystem parasitology*

7.1 Introduction

The previous chapters give an indication of the numerous potential effects of parasites on ecological communities, and effects of communities on parasites. We have met some of the salient interactions in previous chapters:

- Parasites can act as keystone species determining coexistence outcomes for competing species, or as agents of apparent competition (Chapter 2).
- Parasites can alter predator–prey interactions, engaging in resource competition with predators (as parasites of prey), altering predator–prey dynamics (as parasites of predators, prey or both) and altering interaction strengths via behavioural manipulation (by trophically transmitted parasites) (Chapter 3).
- Parasites can alter or engage in intraguild predation (IGP), altering population dynamic and coexistence outcomes (Chapter 4).
- Parasites of plants can have cascading effects on consumers via their impact on the host plant; a wealth of indirect interactions between pathogens, parasites, herbivores and their natural enemies may result from plant defensive signalling (Chapter 5).

Up to this point we have looked at interactions at the level of the module, concentrating on strongly interacting subsets of species. But what are the consequences of such effects for the broader community? Do parasite-mediated interactions ramify across communities to influence food webs, community stability and structure? Perhaps these module-level interactions are 'averaged out' or 'diluted' with the addition of further species and are inconsequential for community-level processes. Indeed, until recently there was very little cross-fertilisation between community ecology and parasitology, a paucity remarked upon by a handful of researchers at the forefront of this now burgeoning research area (Dobson & Hudson, 1986; Marcogliese & Cone, 1997; Thomas *et al.*, 2005b). It has been suggested that the lack of parasites in community ecology studies reflects an underlying belief that their effects could be ignored because of their small size and difficulty of study (Mittelbach, 2005; Lafferty *et al.*, 2006a; Kuris *et al.*, 2008). 'Averaging out', the description of system behaviour in terms of the mean behaviour of its component parts, is a natural consequence of commonly used differential calculus approaches to population dynamics, but it can also result in neglect of important heterogeneity, including that resulting from parasitic infection (Hatcher & Tofts, 2004). We have already encountered examples where averaging out breaks down for the description of one-host–one-parasite systems, as in the case of intraspecific competition of larval amphibians (Section 2.2.2). In that case, disease-induced mortality of conspecifics released uninfected classes from intense intraspecific competition, improving their growth rate and survival (Kiesecker & Blaustein, 1999). To understand population dynamics for the host we must take account of infected and uninfected subclasses and cannot average across them. It would thus seem unreasonable to presume that the effects of parasites can be ignored in the broader community; however, identifying their effects in real systems may prove difficult.

We can distinguish several types of *potential* effects of parasites in ecosystems (Fig. 7.1). Effects of a focal host–parasite interaction may ramify

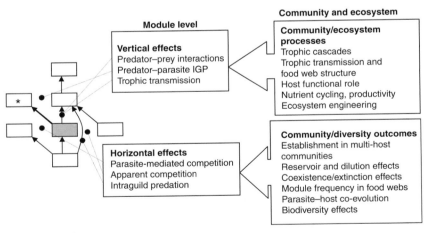

Fig. 7.1. Schematic of possible parasite effects on modules, communities and ecosystems. A parasite (* block) specialising on a single host species (shaded block) can have horizontal and vertical effects in a food web. These local effects can potentially ramify throughout the food web with a variety of conceivable effects on communities and ecosystem processes.

horizontally to other species at the same trophic level as the host, or vertically to trophic levels above or below that of the host. The most obvious vertical ramifications come in the form of trophic cascades (Section 7.2); similar aspects are also covered in the chapter on predation (Chapter 3). In previous chapters we have examined some of the horizontal effects (competition: Chapters 2 and 5; IGP: Chapter 4). Here we extend this analysis to consider further parasite establishment in communities of hosts, and the influence of reservoir and dilution effects (Section 7.3). Perhaps the most compelling arguments to date for the role of parasitism in shaping ecosystems come from food web and bioenergetic studies, examined in Sections 7.5 and 7.6. The effects of parasites may also ramify beyond the food web, influencing ecosystem properties, functions and services. By altering diet preference or feeding behaviour, parasites can influence the functional role of host species (Section 7.5.1). By altering physical or chemical properties of hosts or their environment, parasites can act as ecosystem engineers (Section 7.7). Together, these studies have led to the suggestion that, in general, parasitism and parasite diversity are signs of a well-functioning or 'healthy' ecosystem (Section 7.8). Finally, parasite–host co-evolution may feed back on parasite population dynamics, with ultimate consequences for ecosystem properties (Section 7.9).

7.2 Trophic cascades

Some of the strongest demonstrations of the vertical ramifications of parasitic interactions in communities take the form of trophic cascades, in which indirect top-down regulation of the productivity or biomass of one trophic level is affected by consumers at a higher trophic level (at least one level removed (Table 7.1)). Particularly clear examples showing the potential of parasites to induce trophic cascades come from grazed systems. The general pattern here is that parasite-induced effects on herbivore populations have knock-on effects for the basal resource species. In some cases, trophic cascades are the result of trait-mediated indirect effects (TMIEs), rather than direct effects of parasitism on host density (as seen, for example, in the periwinkle *Littorina littorea*, discussed below). The effects of parasitism can also ramify up the food web, as demonstrated by the indirect effects of plant parasites on herbivore populations (as with many endophytes) and of herbivore pathogens on predators (for example, after rinderpest eradication, discussed below). The almost complete eradication of dominant tree species such as the American chestnut and English elm following pandemic outbreaks of

Table 7.1. *Effects of parasites on trophic cascades*

System	Observation	Reference
Rinderpest in Serengeti ungulates	Rinderpest caused ungulate grazer population crashes; reduced grazing pressure altered plant community structure and succession. Eradication of rinderpest led to increases in some ungulate populations, with knock-on effects for predators and competitors.	Dobson (1995); Thomas *et al.* (2005b); see also Sections 3.3.2 and 7.2
Trematodes of intertidal snails *Littorina littorea*	Trematode infection reduced feeding rate of snails, resulting in increased cover of edible macroalgal species.	Wood *et al.* (2007)
Canine parvovirus in grey wolf *Canis lupus*	CPV induced crash in wolf population; moose (*Alces alces*), freed from predation pressure, undergo boom and bust population cycles linked to climatic oscillations and forage availability.	Wilmers *et al.* (2006); see also Box 3.6
Myxomatosis in rabbits *Oryctolagus cuniculus*	Myxomatosis reduced rabbit populations and so reduced grazing pressure, enabling viable sheep grazing in some areas and shifted plant community structure in others, with knock-on effects for other herbivores and soil-living invertebrates such as ants. Contributory factor in extinction of the large blue butterfly (*Glaucopsyche arion*) in the United Kingdom. Also effects on higher trophic levels; linked to reduction of fox (*Vulpes vulpes*) and buzzard (*Buteo buteo*) populations.	Sumption and Flowerdew (1985); Minchella and Scott (1991); Dobson and Crawley (1994); see also Section 3.2.2

Microsporidia (*Cougourdella* sp.) of herbivorous caddisfly (*Glossosoma nigrior*)	Recurrent parasite outbreaks caused caddisfly population crashes. The resulting reduction in grazing pressure led to increased abundance of attached algae, algae consumers (filter feeders and grazers) and predatory caddisflies and stone flies.	Kohler and Wiley (1997); Kohler (2008)
Acanthocephalan infection of the isopod *Caecidotea communis*	Infection with *Acanthocephalus tahlequahensis* reduced detritus processing by the isopod, reducing the availability of resources to the stream ecosystem.	Hernandez and Sukhdeo (2008a); see Section 7.5.1
Trematode infection of the snail *Hydrobia ulvae*	Parasitism reduced bioturbation by the snail, leading to a reduction in primary productivity and changes in diversity of secondary consumers.	Mouritsen and Haun (2008); see Section 7.7
Trematode infection of the intertidal amphipod *Corophium volutator*	Parasite outbreaks led to a population crash of *C. volutator*, a species which stabilises the sediment through burrowing. Declines in *C. volutator* led to an increase in sediment particle size and an increase in primary production.	Mouritsen *et al.* (1998); see Section 7.7
Endophytes (*Neotyphodium* spp.) of grasses	Infection reduced nutritional quality of plant, reducing vertebrate and invertebrate herbivore pressure; reduced populations of insect herbivores led to reduced populations of predatory spiders and parasitoids.	Omacini *et al.* (2001); Finkes *et al.* (2006); see also Section 5.5

This (non-exhaustive) list focuses on cases where the influence of parasites on documented community change is strongly implicated.

their respective fungal pathogens (chestnut blight and Dutch elm disease) has had clear effects on the landscape (Section 5.1.2), but also results in the loss of associated specialist herbivore guilds with potential knock-on effects for their other competitors and predators; canopy structure and recruitment of other tree species were also affected (Loo, 2009).

7.2.1 Density-mediated trophic cascades

The interaction between herbivore food resources and parasites in determining population densities has been examined theoretically by Grenfell (1992). For 'extensive' ungulate grazing systems, where herbivore numbers are largely naturally regulated (rather than managed), parasites can act as top predators regulating the population density of herbivores, resulting in an increased density at equilibrium for the grazed plant species (i.e. parasites induce a trophic cascade (Fig. 7.2)). This result mirrors the effect of top predators in the classical food chains of producer, consumer and predator (Rosenzweig, 1973). However, more complicated dynamics are also possible. In some cases the system may oscillate between low levels of parasitism (with plant–herbivore cycles at low plant density where vegetation is more or less regulated by the

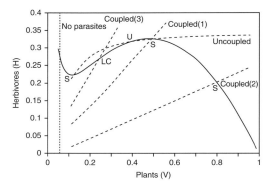

Fig. 7.2. Isoclines for an extensive grazing system where herbivores are hosts to macroparasites. The solid line shows the plant zero-growth isocline; where this crosses the herbivore isocline (dashed lines), the populations are at equilibrium (which can be S: stable; U: unstable; LC: resulting in limit cycles). Parasite transmission can be coupled (dependent on) or uncoupled from herbivore food consumption (dashed lines as labelled, for different parameter combinations). In this example the stable equilibrium for plants in the presence of the parasite is higher than in its absence (where the vertical line intersects the horizontal axis), showing that parasites of herbivores can induce a trophic cascade. From Grenfell (1992).

herbivore) and parasite outbreaks (where plant density shifts to a higher density and herbivores are regulated by the parasite; recall a similar result for forest Lepidoptera in Box 3.3). For the type of parasite modelled by Grenfell (directly transmitted macroparasites such as nematode worms, in which free-living larval stages are consumed along with the vegetation), parasite transmission efficiency is coupled to availability of vegetation: at high levels of forage, larval density is 'diluted', reducing transmission, but when forage is scarce, hosts are more likely to acquire the infection which can exacerbate morbidity or mortality for already nutrient-stressed animals.

There is some evidence for this synergistic interaction between resource supply and parasitism from soay sheep on the island of St Kilda, which undergo boom–bust population cycles over a period of several years; nematode worm burdens and protozoan infections exhibit cycles lagged behind that of their host (Grenfell, 1988; 1992; Craig *et al.*, 2007). For 'intensive' systems, the population density of the herbivore is managed and therefore not regulated by resource availability. Parasitism can also have cascading effects in these systems. For instance, managers may need to take account of potential downsides to anti-parasite treatment of their stock – elimination of parasites can theoretically result in over-grazing by hosts in the absence of the regulatory effects of parasitic infection (Grenfell, 1992).

There are some strong empirical examples of parasite-induced trophic cascades in plant and animal systems (Table 7.1). We discussed plant examples in more detail in Chapter 5. Soil-borne pathogens in particular may influence successional dynamics with community-wide impacts (Section 5.2); some parasitic plants such as the mistletoes can have far-reaching effects on community structure (Section 5.4.2), and a broad range of horizontal and vertical effects across trophic levels have been described for symbiotic endophytes of grasses (Section 5.5).

A series of studies from invertebrate systems (Kohler & Wiley, 1997; Kohler, 2008) examined the effect of parasite-induced mortality of a dominant grazer on the community structure of a series of streams in Michigan. Here, the caddis fly *Glossosoma nigrior* was the dominant grazer, having strong competitive effects on other grazers and filter feeders. Periodic outbreaks of the microsporidian *Courgourdella* sp. led to massive (25-fold) declines in caddis fly density. As a result, periphyton biomass increased. The release from competition by *G. nigrior* also led to an increase (2–5-fold) in the density of grazers and filter feeders, with some species that were rare prior to the epidemic showing large increases in density. Predatory caddis flies and stone flies also increased in density. Thus parasite-induced mortality of a dominant herbivore had effects throughout the trophic levels, affecting both productivity and community structure.

For vertebrate hosts, there is evidence that rinderpest induced trophic cascades in East Africa after its outbreak in the 1890s and its eradication in the 1960s. As rinderpest spread into the wild ungulate populations from infected cattle, it caused massive herbivore mortality, which in turn influenced the composition of the plant community. Reduction in grazing pressure enabled recruitment of tree saplings; in Tanzania, several large areas of even-aged *Acacia tortilis* stands can be identified, which are the result of establishment episodes at the turn of the century (Prins & Weyerhaeuser, 1987). Similar patterns have been reported for *Acacia* establishment in Tanzania after anthrax outbreaks in impala in 1977 and 1983 (Prins & Vanderjeugd, 1993), and also for oak (*Quercus robur*) woodland as evidenced in Silwood Park, England following the arrival of myxomatosis in rabbits in the 1950s (Dobson & Crawley, 1994; Section 3.2.2). When rinderpest was finally eradicated, effects on a broad range of herbivore and carnivore species testified to the impact of this morbillivirus across trophic levels (reviewed in Thomas *et al.*, 2005b). Increases in the abundance of ungulate hosts, particularly wildebeest (*Connochaetes taurinus*) and buffalo (*Syncerus caffer*), led to increases in the population densities of carnivores such as lions (*Panthera leo*) and hyenas (*Crocuta crocuta*). The consequent increased predation pressure is thought to be responsible for subsequent decreases in Thompson's gazelle (*Gazella thomsonii*) populations, and competition with hyenas may have led to declines in African wild dog (*Lycaon pictus*) populations. Wild dogs and lions have also come under pressure from outbreaks of canine distemper, which may itself have increased following eradication of rinderpest (Box 8.10). The dynamics of these closely related diseases may be linked, as evidence suggests that feeding on rinderpest-contaminated carcasses provides cross-immunity to canine distemper (Dobson & Hudson, 1986). Recent statistical modelling suggests that the eradication of rinderpest has led to a long-term change in the trophic balance in the Serengeti, altering the ecosystem's carbon budget. The increase in wildebeest numbers following rinderpest eradication has increased grazing pressure, reducing standing fuel loads and the frequency of fires. This may have transformed the Serengeti into a net sink of carbon instead of a net source (Holdo *et al.*, 2009).

7.2.2 Trait–mediated trophic cascades

In contrast with the density-dependent effects of parasitism described above, a manipulative study by Wood *et al.* (2007) revealed that trait-mediated indirect effects (TMIEs) of parasitism led to a reduction in grazing by the periwinkle *Littorina littorea*. Wood *et al.* (2007) examined the effect of the digenean trematode *Cryptocotyle lingua* on the periwinkle, a dominant algal grazer on Atlantic rocky intertidal shores in North America. As snail density was controlled in the experimental enclosures, effects in this system were largely trait-mediated. Infection reduced feeding rate by 40% in laboratory studies, probably as a result of impaired digestive function (this parasite causes extensive damage to the digestive gland) or reduced energy requirements of infected individuals (infection frequently induces castration, reducing expenditure on reproduction). This effect translated into altered algal community composition in field enclosure experiments. Cages with no snails or stocked with infected snails attained 65% higher cover by edible macroalgae compared to cages stocked with uninfected snails (Fig. 7.3). Possible knock-on effects in communities with high levels of trematode infection could include reduced barnacle and mussel recruitment (both known to be reduced in habitats with a high degree of algal cover); a direct relationship between snail recruitment rates with infection remains to be demonstrated, however.

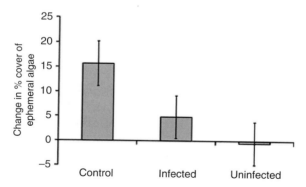

Fig. 7.3. Change in percentage cover of edible macroalgae for experimental treatments for grazing by the snail *Littorina littorea*. After 24 days, percentage cover increased substantially in the control enclosures where snails were excluded and had reduced slightly in enclosures containing one uninfected snail. Cover increased somewhat in enclosures containing one infected snail, demonstrating that grazing was reduced by parasitism but still affected algal growth. From Wood *et al.* (2007).

Parasite-induced reduction in grazing pressure does not, however, appear to be a general pattern, even within snail–trematode systems. In contrast with the systems described above, the trematode *Posthodiplostomum minimum* infecting the freshwater snail *Physa acuta* appears to increase grazing rate in individual laboratory trials, resulting in a 20% reduction in the standing stock of periphyton in mesocosms in which 50% of the snails were infected, compared to mesocosms with uninfected snails or with low (10%) infection rates (Bernot & Lamberti, 2008). This led to dominance of the macroalga *Cladophora glomerata* in the 50% infected treatment, whereas blue-green algae and diatoms dominated the low- or no-infection treatments.

From these studies, it is evident that both density-dependent and trait-mediated effects of parasitism on host grazing can potentially be felt at distant trophic levels. However, relatively few of these studies have actually measured the impact of parasitism beyond the host and its grazed resource, so we still lack information on whether these effects actually propagate to other species. Similarly, we have few detailed pictures of how the effects of parasitism on host population density can influence trophic cascades, but in these few cases, there is compelling evidence for the role of parasites.

7.3 Parasite dynamics in multi-host communities

The vertical ramifications of parasitism through trophic cascades generally concern specialist parasites of a dominant host species. Another way in which parasites can influence, and be influenced by, community structure concerns the effects of generalist parasites. We have examined the population dynamics of a parasite species infecting two host species in the context of competitive interactions (Chapter 2). More generally, alternative host species (for the same life stage of a parasite) may not interact directly at all; they may vary in their 'competence' as hosts (varying in propensity to become infected, mount an immune response or transmit the parasite to other hosts) or they may interact in ways other than resource competition (for instance, influencing alternative host distribution via behavioural effects). Under what circumstances will multiple host species enhance parasite establishment and persistence, and when will they impede it? This question has been of interest to ecologists and parasitologists for some time, but has recently gained new impetus with current concerns over the interaction between parasitism and biodiversity (see Section 7.4).

7.3.1 Baseline model: parasite establishment in multiple host species

A first attempt to investigate these questions in a generalised setting was provided by Holt *et al.* (2003), using a graphical approach. To examine conditions for parasite establishment in two-host communities, Holt *et al.* (2003) developed a graphical isocline model where hosts are treated as resources (this approach can in principle be applied to more than two host species, although the patterns will become increasingly difficult to interpret). For parasites with density-dependent transmission, each host species separately must exceed a threshold population size for parasite establishment (N_T; Box 1.2). For a two-host community, the criterion for parasite establishment depends on some combination of population densities for the two species. If joint densities fall below this level, the parasite cannot invade initially; if the density combination lies above the line, R_0 for the parasite exceeds 1 and the parasite can spread (Fig. 7.4). The shape of these zero-growth isoclines depends on the relative rates of within- and between-species transmission, which in turn will depend on how the hosts interact as well as parasite and host physiology. In many cases the resulting isoclines have patterns similar to those found in classical resource–consumer dynamics (Tilman, 1982), although the underlying mechanisms generating the patterns differ. Although not providing quantitative predictions for any particular system, this approach can be useful in classifying and comparing different systems. For instance, 'non-interactive' hosts with zero cross-species transmission will result in a rectangular isocline requiring at least one of the single host species' N_T to be met for parasite establishment (Fig. 7.4a). Entirely 'substitutable' hosts (with cross-species transmission equal to within-species transmission) result in a linear isocline, with parasite establishment when the weighted sum of the two hosts exceeds a threshold value (Fig. 7.4b). Perhaps the most likely scenario for most two-host systems is depicted in Fig. 7.4c, with between-species transmission lower than within-species transmission, the hosts are partially decoupled and a convex isocline results. In the opposite case, 'complementary' hosts with cross-species transmission exceeding within-species transmission boost the parameter space for parasite invasion with a concave isocline (Fig. 7.4d). However, results from the competition models (Section 2.4.2) suggest that these concave associations are in general not stable in the longer term, with one of the hosts excluded. Holt *et al.* (2003) consider two further cases in this general model: if the parasite requires

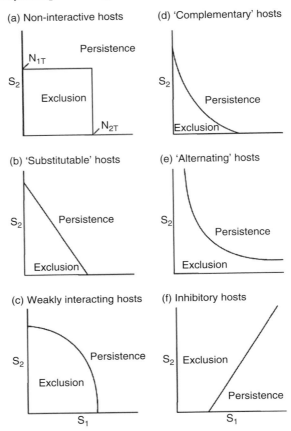

Fig. 7.4. Zero-growth isoclines for the establishment of a parasite in a two-host community. These idealised isoclines take different shapes depending on the pattern of transmission between host species (S_1 and S_2); see text for details. From Holt *et al.* (2003).

alternate hosts (i.e. has an indirect life cycle) no single species can sustain the parasite. Here, between-species transmission does exceed within-species transmission, resulting in a concave isocline (Fig. 7.4e). Finally, one species may inhibit transmission by the other (perhaps because it acts as a sink for one of the infective stages). This will result in an isocline with positive slope, with parasite establishment possible only for sufficiently high population densities of the competent host ameliorating the impact of the inhibitory species (Fig. 7.4f).

Fig. 7.5. Schematic of the relationship between dilution, reservoirs and host competence. As host competence increases, the parasite's basic reproductive number (R_0) increases and, in cases where there is a threshold population size for parasite establishment (N_T), establishment becomes possible at lower population sizes; competent hosts are more likely to act as reservoirs. Hosts of lower competence are more likely to contribute to dilution effects in multi-host communities, and are less likely to sustain the parasite in the absence of other host species (lower R_0 requiring a higher N_T).

7.3.2 Reservoir versus dilution and host competence

The isoclines for two-host systems naturally give rise to the question: what constitutes a 'good' host from the parasites' perspective in multi-host systems? From this perspective, 'good' hosts are those that boost the parameter zone for parasite persistence, shifting the zero–growth isocline leftwards or downwards. Such species will be the most competent hosts and are more likely to act as reservoirs for infection. In contrast, 'poor' hosts will promulgate dilution effects (Fig. 7.5). We have seen several cases of reservoir effects in previous chapters; for instance, barley yellow dwarf virus (BYDV) in grasses (Boxes 5.3 and 6.4) and squirrel pox virus (SQPV) in grey squirrels, which act as reservoirs for infection into red squirrels (Boxes 2.5 and 6.3). Dilution effects have been less explored up to this point, although louping ill (LIV) provides an example (deer act as dilution hosts for the disease through wasted tick bites (Box 2.8)). We review the terminology here:

- *Host competence.* Competent hosts are those that propagate the parasite well, enabling its maintenance and spread. Competence is therefore associated with R_0: all competent hosts result in an $R_0 > 1$ when the parasite is transmitted intraspecifically; more competent hosts result in a higher R_0. In the case of directly transmitted infections, competence is likely to be associated with ability to support a high parasite burden, resulting in higher transmission, or it may involve aspects of host species behaviour (high intraspecific contact rates or living in large free-mixing colonies, for instance). For vector-borne diseases, competence will depend on how well the host propagates the disease *and* the vector: the parasite must replicate sufficiently to enable infection of vectors on taking a blood meal, and vectors must be frequently attracted to feed on that host.

- *Reservoir hosts.* These will be competent hosts, allowing good replication of the parasite, although not necessarily showing strong symptoms of disease. On the contrary, hosts that suffer few virulence effects often support the parasite at a high prevalence, resulting in greater chances of per-encounter cross-species infection. Such is the case for SQPV in grey squirrels: in England, the majority are seropositive for SQPV, but are symptom-free. Reservoir species are key to understanding the processes of 'spillover' and 'spillback' of emerging infectious diseases between populations (Section 8.2.1). Reservoir hosts can enable maintenance of a parasite that cannot persist in the focal susceptible population alone. Consequently, controlling disease in the reservoir host can enable control in susceptible host species; this is particularly important where it would be difficult to target the species at risk directly. For instance, rinderpest was brought under control by vaccination of domestic cattle only (Section 7.2), and outbreaks of canine distemper in lions can be prevented by vaccination of domestic dogs (Packer *et al.*, 1999).

- *Dilution effect.* In contrast to reservoir species, hosts that are relatively inefficient propagators of the disease, or non-hosts that otherwise interfere with disease transmission, can cause a dilution effect. The classic example involves vector-borne diseases: generalist vectors harbouring a parasite may 'waste' an infectious bite on non-host species (the LIV system) or on species that are relatively ineffective hosts of the parasite (Lyme disease, discussed below). Keesing *et al.* (2006) suggest extension of the term to include all net effects of a species that result in decreased disease incidence (hence predators may be responsible for a dilution effect via indirect effects on transmission between prey, for example).

The relationship between dilution and reservoir effects is examined by Keesing *et al.* (2006), who make the point that the pattern seen will depend on whether the most competent host species is already present when a community undergoes change. If the most competent host is already present, increasing species richness will result in a dilution effect; if the current community has relatively incompetent hosts, adding more species, if they include more competent hosts, will tend to increase disease incidence (Fig. 7.5). This pattern is borne out by experimental manipulations of grassland communities showing that reduced species richness tended to increase foliar fungal pathogen load (Mitchell *et al.*,

2002). Variation in outcomes could be attributed to a combination of species composition effects (pathogen load was higher in species-poor communities, provided that the more susceptible grass species were retained) and spatial effects on transmission (transmission of the more species-specific pathogens was more efficient in dense stands of their preferred host).

7.3.2.1 Predation on parasites

Non-host species can also contribute to dilution effects via predation. Recent reassessment of the way in which parasites interact with predators shows that parasites from a variety of ecosystems fall prey to a range of predators (Thieltges *et al.*, 2008b; Johnson *et al.*, 2010). Predation occurs when infected hosts or free-living parasite stages are consumed; eating parasites can contribute substantially to energy budgets for some species. Some predators (such as cleaner wrasses of the genus *Labroides*) specialise in feeding on parasites; in addition, many parasites and parasitoids (i.e. hyperparasitoids – see Section 4.4.2) specifically infect other parasites or parasitoids (or hosts already infected with another parasitoid). Predation on parasites (assuming it is not part of a trophic transmission pathway) has the effect of reducing parasite transmission efficiency (Thieltges *et al.*, 2008b). To the extent that diverse communities contain a greater variety of such predators of parasites, we might expect a greater transmission-blocking effect in these communities; hence diverse communities will be more likely to exhibit dilution effects (Johnson *et al.*, 2010). Furthermore, predation on *hosts* can influence disease prevalence (Section 3.6); this interaction is thought to be partially responsible for patterns of disease risk for Lyme disease, as in Section 7.3.3.

7.3.3 Lyme disease risk, dilution and reservoir hosts

A complex example involving dilution and reservoir effects concerns the spread of Lyme disease across northeastern and central regions of the United States (Box 7.1). This example could equally be considered in the context of emerging diseases (Chapter 8), but we consider it here because this intensively studied system has a wealth of insights to offer concerning dilution and biodiversity–disease relationships. The bacterium responsible for Lyme disease, *Borellia burgdorferi*, is a tick-borne zoonotic infection maintained in a variety of small mammals (and

Box 7.1 Lyme disease

Lyme borrelis, a tick-borne disease caused by the bacterial spiro-chate *Borellia burgdorferi*, has been on the increase in eastern North America since the 1970s (US Center for Disease Control (CDC): http://www.cdc.gov). It is the most widely reported vector-borne disease in the United States, accounting for more than 20 000 cases annually. Lyme disease incidence is also increasing in mainland Europe and the United Kingdom (UK Health Protection Agency: http://www.hpa.org.uk). The host species involved in Europe differ, but similar patterns with respect to habitat and biodiversity changes are likely.

Black-legged ticks, white-footed mice and human disease risk. In most of the US range, the black-legged tick *Ixodes scapularis* is responsible for transmitting the disease. This species has a complex life cycle requiring one blood meal from an appropriate host to complete each develop-mental stage (Fig. 7.6). Adult ticks feed from large mammals including deer and humans; larvae feed on a variety of small mammals which vary in their competence as amplifiers of *B. burgdorferi*. The most effective transmitter of the infection to ticks is the white-footed mouse, *Peromyscus leucopus*; infected mice (which are asymptomatic) transmit the bacterium to larval ticks with 85% probability (LoGiudice *et al.*, 2003). Larvae acquiring the infection will become infective nymphs that quest for their next blood meal from larger mammals, including humans. As the nymphal stage is most likely to transmit the

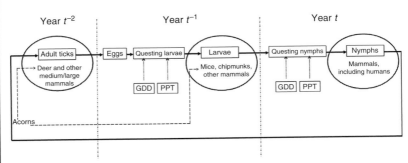

Fig. 7.6. Life cycle of the black-legged tick, *Ixodes scapularis*. The frequency of each life stage (egg, larva, nymph, adult) is influenced by various abiotic (GDD: temperature; PPT: precipitation) and biotic (acorns, hosts) factors. Year *t* is the focal year for disease risk to humans, as it is then that nymphal ticks search for hosts, including humans. From Ostfeld *et al.* (2006).

Box 7.1 (*continued*)

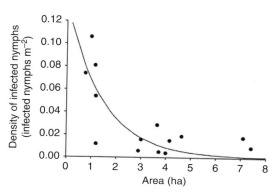

Fig. 7.7. Lyme disease risk in relation to habitat fragmentation. The density of *Ixodes scapularis* nymphs infected with *Borellia burgdorferi* declines as an exponential decay as forest fragment area increases, near the epicentre of infection in Duchess County, New York. From Allan *et al.* (2003).

infection to man, the density of infected nymphs per unit area provides a good estimate of disease risk (Ostfeld & LoGiudice, 2003). Regression analysis of 13 years of data for forest sites near the centre of the endemic range in the state of New York found that the best predictors of disease risk in the current year were the prior year's abundance of *P. leucopus* and chipmunks *Tamias striatus*, and abundance of acorns (a critical food resource for these hosts) two years previously (Ostfeld *et al.*, 2006).

Habitat fragmentation and disease risk. Sampling across sites at the centre of the endemic zone indicates a strong relationship between habitat size and density of infected nymphs (Allan *et al.*, 2003). The relationship is non-linear, with small patches (under 2 ha) having particularly high infected nymph and hence risk levels (Fig. 7.7). This pattern seems to occur because small habitat fragments have reduced biodiversity but retain resilient rodent species such as *P. leucopus*. It has been suggested that mice (and other adaptable small mammals, see below) prevail in these patches as they incur reduced predation pressure, larger predators often being the first species to disappear in fragmented habitats (Ostfeld & Holt, 2004). Data on host competence, species frequency and tick burdens have been used

Box 7.1 (*continued*)

to parameterise simulation models of disease risk (LoGiudice *et al.*, 2003; Ostfeld & LoGiudice, 2003). Recent analysis of 40 habitat fragments provides a statistically significant correlation between predicted and observed risk, although the models consistently overestimate disease risk (LoGiudice *et al.*, 2008).

Multiple reservoir species. Recent genetic analysis of the bacterial strains in hosts reveals that two species of shrew are also important reservoir hosts (Brisson *et al.*, 2008). Transmission of the infection to ticks feeding on short-tailed shrews, *Blarina brevicauda*, and masked shrews, *Sorex cinereus*, is moderate (42–57%), but these species feed a substantial proportion of ticks. At the site investigated, white-footed mice fed 10% of all ticks (contributing 25% of the infected tick population), whereas the two shrews together fed 35% of all ticks, contributing 55% of those infected. Chipmunks contribute a further 10% of the infected tick population (Brisson *et al.*, 2008). The occurrence of multiple reservoir species has implications for disease control strategies (see below).

---

possibly birds); larger mammals that do not propagate the bacterium but support the adult tick are also relevant. Anthropogenic change in the form of habitat fragmentation appears to be responsible for the increased incidence in Lyme disease, via its positive impact on the most competent host species for the disease, the white-footed mouse *Peromyscus leucopus*.

The lessons from this system for disease control strategists are:

- *Habitat fragmentation increases the risk of human disease.* Small habitat fragments (under 2 ha) are particularly vulnerable to increased risk of disease spread to humans, because biodiversity and predator loss in these patches allows proliferation of the most competent host species (mice and shrews). Ticks in these patches are more likely to feed on an infected host, and to take up the infection themselves. Fragmented habitats usually occur in close association with urban or suburban development, further exacerbating the effect as humans are more active in these fragments.
- *Control should target several hosts.* Targeting the white-footed mouse as the sole reservoir for the disease is unlikely to be effective: vaccination or culling of this species (assuming 100% efficiency) would only result in a

25% reduction in disease incidence if the recent work on shrews as reservoir species is generally applicable. If the four most competent species are targeted, up to 90% disease risk reduction can be achieved (Box 7.1).

- *Infection risk could be mitigated by public health warnings.* The strong correlation between disease risk and acorn production two years previously suggests the use of public health warnings of increased disease risk suitably lagged after mast years.

7.4 Biodiversity and disease

Lyme disease highlights current research concerns over the effect of biodiversity on disease risk. With particular emphasis on the anthropo-genic changes that are leading to reduced biodiversity and degraded communities, the question arises: to what extent does biodiversity buffer organisms from disease, and under what circumstances does increased biodiversity enhance the spread of disease?

7.4.1 Determinants of disease spread

An examination of the behaviour of one-host models (Chapter 1) allows us to identify several biodiversity-related factors that can alter the rate of infection of the host (Keesing *et al.*, 2006). Adding non-host species can reduce disease persistence in a focal species through a number of processes:

- reduced rates of encounter between the susceptible (S) and infectious (I) subclasses – for example, through changes in social or predator-avoidance behaviour;
- decreased density of susceptibles through enhanced interspecific com-petition or predation pressure;
- increased recovery rate of infected individuals (recovery augmenta-tion) – for example, the addition of a resource species may enhance a host's immune response, speeding its recovery. This reduces the time that individuals are infectious, thus reducing S–I encounter likelihood;
- increased death rate of infected individuals – increased predatory pressure or interspecific competition may differentially affect infected individuals.

Opposing scenarios can also be envisaged for each of these cases whereby additional non-host species enhance parasite persistence and spread (Keesing *et al.*, 2006).

These arguments focus on effects for disease risk in a single species; in multi-host systems, the consequences for disease risk to other organisms require further elaboration. Disease risk (usually referred to in the context of risk to sensitive wildlife species, domestic stock or humans) will depend on how the above factors influence parasite prevalence in the (reservoir) host(s), plus other factors that affect contact and transmission between reservoir hosts and the focal host of economic, conservation or health interest. For instance, in the case of Lyme disease, although white-footed mice are the most efficient transmitter of the infection to ticks, they are not the most frequently encountered or preferred host for ticks; it now appears that two species of shrews are equally important in determining disease risk as they are reasonably effective propagators of the infection to ticks and incur a high proportion of tick bites (Box 7.1).

7.4.2 Transmission models and biodiversity relationships

General models for the establishment of parasites in multi-host communities suggest that the transmission dynamics of disease will play an important part in determining the relationship between biodiversity and disease. Parasite transmission can be regarded as either density dependent or frequency dependent, depending on the types of contact expected between susceptible and infected hosts (Box 1.4). These concepts, usually applied to within-species transmission, require extension to the between-species case in the multi-host context. Dobson (2004) has used a matrix approach (Box 7.5) to characterise within- and between-species infection dynamics, allowing derivation of the conditions for initial spread of a directly transmitted disease in multiple host species. If transmission within and between species is density dependent, increasing the number of host species increases the range of conditions for disease spread (where $R_0 > 1$), so parasite establishment is enhanced with increasing biodiversity. However, if transmission between species is frequency dependent, R_0 decreases with additional host species, and biodiversity is predicted to inhibit establishment of disease. This occurs because, for a parasite with frequency-dependent transmission, each additional host species causes a dilution effect by reducing the relative frequency of each host species in the community. If transmission is density dependent, additional hosts simply add to the summed densities contributing to transmission (even if they are less competent hosts for the parasite, they nevertheless provide an increased 'resource space' for the parasite, without detracting from available 'high-quality resources'). Similarly, Rudolf and Antonovics (2005) found a dilution effect in a two-host model with frequency-dependent transmission. An interesting implication of these results is that, in contrast to apparent competition models with density-dependent transmission (Chapter 2), frequency-dependent transmission induces apparent mutualism (i.e. a ++ interaction) between the host species (Abrams & Matsuda, 1996), whereby the presence of one host species buffers another host from parasite-induced extinction (Rudolf & Antonovics, 2005). Both of these models assume no interaction between the host species other than that produced by shared parasitism, so no competitive or predatory interactions affect the densities of hosts in the community. Whether this simplification can realistically be applied in real-world instances, and the consequences of relaxing this assumption, remain to be explored. In a single-species context, density-dependent transmission is the usual pattern expected for directly transmitted infectious diseases (such as viral colds and flu in humans), or for diseases attained from a free-living stage (such as helminths ingested by grazing herbivores). Frequency-dependent transmission is the form taken by most sexually transmitted diseases and is also the more likely pattern for many vector-borne diseases (Chapter 1). Hence, in one-host–one-parasite systems, density dependence is often assumed to be the norm, but there is no *a priori* reason to expect this pattern to apply to multi-host communities, which ideally should be examined individually for their characteristic transmission properties.

The implications of these diversity–transmission relationships are interesting for biodiversity conservation. In the case of frequency-dependent transmission, positive feed back between reduced diversity and parasite-induced extinction could set off an extinction cascade (Fig. 7.8), whereby reduced biodiversity (perhaps resulting from anthropogenic influences) would increase disease incidence in remaining species, potentially resulting in further extinctions in a cascading process reminiscent of the habitat destruction–extinction cascades proposed by Tilman *et al.* (1994). A parasite-induced cascade requires that hosts suffer reduced fitness from parasitism, but in some multi-host systems many host species are asymptomatic (which is why they are good reservoir hosts). For instance, Lyme disease (Box 7.1) exhibits a negative relationship between biodiversity and infection prevalence, and most of the natural community of hosts in these systems suffer limited or no pathology; it is only humans that need fear

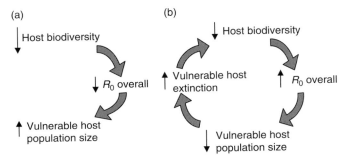

Fig. 7.8. Schematic of host biodiversity–disease relationships in multi-host communities. (a) If parasite transmission is density dependent, a reduction in host biodiversity results in a reduced total R_0 for the parasite in the community; as this results in reduced parasite prevalence in the more vulnerable host species, its population size will increase and parasitism will not contribute to further extinctions. (b) If transmission is frequency dependent, a reduction in host biodiversity results in increased R_0 across the community; the most vulnerable host population will decline as a result of increased parasite prevalence, making it more vulnerable to extinction, setting up a cycle of positive feed back with successive extinctions of more vulnerable species.

the consequences of this particular diversity–disease relationship. However, for West Nile virus (WNV), there is also a negative relationship between biodiversity and human disease risk, but many bird species are also affected (Box 8.8). Nevertheless, parasite-induced extinction of a keystone species could set off an extinction cascade driven by other demographic processes (de Castro & Bolker, 2005). Conversely, parasites may make ecosystems more robust to secondary extinction, as we discuss in Section 7.5.3. Recent evidence from food web studies also suggests that parasites themselves can be particularly vulnerable to extinction and their loss can have important consequences for ecosystem properties (Section 7.8).

7.5 Parasites in the food web

Despite some earlier suggestions that parasites should be included in food web studies (Price *et al.*, 1986; Minchella & Scott, 1991), it was well into the new millennium before this area of research started to show promising growth (Table 7.2). We now know that parasites can influence the functional role of hosts in ecosystems (Section 7.5.1), and it is well known that they can alter the strength of interaction between the host and other species (for instance, via behavioural manipulation in trophic transmission (Section 3.5.1)), so it seems likely that they could have important effects on food web properties (Section 7.5.2). We discuss how these studies shed light on the role of parasitism for community properties in Section 7.5.3.

7.5.1 Functional role and interaction strength

Parasitism often leads to changes in food consumption by the host. In many cases, parasite-induced morbidity may lead to decreased consumption. Less frequently, parasitism may cause increased food intake by the host. For example, sticklebacks with early-stage infection by *Schistocephalus solidus* showed increased food intake (Wright *et al.*, 2006). Whilst evidence for altered foraging and consumption in infected hosts is widespread, there have been fewer attempts to look for wider community-level effects of parasite-induced changes in consumption, although there are examples suggesting that parasites affect energy transfer through communities via trophic cascades (Section 7.2.2). Such parasite-induced changes in resource utilisation may therefore impact the functional role played by hosts in energy transfer through the ecosystem.

In many stream ecosystems, the primary basal energy resource is provided by leafy detritus deposited in the water. Shredding and detritivorous macroinvertebrates play a key role in processing this detritus, converting large particles into fragments which are then available to filter feeders and collector/gatherer invertebrates. Hernandez and Sukhdeo (2008a; 2008b) examined the impact of parasitism on detritus processing by the isopod *Caecidotea communis*, the dominant detritivore in their study stream in North America. Isopods infected by the acanthocephalan parasite *Acanthocephalus tahlequahensis* showed a reduction in feeding. The authors went on to estimate the impact of this on the ecosystem using measurements of isopod and parasite densities over a two-year period. Both isopod and parasite showed seasonal change in density. Parasites were predicted to have the greatest impact on detritus processing in the autumn, when the bulk of leaf litter is deposited in the stream, reducing the amount of detritus processed by up to 47%. These studies suggest that parasitism can impact the functional role of isopods as detritus processors in a seasonally dependent manner.

Parasites can also affect the predatory strength of their hosts; examples are considered in Section 3.4.2. We have investigated the impact of

Table 7.2. *Food web studies including parasites*

System	No. free-living species (F) and parasites (P)	Findings: how did inclusion of parasites alter web metrics?	References
Ythan (Y) and Loch Leven (L) estuaries, Scotland	Y: 94F, 41P (78T*) L: 22F, 30P (43T*)	Chain length increased; linkage density slightly increased; % omnivory generally increased; depends on how parasites are included. Parasites were included as species (P) or as separate feeding stages (*trophospecies, T).	Huxham *et al.* (1995)
Ythan estuary, Scotland	As above	Linkage density and connectance reduced slightly, intervality reduced.	Huxham *et al.* (1996).
Scotch broom, *Cytisus scoparius*, Silwood Park, England	85F, 69P (3 pathogens and 66 parasitoids)	Predators were more generalist than parasitoids and thus contributed more to connectance than parasitoids. Parasitoids were generally smaller than their host, violating cascade model of food chains.	Memmott *et al.* (2000)
Company Bay intertidal mudflat, New Zealand	67F, 9P	Greatly increased mean and maximum food chain length; linkage density increased slightly; connectance reduced slightly.	Thompson *et al.* (2005)
Muskingham Brook, New Jersey Pinelands, USA	28F, 10P	Increased linkage density; decreased nestedness; 4% increase in connectance (when adjusted to discount illogical links).	Hernandez and Sukhdeo (2008b)
Carpinteria Salt Marsh, California USA	87F, 47P	Increased linkage density, nestedess and connectance. Effects strongest when all logically possible parasite links are counted in the data and included in the model (connectance increased by 93%, nestedness by 439%. Muskingham, Company Bay and Ythan re-examined showing similar but weaker effects.	Lafferty *et al.* (2006a); Lafferty *et al.* (2006b)

System	Size	Findings	Reference
Flea–mammal webs worldwide; freshwater fish webs, Canada	33 flea webs each with P ≥ 10; 7 fish webs with P ≥ 25	Decreasing relationship between species richness and connectance; additional potential links not realised in host species-rich assemblages; host–parasite webs are close to interval with dietary gaps most likely due to phylogenetic constraints.	Mouillot et al. (2008a); Mouillot et al. (2008b)
Takvatn lake pelagic zone, Norway	37F, 13P	Increased food chain length, linkage density, connectance and omnivory due to high proportion (>50%) trophically transmitted parasites. More connected host species supported more trophically transmitted parasite species.	Amundsen et al. (2009)
Carpinteria, Company Bay and Ythan webs	As above	Network analysis showed parasites preferentially exploit highly connected host species at higher trophic positions.	Chen et al. (2008)
Aphid-natural enemy subwebs, Silwood Park, England	40 aphid species; 30 parasitoids; 5 pathogens; 13 predators	Quantitative webs used to assess likelihood of apparent competition in structuring guilds; highest linkage density and connectance in predator subweb; intermediate for pathogens; least for parasitoids.	Van Veen et al. (2008)

Note the relatively high numbers of parasites and parasitoids in comparison to other species. The significance of the findings for web metrics is discussed in the text.

parasitism on predation by a number of aquatic crustaceans and find examples of decreased and increased predatory strength as a result of infection. These changes may influence the status of the host as keystone predators (Box 7.2) or have knock-on effects for competition and the outcome of biological invasions (Boxes 6.2 and 6.5).

Box 7.2 Parasitism, predation and invasions in aquatic communities

In Northern Ireland, *Gammarus pulex* is an invasive species that has displaced the native *G. duebeni celticus* throughout many river systems. Studies of the functional responses of the two species (in the absence of parasitism) have revealed that *G. pulex* is a stronger predator than *G. duebeni celticus* (Dick, 2008). Parasitism also plays a role in this species displacement. The microsporidian *Pleistophora mulleri*, which is specific to the native species, reduces the predatory strength of its host. We offered infected and uninfected *G. duebeni celticus* individuals *Asellus aquaticus*, a common prey item, at increasing densities. The functional response of a predator is the relationship between prey density and prey consumption. For both infected and uninfected individuals, we found a typical Type II functional response; prey intake increased with prey density, reaching a plateau as prey handling time limits prey intake. However, the curve for infected individuals reached a plateau at a lower prey density, indicating a weakened predatory impact by infected hosts. *P. mulleri* infects the abdominal musculature of the host and causes reduced activity in the host (MacNeil *et al.*, 2003a; 2003e). The reduced predatory strength is likely to reflect increased prey handling time by individuals that are weakened by the infection.

In contrast, we found that the acanthocephalan parasite *Echinorhynchus truttae* increased the predatory strength of *G. pulex* (Dick *et al.*, 2010; Fig. 7.9). The functional response of infected individuals rose more steeply and had a higher asymptote compared with uninfected individuals. This increase in the predatory functional response may result both from the direct metabolic demands of the parasite (which comprised 17% of the mass of infected individuals) and from the manipulative effect of the parasite. Infected *G. pulex* show increased activity (increasing the risk of predation by the definitive fish host), which is likely to increase the nutritional demands on the host.

Box 7.2 (*continued*)

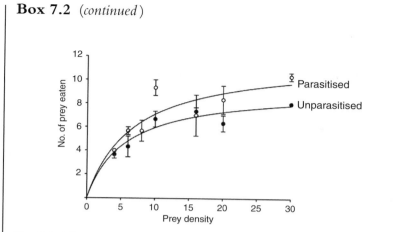

Fig. 7.9. The predatory functional response of healthy versus *Echinorhynchus truttae*-infected *Gammarus pulex*. From Dick *et al.* (2010).

In the native–invader *Gammarus* system described in Box 7.2, both parasites studied facilitate the displacement of the native species by the invader: *P. mulleri* by reducing the predatory strength of the native host; and *E. truttae* by increasing predation by the invader. The effects are likely to be felt throughout the community. *G. pulex* invasion has been found to alter the structure of the invaded communities in the field. The diversity and richness of macroinvertebrates was found to be lower in invaded river reaches than in reaches where the native species occurred and the abundance of several prey species was also reduced (Kelly *et al.*, 2006). *E. truttae* can reach high prevalence in *G. pulex* in Northern Ireland. By increasing the predatory strength of its host, the parasite may increase the impact of *G. pulex* invasion on the freshwater community structure.

7.5.2 Parasitism and food web topology

Food webs are a basic tool in ecology, describing ecological communities as a network of trophic relationships. Despite their long history of use, food web studies have only recently begun to include parasites. This is particularly surprising since food webs focus on the flow of energy through feeding relationships, but parasitism, a key trophic relationship and possibly the most common consumer strategy (Kuris *et al.*, 2008), has been largely ignored. Food webs have proved useful tools in ecology from a descriptive and categorical perspective. A number of different metrics can be calculated, enabling comparison between webs; theoretical work has demonstrated how these measures relate to community stability properties such as robustness to extinction or invasion. Recently, the inclusion of parasites in food web studies has revealed that parasites can have a dramatic influence on food web metrics and web topology in general (Table 7.2).

7.5.2.1 Web metrics

- *Weighting of trophic levels.* The proportion of top predator species, intermediate species and basal resource species gives some indication of web complexity and community function. The relative biomass of species at each trophic level is another key measure: classic community composition theories often assume that species can only consume organisms smaller than themselves; this is clearly contradicted by parasites and many parasitoids.
- *Food chain length.* The maximum number of trophic links between basal resource species and top predator species is the maximum food chain length, giving some indication of the complexity of the food web (Fig. 7.10a). It had been thought that longer food chains indicate

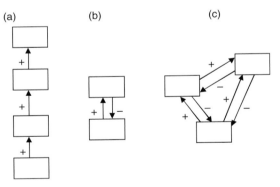

Fig. 7.10. Chain length and loop length in food webs. (a) A simple food chain with four trophic levels has a maximum chain length of 3; only the bottom-up effects (i.e. the positive interaction effects accruing to the consumer) are shown. (b) A simple consumer–resource module has one loop of length 2, tracing the positive and negative interactions from one species back to itself. (c) An intraguild predation module has three loops of length 2 (each of the three consumer–resource modules), and two loops of length 3; starting from any one species, one can travel clockwise or anticlockwise back to the same species – each route involves three links.

a more productive ecosystem as more resources are required to support more trophic levels, given the inefficiency of energy conversion at each stage (Pimm, 1982). Not surprisingly, the inclusion of parasites increases maximum food chain length; what is surprising, however, is the extent to which it can increase chain length; for instance, including parasites in the Company Bay ecosystem doubled maximum chain length from three to six (Thompson *et al.*, 2005). This is because there are parasites for every free-living species in the web, and their consumption by higher trophic levels means that many parasite species can occur in any given chain.

- *Loop length*. A loop describes the path from a given species back to itself without visiting other species more than once, via the bottom-up (positive) and top-down (negative) trophic links. A simple predator–prey association has one loop of length two, whereas intraguild predation (IGP) has two loops of length three and three loops of length two (Fig. 7.10b, c). Loop length provides a measure of food web complexity that can be used to infer stability properties (Section 7.5.3). It follows from analysis of food chain length, and the propensity of parasites to be involved in IGP (Chapter 4), that the inclusion of parasites will likely increase the average loop length of food webs.

- *Connectance* is defined as the proportion of possible links that are realised in the food web. If all species can potentially eat each other, with S species in the web there are S^2 potential links. In this case, connectance is L_o/S^2 where L_o is the number of trophic links observed. Earlier studies suggested that including parasites led to a slight reduction in connectance; however, recent studies suggest that parasites can boost connectance substantially (Box 7.3; Table 7.2). Theoretical studies have also conflicted over whether connectance increases or reduces food web stability (Section 7.5.3).

- *Nestedness* describes an aspect of how links are organised in a network. In a perfectly nested network, each species interacts with a strict subset of other species in order of increasing generality. Hence, species can be arranged along the rows and columns of a matrix in such a way that the interactions will form a triangle (Fig. 7.11a). In such communities, there will be a core community of strongly interacting species around which other less connected species associate. A well-nested consumer–producer system has the property that specialist consumers use generalist producers, and specialist producers are utilised by generalist consumers. Nestedness has implications for community robustness (Section 7.5.3).

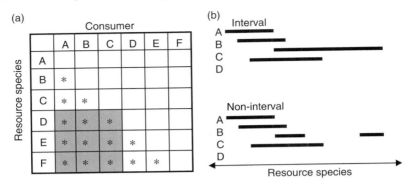

Fig. 7.11. Nestedness and intervality in food webs. (a) A community is perfectly nested if the consumer and resource species can be arranged so that each species interacts (denoted by *) with a successively greater subset of species. This arrangement has a core of strongly interacting species (shaded), around which less connected species are attached. (b) A food web is perfectly interval if the resource species (not labelled, arranged horizontally) can be arranged so that the diets of each consumer (species A–D) are continuous (top panel); if no such arrangement exists, at least one species has a 'dietary gap' and the web is non-interval. A perfectly nested web will be interval, but not all interval graphs are perfectly nested.

- *Intervality* is linked to nestedness, and describes diet breadth in a trophic web. A food web is described as 'perfectly interval' if the species can be arranged in such a way that each consumer utilises a continuous portion of the resource gradient (Fig. 7.11b). Resource use by different consumers can overlap, but if the graph is interval (i.e. has no internal gaps within a consumer's diet), it suggests that some aspects of community assembly can be described using a single dimension of resource quality. Non-intervality (i.e. existence of dietary gaps) may mean that other factors such as phylogenetic constraints are important in structuring communities (Mouillot *et al.*, 2008a).

- *Stability* has many definitions in community ecology. Dynamic stability refers to the theoretical properties of population or community dynamics: a system is considered stable if it returns to equilibrium after a small perturbation. Usage has been extended to describe properties of populations or communities which may not be at equilibrium: 'general stability' refers to systems where the population densities are bounded further away from zero and are therefore less prone to extinction (McCann, 2000). Communities and ecosystems may also be described as stable if they are resilient to invasion or robust to extinction (see below).

- *Resilience and resistance.* A resilient system returns quickly to its former state following a perturbation; the faster the return, the more resilient the system. Resistance implies little change in the properties of a system following a perturbation. Communities are described as resilient if structural properties, species identity and abundance remain relatively unchanged following disturbance. In the context of biological invasions, communities are described as resistant if the invader fails to become established, or if it has little impact once established. 'Equilibrium' and 'general' resilience can be defined depending on the assumption of equilibrium or non-equilibrium dynamics respectively, as for stability (McCann, 2000).
- *Robustness.* In the food web literature, this term usually refers to buffering against secondary extinctions. Robustness is defined as the proportion of species that must be removed in order that 50% of the species in the web go extinct because they lack resource species. For a given community, robustness is maximally $S/2$ (where $S =$ initial number of species), and minimally $1/S$ (Dunne *et al.*, 2002b). Robustness can be estimated by repeated simulation of the knock-on effects of removal of a species chosen by some algorithm (for instance, a randomly chosen species, the species with most or fewest links to others in the web or one from a particular trophic level).

Recently, studies looking at the effect of parasite inclusion on network metrics have shed new light on the role of parasitism in communities. Key among these studies is one by Lafferty *et al.* (2006a), detailing the food web for a salt marsh ecosystem along coastal California: Carpinteria Salt Marsh (Box 7.3).

The Carpinteria study in Box 7.3 is key because it overturns previous work that tended to suggest a fairly minor effect of parasites on food web connectance. Lafferty *et al.* (2006a) argued that previous studies, by ignoring parasite–parasite interactions and incidental consumption by predators, had used an inappropriate calculation for connectance; once corrected, it becomes apparent that the inclusion of parasites can dramatically increase linkage density and connectance. Re-analysis of other food webs showed, in agreement with the Carpinteria study, that parasites enhance web connectance and nestedness (Lafferty *et al.*, 2006a). Other studies have since corroborated this finding, albeit to a lesser extent than found at Carpinteria (Table 7.2). Connectance and nestedness have implications for community stability as discussed in Section

Box 7.3 The food web for Carpinteria Salt Marsh

Carpinteria Salt Marsh comprises 93ha of wetland area east of Santa Barbara, California. This site has also been studied in the context of plant parasites; parasitic dodders are important in determining competitive outcomes between plant species, ultimately influencing their zonation at higher elevations in the salt marsh (Box 5.4). Lafferty *et al.* (2006a; 2006b) focused on interactions in the tidal mud and vegetation of Carpinteria (including the dodder zone), documenting trophic interactions for free-living species and parasites from published literature combined with unpublished data from their own research over 20 years.

The complete food web for Carpinteria included 47 parasite species and 87 free-living species. Parasites dominated food web links, with 78% of all links involving a parasite. This suggests that in most food web studies where parasites are ignored, the number of realised links is greatly underestimated (Fig. 7.12). In the most complete version of the web (taking account of all possible interactions involving parasites), inclusion of parasites more than doubled linkage density and maximum chain length, and nestedness was increased more than four-fold. Other aspects of trophic structure were affected by the addition of parasites: while lower trophic levels were more vulnerable to predators, higher trophic levels were more vulnerable to parasites; taken overall, intermediate trophic levels were most vulnerable to natural enemies.

Fig. 7.12. Three-dimensional visualisation of the Carpinteria Salt Marsh food web, showing how parasites add to web complexity. Dark balls represent free-living species, light balls are parasites; lines connecting balls represent trophic links. Basal resources are at the bottom of the picture, with successively higher trophic levels above. The majority of parasite trophic links are to higher trophic levels. From Lafferty *et al.* (2008a).

Box 7.3 (*continued*)

Analysis of the subwebs making up the complete Carpinteria food web yielded further surprising properties (Fig. 7.13). Both the parasite–parasite and predator–parasite subwebs were rich in trophic links, resulting mainly from IGP (in the parasite–parasite web) and accidental consumption or trophic transmission (in the predator–parasite web). Full inclusion of parasites and their links increased connectance in the entire web by 93%, and high measures of connectance in the parasite–host and predator–parasite subwebs revealed the importance of these subwebs in driving linkage properties for the entire web.

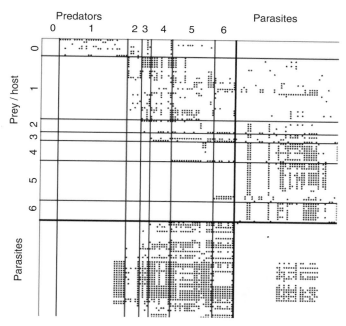

Fig. 7.13. Carpinteria food web, showing the four component subwebs. All species are listed as potential consumers (columns) or prey (rows). A dot in column *i*, row *j* indicates that species *i* consumes species *j*. Numbers indicate trophic levels. The density of dots gives an indication of connectance in the four subwebs: predator–prey (top left, connectance 6.7%), parasite–host (top right, 15%), predator–parasite (bottom left, 25%) and parasite–parasite (bottom right, 7.8%). From Lafferty *et al.* (2006).

7.5.3. In addition, recognition of the importance of other (non-parasite–host) subwebs involving parasites highlights another role for parasites in ecosystems. The predator–parasite subweb is a case in point: recent work shows that parasites are frequently consumed as prey items and can contribute substantially to energy budgets for some organisms (Johnson *et al.*, 2010; Section 7.3.2).

These analyses highlight conceptual and methodological problems with the inclusion of parasites in food webs (Lafferty *et al.*, 2008a; Byers, 2009). For instance, how should we incorporate parasites with complex life cycles into food webs? Each life stage may specialise on particular hosts, and including such parasites as a single species node will inflate their generality. An alternative is to include each life stage as a separate 'trophospecies' (see Huxham *et al.*, 1995 for an example), but this inflates species richness, and links between trophospecies are not logically equivalent to trophic links between real species. As with all food webs, there are issues over completeness, with some parasite taxa likely to be consistently neglected. Also, webs focusing on particular host–parasite assemblages may neglect important interactions involving other species. As seen above, webs that include parasites may underestimate non-parasitic interactions involving parasites, resulting in artificially low estimates of linkage and connectance. Finally, and ideally, the relative strength of interaction should be considered in the calculation of web metrics. Interaction strength will vary greatly between different associations, depending on relative abundance, contact rates, prey/host preference and virulence. Most parasites do not kill their host and so will not have the same instantaneous effect as a predator; however, they may have substantial long-term consequences for host biomass and reproduction (see, for example, Section 7.6). Incorporation of interaction strength is an ongoing area of research in food web ecology and very few webs including parasites have taken this approach (see Van Veen *et al.*, 2008 for a recent exception; Table 7.2; Section 7.9).

7.5.3 Implications for community stability

The role of food web complexity (the number of species involved and how they are linked) for stability has undergone two paradigm shifts since the seminal works of 1950s ecologists. Odum (1953), MacArthur (1955) and Elton (1958) all regarded more diverse communities as more stable, reasoning that they would be less prone to strong fluctuations in population densities of the component species. This reasoning was based, in part, on observations that simple experimental or agricultural communities were more prone to catastrophic extinctions or invasions. In more complex communities, it was argued, variety in natural enemies buffered prey populations from dramatic variation (if one species was strongly perturbed, the actions of another species would compensate for it). This view was contradicted by the theoretical work of May (1973), who demonstrated that in randomly constructed community networks, greater diversity tended to be destabilising. In May's models, adding more species with more links to the community matrix decreased the chances that the matrix would have dynamics tending towards a stable equilibrium (see also Box 7.5). This pioneering work led to a general perspective held throughout the 1970s and 1980s that complex communities and more complex interactions within communities (such as omnivory) were inherently more unstable than simple communities (Pimm & Lawton, 1978; Section 4.1.2). However, some of the simplifying assumptions used by May concerning the distribution and magnitude of interactions between species now look to have been uncharacteristic of real populations and responsible for the mathematical instability. Real communities appear to be characterised by many interactions between species; a few of the interactions are strong and many are relatively weak (Dunne et al., 2002a; Neutel et al., 2002). This skewed distribution towards weak interactions appears to help stabilise real-world food webs. Several influential theoretical studies have now demonstrated a positive relationship between connectance and food web stability. General stability and community persistence is discussed in McCann et al. (1998). Robustness to secondary extinctions is considered by Dunne et al. (2002b). These works return us to the original view that complexity generally enhances stability. Within the current paradigm, higher connectance in real food webs enhances community stability because the many weak interactions enjoyed by a focal species serve to dampen the strong fluctuations (from strong negative covariance (McCann, 2000)) caused by the few strongly interacting species (Fig. 7.14). More complex trophic structures (such as those caused by intraguild predation (IGP) – Fig. 7.10) generate longer loops within food webs; this again dampens the feed back between strongly interacting species provided a strong interaction is counteracted by other weak interactions (Neutel et al., 2002).

Although the effects of parasitism were not considered explicitly in the theoretical literature discussed above, extrapolating from the current paradigm suggests that parasitism could potentially provide an important service to the ecosystem in enhancing community stability (see also

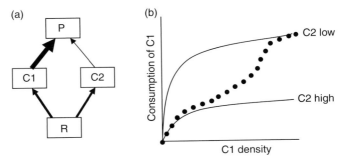

Fig. 7.14. Feed-back dampening by weak interactions. (a) Consumers C1 and C2 compete for a shared resource R, so their densities covary negatively. Parasite (or predator) P consumes C1 and C2 with a strong preference for C1 (depicted by the bold arrow). (b) Consumption of C1 depends on both C1 and C2 density (when C2 is common, C1 consumption is weakly density dependent, depicted by the shallow curve; when C2 is scarce, C1 consumption is strongly density dependent). Because C1 and C2 negatively covary, the actual consumption rate (dots) will be closer to the C2 high curve at low C1 densities and the C2 low curve at high C1 densities. This effectively weakens the impact of the parasite/predator at low host/prey densities and dampens population oscillations. Adapted from McCann (2000).

Section 7.8). In particular, weak interactions in long loops help to stabilise communities, and parasites can potentially contribute both to loop length and the relative frequency of weak interactions:

- *Loop length.* Parasites are often consumed by predators of the host (either incidentally or as part of a trophic transmission strategy); as such they entail omnivory or IGP. Different parasitoids competing for the same host also often engage in IGP (Van Veen *et al.*, 2008). Each such IGP association increases the average loop length of the food web, enhancing stability provided the interactions are generally weak (Neutel *et al.*, 2002).
- *Interaction strength.* Interactions between hosts and parasites may frequently be relatively weak (compared, for instance, to predatory interactions), as effects on individual hosts are usually sublethal; the addition of many weak interactions offsets the destabilising effects of fewer strong interactions.
- *Connectance.* Recent analyses suggest that adding parasites to food webs increases connectance; provided links are skewed towards weak interactions, such increased connectance stabilises food webs by increasing robustness to secondary extinctions (Dunne *et al.*, 2002b).

- *Nestedness.* Food webs involving parasites and mutualists appear to be well nested (Bascompte *et al.*, 2003; Lafferty *et al.*, 2006a). Strong nestedness can enhance robustness to secondary extinction because generalist species consume many specialists (so are not exposed to catastrophic resource loss if a single resource species is lost); if a generalist resource species is lost, its specialist consumers will be lost, but these will be a subset of the consumer community.

These arguments generally suggest that parasites are good for community stability, promoting dynamic stability by counteracting strong feed-back relationships and enhancing robustness to secondary extinction. However, we must caution that this is still an ongoing area for further research, and stability relationships will depend on the precise make-up of parasitic relationships within communities (see also Section 7.8.2).

7.6 Bioenergetic implications of parasitism

One reason why parasites may have been ignored up to now in food web studies is the widespread belief that because most parasites are small, they will account for little energy transfer within most ecosystems. However, recent studies of estuarine systems now challenge this view, demonstrating that parasites can account for a substantial proportion of both the biomass and productivity in these ecosystems.

7.6.1 Parasite biomass

Kuris *et al.* (2008) compared biomass over a five-year period for parasitic and free-living taxa in three Californian estuaries, including the Carpinteria Salt Marsh (Box 7.3). Overall, they recorded 199 species of free-living animals, 15 free-living vascular plants and 138 parasitic species (including the parasitic dodder plant *Cuscuta*, discussed in Box 5.4). In these estuaries, parasite biomass comprised 0.2–1.2% of all animal biomass. This may seem rather an inconsequential figure, but it exceeds the combined biomass of the vertebrate top predators in these systems; large, visible top predators are commonly regarded as key to ecosystem energetics and we would not expect them to be ignored in community studies! Parasitic helminths, particularly trematodes, were abundant at all three sites, having a biomass comparable to the combined biomass of birds, fish and burrowing shrimps and

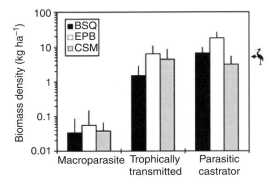

Fig. 7.15. Biomass density (kg ha^{-1}) of parasite functional groups in three Californian estuarine ecosystems. The reference arrow at the bird icon marks the mean winter bird mass density (which is much higher than the summer density). The three estuaries are part of the Baja California system (BSQ: Bahia San Quentin; EPB: Estero de Punta Banda; CSM: Carpinteria Salt Marsh). From Kuris *et al.* (2008).

worms (Fig. 7.15). In terms of biomass, the California horn snail *Cerithidea californica* was the dominant invertebrate at two of the three sites, and is host to 18 species of trematode. Trematodes averaged 22% of the soft-tissue weight of infected snails. Furthermore, many of these trematodes castrate their host, thereby subverting host metabolism away from reproduction and towards the production of parasitic free-swimming stages (cercariae). Castrated snails can be seen as an extended phenotype of the parasite (Dawkins, 1982). Together, castrating parasites and their extended snail phenotype accounted for up to ten times the high winter bird mass across the three estuaries (Fig. 7.15).

7.6.2 Parasite productivity

In addition to effects on the standing crop biomass in food webs, some parasites contribute directly and substantially to system productivity. The parasitic mode may offer substantial opportunities for enhanced reproductive potential, partly because parasites are freed from many of the defence constraints placed on free-living species. The trematode castrators of the snail *C. californica* are a case in point. These parasites release large quantities of free-swimming stages (cercariae), some of which go on to infect the next intermediate (fish) host; those that do not succeed

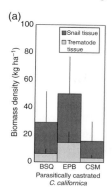
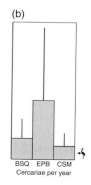

Fig. 7.16. Standing crop biomass and cercarial productivity of parasitic castrators in Californian estuaries. (a) Biomass density of castrated host snail (*Cerithidea californica*) and trematode parasite (together, the extended phenotype of the parasite); (b) Biomass density of the free-living cercarial stages (i.e. parasite productivity). The three sites are as detailed in the previous figure. From Kuris *et al.* (2008).

in infecting their target host can be an important dietary constituent for many planktonivorous species. Kuris *et al.* (2008) found that the annual cercarial production from infected *C. californica* was about three times the standing stock biomass of trematodes within snails and 3–10 times the winter biomass of birds (Fig. 7.16).

High parasite productivity does not appear to be confined to the Carpinteria system. Cercarial biomass density is similarly high for parasites from a wide range of snail hosts in different geographic localities (Thieltges *et al.*, 2008a). These findings have interesting implications for the role of parasitism in energy flow within ecosystems:

- Parasites may comprise a significant biomass in the system.
- Parasite productivity may be high for some life-forms, contributing directly to energy flow in the system.
- For trophically transmitted parasites, behavioural manipulation may enhance consumption rates of infected hosts (Section 3.5.1), thereby influencing conversion of biomass from one form (the infected host) to another (predators or scavengers).

Many parasitic plants play similar roles, speeding up nutrient cycling and litter production via their own productivity or via effects on mortality and morbidity of long-lived host plant species (see mistletoe and rattle examples, Section 5.4). Furthermore, these findings underline the importance of trophic links involving parasites in food webs (Section 7.5); not only do parasites add many extra links, some of these links can be strong.

7.7 Ecosystem engineering

Many organisms play a role in creating, modifying and maintaining habitats. Jones *et al.* (1994) coined the term 'ecosystem engineer' to describe an organism that modulates, either directly or indirectly, the resources available in an ecosystem by causing changes in the physical state of either non-living or living material. Although these changes do not involve direct trophic interactions between species, they play an important role in determining community structure. Allogenic engineers transform materials (living or non-living) from one physical state to another. Terrestrial examples include nest construction by termites and ants. In aquatic habitats, burrowing invertebrates redistribute sediments and affect bioturbation. For example, the burrowing activity of the amphipod *Corophium volutator* redistributes and stabilises the sediment of intertidal mud flats (Table 7.1; Mouritsen *et al.*, 1998). Autogenic engineers change the environment with their own physical structures. For example, in terrestrial habitats, leaf litter from plants affects soil drainage and surface structure; in aquatic habitats many molluscs and crustaceans provide habitat for the attachment of epibionts (Box 7.4).

Thomas *et al.* (1999) pointed out that parasites can modify the traits exhibited by species that act as ecosystem engineers or may act themselves as ecosystem engineers (Fig. 7.17). For example, aquatic molluscs provide substrate for the various epibionts that attach to mollusc shells. Trematode infections can lead to gigantism in molluscs, with larger shells offering a greater area of substrate for colonisation. Parasites may act as ecosystem engineers themselves if the phenotypic changes induced in the host modify the habitat of other species that inhabit the host. For example, crabs do not normally provide a substrate for epibionts as these organisms are lost during moult. However, the parasitic barnacle *Sacculina carcini* prevents its crab host from moulting, thereby creating a new habitat for colonisation by epibionts. As a result, the carapaces of sacculinid-infected crabs harbour a higher abundance of barnacles and serpulid worms than do the carapaces of uninfected individuals (Mouritsen & Jensen, 2006). Parasite-induced changes in host defences can also be thought of as ecosystem engineering as infected hosts may become more susceptible to infection by other parasites. For example, humans carrying

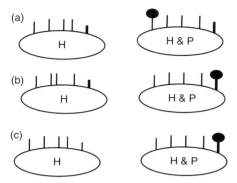

Fig. 7.17. Traits affected by parasites and traits involved in ecosystem engineering. (a) Host traits that the parasite affects are not those traits that are involved in ecosystem engineering by the host; (b) the parasitism alters the ecosystem engineering traits of the host; (c) parasitism leads to the emergence of new engineering functions in the host (H: host; P: parasite). Adapted from Thomas *et al.* (1999).

the human immunodeficiency virus (HIV) are more susceptible to a range of opportunistic infections, whilst the sacculinid described above also reduces the antifouling defence of host burrowing (host susceptibility and coinfection are also considered in Section 2.7).

The role of parasites in ecosystem engineering has been extensively examined by Poulin and colleagues in a series of studies of intertidal communities in New Zealand (Box 7.4). In these soft-sediment intertidal communities, burrowing cockles play an important role in sediment bioturbation and by acting as substrate for epibionts. Trematode parasites manipulate the burrowing behaviour of cockles. Communities affected by the parasite show changes in the distribution of predators and epibionts of cockles. Furthermore, this manipulation has ramifications throughout the community; there is strong evidence that parasite manipulation of cockle behaviour drives changes in diversity and productivity throughout the intertidal community.

Mouritsen and Haun (2008) also reported modification of an ecosystem engineer as a result of parasitism. The snail *Hydrobia ulvae* dominates energy flow in the intertidal zone in Denmark, where it can make up 80% of the macrofaunal biomass. The snails are herbivores but they also play an important role as ecosystem engineers, causing bioturbation which may release nutrients from the substrate as well as exposing the dominant epipsammic algae to light. *H. ulvae* is the intermediate host for several species of trematode, and infection leads to reduced snail activity.

Box 7.4 Ecosystem engineering by parasites in the intertidal zone

The cockle *Austrovenus stutchburyi* is a dominant species in the intertidal zone of soft-sediment shores in New Zealand. These bivalves burrow just below the surface of the mud and play a key role in providing a substrate for attachment of a range of epibionts, including anemones, limpets, barnacles and polychaetes (Poulin, 1999). The trematode *Curtuteria australis* (subsequently described as a coalition of related species (Leung *et al.*, 2009)), uses this cockle as its intermediate host. The parasite has an indirect life cycle and is trophically transmitted to the definitive host, the oyster catcher (*Haematopus* sp.; Fig. 7.18). Eggs passed out in the faeces hatch into miracidia that go on to infect the first intermediate host, the whelk *Cominella glandiformis*. Cercariae are released from the whelk and penetrate cockles via their inhalant siphon. Development in the cockle leads to the formation of metacercariae in the host foot. The life cycle is completed when an infected cockle is preyed upon by the definitive host. The parasite manipulates the behaviour of its cockle host, reducing its burrowing activity. As a result, infected cockles suffer an increase in predation by the definitive host. This manipulation has ramifications throughout the community, modulating the availability of resources in the community, leading to changes in species diversity and distribution, as well as in overall productivity.

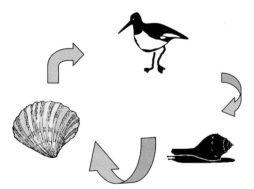

Fig. 7.18. Life cycle of the trematode *Curtuteria australis.*

Epibiont distribution. Cockles provide sites of attachment for the anemone *Anthopleura aureoradiata* and the limpet *Notoacmea helmsi.* The two species compete for habitat and anemones may prey upon

Box 7.4 (*continued*)

limpets, leading to a negative correlation between the abundance of anemones and limpets across different sites (Poulin, 1999). In sites affected by the trematode, manipulation of the cockle behaviour has led to two subpopulations of cockles (exposed, infected individuals and uninfected cockles with normal burrowing behaviour). Studies of colonisation revealed anemones were more abundant on buried cockles, reflecting their vulnerability to desiccation when exposed. In contrast, limpets preferred surfaced over buried cockles. By altering the cockle behaviour, the parasite creates a new habitat/substrate of exposed cockles, decreasing limpet/anemone competition and allowing limpets to coexist (Poulin, 1999).

Predator distribution. The zonation of two predators was also modified in habitats affected by the trematode. The whelk *Cominella glandiformis* showed a change in distribution towards the upper intertidal in response to the availability of exposed (infected) cockles (therefore engaging in both predation on cockles and intraguild predation on the trematode). The distribution of the fish *Notolabrus celidotus* also shifted towards the upper intertidal where the fish exhibited non-lethal predation by cropping the foot of exposed (parasitised) cockles.

Community productivity. Field manipulations (Mouritsen & Poulin, 2005; Mouritsen & Polml, 2006) demonstrated that parasite-induced modification of cockle behaviour affected the whole community, leading to changes in the abundance of primary producers and the biomass of secondary producers. In the first experiment, parasite load was manipulated, leading to an increased abundance of surfaced cockles. The abundance of macroinvertebrates increased as did the zoobenthic biomass, reflecting a reduction in bioturbation as there was less burrowing activity. The second experiment investigated the effect of surfaced cockles on hydrodynamics and sediment conditions. Here, mimic cockles were added to the mud surface. This experiment had an even greater impact, leading to a decrease (8–22%) in benthic primary production and an increase (up to five-fold) in secondary production of macroinvertebrates from diverse taxa. These experiments provide strong evidence that, by modifying the behaviour of a single host species, parasitism can indirectly affect overall ecosystem productivity.

For more discussion of this system, we direct the reader to a forthcoming book in this series, *Parasitism and the Structure of Communities: Intertidal Habitats as Model Systems*, by Mouritsen and Poulin (2011, forthcoming).

Using a caging experiment, Mouritsen and Haun demonstrated that increases in snail density led to an increase in primary production, resulting from their role as ecosystem engineers. However, parasitism reduced the role of the snail as an ecosystem engineer and led to a decrease in primary production. The density and richness of various zoobenthic species was also altered (in some cases positively, in others negatively), with an overall increase in the diversity of zoobenthic secondary producers.

7.8 Ecosystem health

The importance of food web and bioenergetic studies for our understanding of ecosystem processes has yet to be evaluated, and incorporating parasites presents real research problems for the mathematical analysis of food webs (Lafferty et al., 2008a). In particular: it is unclear how we should represent parasite species with complex life cycles that specialise on different host species at each life stage; relative interaction strength has largely been ignored; and few, if any, studies comprehensively include a fair representation of all the parasite taxa associated with an ecosystem. Nevertheless, the recent foray into community ecology perspectives for parasitology – which we term *ecosystem parasitology* – does highlight some important questions. Here, we consider recent advances into how to represent such complicated systems mathematically (Section 7.8.1), and discuss what the evidence so far suggests for the role of parasites in maintaining community structure and ecosystem function (Section 7.8.2).

7.8.1 Integrating population and community approaches to the study of ecosystems

Given the complexity of interactions revealed in these studies, how are we to proceed with population dynamic modelling of the systems with a view to predicting community stability, efficacy of management strategies or the effects on anthropogenic change on communities? One promising approach developed by Dobson (2004) is the use of Who Acquires Infection from Whom (WAIFW) matrices, originally developed for diseases of one host (or host and vector) with heterogeneous infection between host age or stage categories (Box 7.5).

Matrices are a succinct representation for community–parasitology systems, and have proved useful in elucidating some key concepts such as

Box 7.5 Who acquires infection from whom (WAIFW) matrix

In a multi-host species context, the WAIFW matrix (Anderson & May, 1985) allows succinct representation of all the different components of transmission from individuals of one host species i to host species j (Eqn 7.1).

$$W = \begin{pmatrix} \tau_{i,i} & \tau_{j,i} & \tau_{k,i} \\ \tau_{i,j} & \tau_{j,j} & \tau_{k,j} \\ \tau_{i,k} & \tau_{j,k} & \tau_{k,k} \end{pmatrix}, \tag{7.1}$$

$$G = \begin{pmatrix} \dfrac{\beta_{i,i}p_{i,i}}{(\alpha_i+\delta_i+d_i)} & \dfrac{\beta_{j,i}p_{j,i}}{(\alpha_i+\delta_i+d_i)} & \dfrac{\beta_{k,i}p_{k,i}}{(\alpha_i+\delta_i+d_i)} \\ \dfrac{\beta_{i,j}p_{i,j}}{(\alpha_j+\delta_j+d_j)} & \dfrac{\beta_{j,j}p_{j,j}}{(\alpha_j+\delta_j+d_j)} & \dfrac{\beta_{k,j}p_{k,j}}{(\alpha_j+\delta_j+d_j)} \\ \dfrac{\beta_{i,k}p_{i,k}}{(\alpha_k+\delta_k+d_k)} & \dfrac{\beta_{j,k}p_{j,k}}{(\alpha_k+\delta_k+d_k)} & \dfrac{\beta_{k,k}p_{k,k}}{(\alpha_k+\delta_k+d_k)} \end{pmatrix}. \tag{7.2}$$

Eqn (7.1) shows the WAIFW matrix for three species, W. Each element τ_{ij} combines all the processes contributing to transmission from species i to species j. Eqn (7.2) shows the G matrix version of W; β_{ij} = transmission efficiency from species i to species j; p_{ij} = a function describing whether that species pair has density- or frequency-dependent transmission; α_i = pathogen-induced mortality rate; d_i = natural mortality rate; δ_I = recovery rate for species i. Each element now corresponds to the transmission rate from i to j, multiplied by the duration for which individuals of species i remain infectious.

Calculating R_0. A modified form of this matrix (now called the 'next-generation matrix', G) can be used directly to calculate R_0 (the basic reproductive number) for the shared parasite (Diekmann *et al.*, 1990; Dobson, 2004). To obtain G, we expand each term in W so it corresponds to the rate of transmission times the average lifetime of an infectious individual (Eqn 7.2). R_0 is now equal to the dominant eigenvalue for G, a measure found by standard algebraic or numerical methods. $R_0 > 1$ tells us that the parasite will increase initially in a community described by G; so it provides a useful guideline when considering the initial spread of emerging infectious diseases or pathogen spillover into new populations (Dobson & Foufopoulos, 2001; Chapter 8).

Transient dynamics and coexistence. Dobson (2004) used further analyses to show that the short-term (transient) dynamics of

Box 7.5 (*continued*)

parasitism in multi-host systems depend on the relative magnitude of within- and between-species transmission (paralleling the results from two-host systems – Section 2.4.2). If interspecific transmission is rare, each species exhibits independent epidemic outbreaks of disease; as interspecific transmission is increased, outbreaks become buffered and the system settles into a stable pattern of constant abundance (parasite-mediated coexistence). However, if interspecific transmission is increased still further, the species best able to recover from the impact of the pathogen will exclude the other host species. Hence, the reservoir species acts as a source for repeated infection into the more vulnerable sink population.

Force of infection. Another important result can be obtained directly from the next-generation matrix. If we sum the elements of row *j*, we obtain the force of infection (Box 1.2) *experienced* by species *j*; summing the *j*th column elements gives the force of infection *exerted* by species *j*. By calculating this latter expression, we can identify the species that makes the largest contribution to epidemic outbreaks. If one species produces a substantially greater force of infection than the others, control strategies focused against this species should yield the best results (Dobson, 2004).

Stability of food webs. The matrix approach has been used by Dobson *et al.* (2008) to demonstrate that generalist parasites in communities of competitive species help to stabilise the food web (see also Section 7.5.3). In May's (1973) version, all the species in the matrix were potential competitors, so pairwise interactions were always 0 or (−). Matrix stability is determined by the dominant eigenvalue, found by summing the cross-products of elements in the matrix; a negative dominant eigenvalue is required for dynamical stability. Since − − multiplication yields a positive result, adding more species results in more positive cross-products, reducing the chance that the system will be stable. When we introduce parasites, we introduce +− host–parasite interactions, especially if the parasite is a generalist infecting several species. The inclusion of these negative terms reduces the value of the dominant eigenvalue, increasing the likelihood of system stability. As species richness *n* is increased for the matrix, potentially destabilising competitive interactions increase at a rate $(n^2 - n)/2$, but stabilising shared parasite interactions increase at the faster rate of n^2 (Dobson *et al.*, 2008).

R_0 and the force of infection in the context of multiple hosts (Box 7.5). However, they are not immune to the drawback faced by other approaches: once the system has more than a few host species, mathematical analysis becomes intractable. For systems of more than four simultaneous equations (an irreducible 5×5 matrix or larger), numerical or simulation-based approaches are required. This reduces the general applicability of results to alternative systems, and the reliability of model predictions stands or falls on the confidence with which key parameters can be estimated. One way to reduce the complexity of multi-host systems is to use allometric body size scaling rules so that species with similar life history traits and dynamics can be represented together (Cohen *et al.*, 2003). Work in this area has begun to identify how disease establishment and patterns of disease dynamics scale with host body size (Bolzoni *et al.*, 2008). Some simplifications do exist for the analysis of sparse matrices (ones in which many elements are zero); some emerging infectious diseases have largely disconnected species with strongly asymmetric transmission terms and may lend themselves to these techniques (Dobson & Foufopoulos, 2001; Dobson, 2004).

7.8.2 Are parasites indicators of healthy ecosystems?

In an important state-of-the-field review, Hudson *et al.* (2006) proposed (contrary to traditional intuition) that a healthy ecosystem is one that is rich in parasites. A healthy ecosystem was defined in terms of ecosystem functioning as one that was productive, diverse and resilient to change. Hudson *et al.* examined several lines of evidence supporting their claim: the keystone effects of parasites can enhance coexistence (Chapter 2; Box 7.5); trophic interactions involving parasites tend to stabilise food web dynamics through increased connectance and loop length (Section 7.5.3; Box 7.5); and enemy release, the process whereby invasive species freed of their native parasites are more successful in a novel habitat (Section 6.3). Mechanistically, the processes underlying Hudson *et al.*'s 'healthy ecosystems' hypothesis may be the result of affirmative answers to at least one of the following questions: does parasite diversity beget host diversity? Or, does host diversity beget parasite diversity?

7.8.2.1 Does parasite diversity beget host diversity?

The majority of the material mustered in this chapter (and by Hudson *et al.*, 2006) indicates a potential for parasites to influence diversity. What is less clear is whether parasitism will drive diversity upwards or downwards. Certainly, a number of keystone effects of parasitism can help maintain diversity. For instance, parasite-mediated competition can enable coexistence of an otherwise 'inferior' competitor (Chapters 2 and 5), and similar processes can apply in the case of intraguild predation (IGP; Chapter 4). However, for each example there is a counter-example: parasite-mediated competition or IGP can also result in exclusion of the more vulnerable species. Parasites can also interact with predation to either enhance or decrease host/prey population viability (Chapter 3). In Chapter 6, we noted that species may become invasive (in part, at least) because of parasites they leave behind (the many instances of enemy release), or because of the parasites they bring with them (for instance, the grey squirrel and SQPV). Endophytes of grasses provide an extreme example of the latter process (Section 5.5.1); by enhancing the competitive ability of infected plants, they obscure the normally positive relationship between diversity and resilience to invasion documented for plant communities (Rudgers et al., 2005).

Network analysis provides a novel perspective on the problem: the addition of parasites to food webs generally appears to be stabilising, thus suggesting that parasites are important contributors to, at least, a dynamically 'healthy' ecosystem (Section 7.5.3). That said, counter-examples do exist: the inclusion of parasites in one salt marsh food web reduced robustness to secondary extinction (Lafferty & Kuris, 2009). This occcured because the parasites, 17 trematode species, were all specialist on the California horn snail (*Cerithidea californica*), the dominant intertidal grazer in the system. This species is vulnerable to replacement by the exotic Japansese mud snail (*Battilaria attramenteria*), which harbours only one trematode species. Under computer-simulated invasion scenarios, extinction of the horn snail led to secondary extinction of all the specialist trematodes, resulting in dramatically reduced connectance with the removal of many trophic links.

7.8.2.2 Does host diversity beget parasite diversity?

This would appear to have a clear answer in the affirmative; the more host species present in a community, the more parasite species should, logically, be present. There are several provisos, however; greater host diversity will only result in greater parasite diversity if:

- there are more specialist parasite species than generalists (generalists may do better on one abundant host species if within-species transmission exceeds between-species transmission);
- links between parasite and host are sufficiently strong that other intervening factors (biotic and abiotic) do not obscure the pattern;

- if specialist parasites are scarce, generalists should be more efficiently transmitted between species than within species (usually regarded as unlikely, but trophically transmitted parasites tend to fall into this category);
- the populations of potential host species are each sufficiently large to support their specialist parasites (they exceed the threshold for establishment – Box 1.2);
- the spatial distribution or interactions between potential host species do not confer too strong a 'dilution effect' (Section 7.3.2) on parasite transmission and establishment; for instance, increasing host diversity will decrease parasite frequency if alternative host species are of low competence.

Hechinger and Lafferty (2005), in a direct test of the 'host diversity begets parasite diversity' idea, found that trematode parasite diversity in snails (the 'downstream' intermediate host) increased with the diversity of birds (the final 'upstream' host) at sites in the Carpinteria Salt Marsh system. The positive relationship between host and parasite diversity can have some useful applications in conservation monitoring. In some systems, it may be easier to monitor and quantify parasites than hosts; here, parasites might be used as indicators of ecological condition.

Huspeni and Lafferty (2004) used trematode parasites of the snail *C. californica* to assess the success of a salt marsh restoration project. Six years after the restoration, trematode abundance and species richness had increased significantly, suggesting that the project had been successful at attracting increased numbers and variety of bird species to the area. Parasite species richness also correlated with ecosystem health in a comparative study of pacific marine ecosystems (Lafferty *et al.*, 2008b).

At the relatively pristine Palmyra Atoll, reef fish supported significantly higher richness of parasitic nematodes, helminths and arthropods than at heavily fished Kiritimati Island (Fig. 7.19); this pattern was repeated across each broad group of parasites separately, with the exception of monogenean helminths. For all five fish species sampled, Palmyra fish also had higher parasite prevalence and burden than those at Kiritimati. Research into parasites as bioindicators is still in its infancy, however; patterns are not repeated across all ecosystems and host or parasite groups (Hechinger *et al.*, 2007), and the criteria for choosing particular parasites as bioindicators requires further examination.

Interestingly, a recent large-scale biodiversity monitoring programme in Ireland (Ag-Biota) provided evidence for the applicability

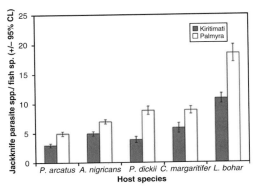

Fig. 7.19. Parasite species richness estimates for five fish species at Palmyra Atoll (unfished) and Kiritimati Island (fished). Paired *t*-tests indicated a significant difference in parasite species richness between sites. From Lafferty *et al.* (2008b).

of parasites as bioindicators in terrestrial ecosystems. In samples from 50 commercial farms, both the abundance and genera richness of parasitoid wasps was strongly correlated with that of all other arthropods, leading to the suggestion that parasitoids would make good 'front-line' bioindicators of the effects of farming practices on lower trophic levels (Anderson *et al.*, 2011). Parasitoids fit a number of the criteria listed above for effective indicators of diversity: most are specialists; their population dynamics are tightly coupled to those of their hosts; and they are mobile so dilution effects resulting from failure to locate rare hosts are less likely to apply. They are also a diverse taxon (over 50 genera were identified from samples at some farms in the study, with other arthropods accounting for up to 90 genera); this is an important property for bioindicators as it results in a high-resolution index.

7.8.2.3 A tentative model for parasite and host biodiversity

The question of how parasite diversity relates to community diversity is yielding some interesting results, but no general patterns have emerged as yet. It is interesting to speculate that patterns might be linked to parasite transmission strategies, taking into account the mathematics of parasite establishment in multi-host communities. We previously drew attention to possible consequence of such processes for host biodiversity (Fig. 7.8). These ideas can be extended to consider the effects on parasite diversity (Fig. 7.20), and perhaps indicate where we should look for general patterns that underpin the relationship in different ecosystems. Extrapolating the effects on parasite populations, we would predict that in ecosystems where density-dependent parasite transmission is the norm, parasite diversity will likely be positively related to host diversity (Fig. 7.20a). This occurs because any reduction in host diversity reduces R_0 for such parasites and increases the minimum population size for parasite establishment (N_T). All other things being equal, parasites close to their establishment threshold may be taken below it to deterministic extinction by reduced host diversity. The relationship here is only causal from the direction of hosts to parasites: parasite diversity itself does not feed back onto host diversity. In contrast, in communities where frequency-dependent transmission dominates, positive feed back between host and parasite diversity may lead to an extinction cascade (Fig. 7.20b). Here, reduced host diversity enhances infection prevalence of existing parasites in

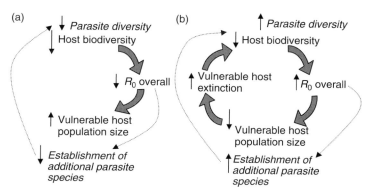

Fig. 7.20. Hypothetical relationships between host and parasite diversity in multi-host communities. (a) If parasite transmission is density dependent, a reduction in host biodiversity results in a reduced total R_0 for the parasite in the community, reducing parasite prevalence possibly to the point of extinction. Because of the positive relationship between R_0 and host diversity, parasite species with density-dependent transmission are less able to establish in less-diverse communities, hence parasite diversity will covary positively with host diversity. (b) If transmission is frequency dependent, a reduction in host biodiversity results in increased R_0, increased parasite prevalence and population declines for more vulnerable host species, potentially to the point of host extinction. Because of the negative relationship between host diversity and R_0, parasites with frequency-dependent transmission are more able to establish in less-diverse communities, and parasite diversity will covary negatively (and feed back upon) host diversity.

vulnerable hosts at the same time as increasing the opportunities for establishment of additional parasite species. In frequency-dependent transmission systems, we would thus expect parasite diversity to be inversely linked to host diversity, with each potentially influencing the other. We should caveat here that this simple extrapolation makes some potentially unrealistic assumptions (most ecosystems probably have a mix of parasite transmission patterns, for instance). It remains to be seen whether real systems reflecting these idealised patterns exist, and whether detailed mathematical analyses bear out the processes outlined (Hatcher *et al.*, forthcoming).

7.9 Evolutionary considerations

It must be remembered that food webs are not static entities; the strength (and type) of interactions, and even the identity of the players are subject to change, in part the result of adaptation and co-evolution between interacting species. There is now good evidence that evolutionary dynamics can occur on a timescale similar to that of population dynamics (rapid evolution; see Section 3.3.3), particularly for 'r-selected' organisms with short generation times and high reproductive rates (Yoshida *et al.*, 2007; Duffy & Hall, 2008). It therefore seems reasonable to suppose that food webs will also be in a state of flux reflecting current population dynamics and recent as well as long-term evolutionary processes.

There is evidence that the distribution of parasites in food webs is influenced by parasite transmission strategy, and this distribution feeds back on food web topology. In lake Takvatn (northern Norway), more than half the parasite species in the pelagic zone were trophically transmitted (Amundsen *et al.*, 2009). There was a positive correlation between the diversity of trophically transmitted parasites in a host species and the number of predator–prey links around the host, suggesting that these parasites tended to utilise the most well-connected species in the web. In contrast, the diversity of non-trophically transmitted parasites was independent of host connectivity. This pattern is in accord with evolutionary explanations as the use of highly connected hosts should enhance transmission for trophically transmitted parasites (Parker *et al.*, 2003). In addition, well-connected hosts may be more exposed to trophically transmitted parasites selecting for adaptation by the parasite to suites of strongly connected species.

In the Takvatn study, the trophically transmitted parasites were more likely to be generalists, having several alternative hosts at each life stage, proceeding along the food chain from one host to its consumers at successive life stages (Amundsen *et al.*, 2009). Such parasites were therefore important contributors to the frequency of omnivory in this food web. As discussed in Section 7.5.3, IGP or omnivory is likely to contribute to stability in food webs via the introduction of more, weaker interactions between species.

This pattern appears to generalise to several other salt marsh and estuarine systems. Chen *et al.* (2008) compared the distribution of parasites in several of these food webs (Table 7.2) to the distributions in randomised versions of the webs, finding that the number of parasite species harboured by a host species was related to the host's network position. Species with many predators or involved in many food chains were important hosts for trophically transmitted parasites. In addition, species with a high diversity of parasites were efficient predators: they tended to be highly connected, in trophic positions close to many prey species or better able to attain resources from lower trophic levels. There is also evidence that parasites can drive the evolution of the host's life history traits, altering interaction strengths over the long term for key species in food webs. Perch (*Perca fluviatilis*) are under opposing selective pressures on growth rate driven by parasitism and predation; this alters the relative strength of predator–prey versus competitive interaction with pike (*Esox lucius*) (Edeline *et al.*, 2008; Section 3.3.3).

Parasites may also mediate indirect interactions such as apparent competition (Chapter 2), which can be important in structuring communities. In some phytophagous insect communities, few species compete directly for resources (most specialise on particular host plant species). In these situations, apparent competition mediated through predators and parasites may play a key role in community structure. Van Veen *et al.* (2008) used a quantitative food web approach where the relative strength of interactions was included to assess how different consumer groups were involved (Table 7.2). The scope for apparent competition was greatest for predators (there was higher connectance in the predator subweb, with more generalist predators), intermediate for pathogens and least for parasitoids (which were mostly specialists, with a low subweb connectance). This quantitative approach allows identification of key interactors; for instance, several fungal pathogens were generalists attacking many aphid species, and one aphid species (the nettle aphid *Microlophium carnosum*) was identified as the chief reservoir for fungal spores.

7.10 Conclusions

We have come a long way from the early studies on apparent competition suggesting a role for parasites in structuring communities (Park, 1948). However, progress in theory (and speculative reasoning) probably exceeds empirical confirmation in the majority of areas that we might now study under the umbrella of ecosystem parasitology.

This rather pessimistic tone is not meant to be doom-laden, but to provide impetus for further study in an area that is just beginning to come of age. For instance, there is a theoretical expectation that parasites can induce trophic cascades (Section 7.2). There is some empirical evidence (largely anecdotal or historical) tracing the effects of parasitism through successive trophic levels, and far more evidence of limited trophic effects; further controlled studies quantifying effects throughout food chains and webs would be instructive. In multi-host communities, the mathematics has moved on from two-host models, but further insight is held up by the intractable nature of equational systems with many variables (Sections 7.3 and 7.4); there has been progress with the use of matrix approaches (Box 7.5), but perhaps we are still searching for the right representational tool for these complicated systems.

The recent food web, bioenergetic and ecological engineering studies (Sections 7.5–7.7) have produced exciting evidence for multiple, significant roles for parasites in community processes. However, there is a legitimate complaint that these studies merely reflect the extremes of researcher bias (nobody studies ecoparasitology in systems where parasites don't appear to be important, after all). These studies predominantly involve estuarine and intertidal systems, which seem to harbour a preponderance of trophically transmitted or castrating helminth parasites; we do not know if the strong patterns described are unique to these parasite groups. There are some notable exceptions reported in this chapter – plant parasites and endophytes, crustaceans and their microsporidia and viral infections of mammals – but none are direct or comprehensive food web studies.

Recent theoretical analyses of networks give hope that some general principles will emerge – for instance, with regard to community stability and connectance – but we are a long way from generality (we need more basic studies of the relationship between parasite diversity and connectance, for example). Finally, we turn to the relationship between ecosystem health and parasitism. Again, we need data from a greater variety of systems. But at this point, it would seem fair to conclude in favour of parasite diversity being positively associated with healthy, diverse and robust ecosystems. Indeed, the current focus on so-called beneficial players such as pollinators in studies of ecosystem health may be simplistic, and we call for a greater consideration of the impact of parasites on ecosystem health.

So, are the studies here that suggest a strong role for parasites just the result of observer bias? We do not expect so, partly because relationships between parasitism and the metrics of ecosystem health are now being found in systems where the research questions were *not* focused primarily on the ecological role of parasites (such as the relationship between parasitoid diversity and arthropod diversity in agroecosystems (Anderson *et al.*, 2011; Section 7.8.2)). Parasites also impact on populations and communities when they emerge *de novo* or increase in prevalence; these emerging infectious diseases (EIDs) are the subject of the next chapter. EIDs are experiments that have 'chosen' themselves; they are not the result of researcher preference. The examples we have to date demonstrate that parasites can play significant roles across population boundaries and out into the ecological community as a whole.

8 · Emerging diseases in humans and wildlife

8.1 Introduction

Emerging infectious diseases (EIDs) such as human immunodeficiency virus (HIV), severe acute respiratory syndrome (SARS), malaria and bovine tuberculosis have significant social and economic costs, threatening human health, wildlife conservation, biodiversity and sustainable agriculture. An EID can be broadly defined as a disease that is increasing in range, incidence or impact. This inevitably means that there is some contingency in deciding what classifies as an EID. Pathogens may be classed as emerging for a number of reasons:

(1) increasing incidence of infection in a host population;
(2) increasing severity of infection (virulence) in a host population;
(3) spread to novel populations;
(4) spread to novel species;
(5) a recently evolved novel pathogen or strain, including drug-resistant strains.

Some researchers confine their definitions to a subset of these processes – for instance, Woolhouse et al. (2005) are generally concerned only with categories 4 and 5 above, whereas others (Ostfeld et al., 2005; MacDonald & Laurenson, 2006) take a broader view. The term 're-emergence' is sometimes used with reference to long-established infectious diseases, such as malaria and bovine tuberculosis, which are increasing in incidence. Partly resulting from differences in classification, estimates for the number of EIDs vary. Childs et al. (2007b) suggest a rule of thumb of about 30 zoonotic human EIDs in the last 30 years, whereas Jones et al. (2008) identify 335 human EID first records between 1940 and 2004; this latter study includes many novel antibiotic-resistant strains of bacteria and novel opportunistic infections associated with the pandemic spread of HIV and AIDS. Dobson and Foufopoulos (2001) identify 31 wildlife EIDs in North America over a two-year period, collating reports from the ProMED disease database (see Section 8.6.1).

8.1.1 Emerging approaches to the problem of EIDs

Although diseases have been emerging in human societies and wildlife throughout evolutionary history (Diamond, 1997; Wolfe *et al.*, 2007), there is a strong perception that the frequency of emergence has increased dramatically in recent years and will continue to increase as a result of increasing effects of anthropogenic activity on climate and habitat (evidence for this is discussed in Section 8.4). This perception has driven a number of global collaborative initiatives (in addition to efforts by the World Health Organisation (WHO) and US Center for Disease Control (CDC) – Section 8.6) to improve interdisciplinary research and delivery in human health, animal health and conservation (Table 8.1) which recognise the importance of the community ecology approach to parasitism and encourage research in this area.

8.1.2 What are the problems caused by EIDs?

Emerging diseases can have potentially global impacts socially, ecologically and economically, for public health, conservation, veterinary science and agriculture. In this chapter we consider EIDs from a community ecology perspective. This subject warrants greater depth than we can afford here and we direct the interested reader to two recent volumes that examine community ecology issues for some of the more important human and wildlife EIDs. Childs *et al.* (2007a) contains current concerns to human health from zoonotic EIDs, include comprehensive discussion of the factors influencing emergence and potential control and monitoring strategies; Collinge and Ray (2006) provide further ecological and monitoring perspectives on key wildlife (and some human) systems. Most EIDs have a community ecology component at some point because they usually involve parasites of more than one host species (Fig. 8.1). For instance, of the 1415 pathogens recorded as infecting humans by the turn of the century, 62% were known to be zoonotic in origin (Taylor *et al.*, 2001). At its simplest, an EID system can consist of repeated cross-species infection events (spillovers) of a pathogen from a reservoir host to a novel susceptible host in which intraspecific transmission cannot occur (Fig. 8.1a) (Hendra virus is an example – see Box 8.5). Alternatively, involvement of multiple hosts may occur for a relatively brief (in evolutionary time) period during the evolution of intraspecific transmission in the novel host (a host-species jump (Fig. 8.1b) – HIV, for example, discussed in Box 8.2). Many EIDs involve a combination of multiple host species, multiple vector

Table 8.1. *Conservation and healthcare initiatives that promote research and public understanding of EIDs*

Initiative	Aims	Website
Consortium for Conservation Medicine	Collaborative research in conservation, the link between human and animal health and biodiversity (partners include the Wildlife Trust, academic institutions and government agencies)	http://www.conservationmedicine.org
One Health Initiative	Recognise links between human and animal health, foster collaboration between medical and veterinary science	http://www.onehealthinitiative.com
One World, One Health	Recognise and research links between human, wildlife, domestic animal health and functioning ecosystems (initiative of the Wildlife Conservation Society)	http://www.oneworldonehealth.org
Global Health	Membership alliance of non-governmental organisations, government agencies, academic institutions and industry for improving global human health through research, publicity and health lobbying	http://www.globalhealth.org
EcoHealth	International Association for Ecology and Health. Mission to promote sustainable health of people, wildlife and ecosystems; student association and publication of 'EcoHealth' peer-review journal	http://www.ecohealth.net

species or multiple pathogen strains and may have varying degrees of within- and between-species transmission (Fig. 8.1c, 8.1d – West Nile virus, for example, discussed in Box 8.8).

In Section 8.2 we consider the stages leading to emergence of a new disease, and examine the population dynamics of newly emerging diseases. We then examine in more detail how evolutionary factors contribute to emergence (Section 8.3), before examining the broader patterns of

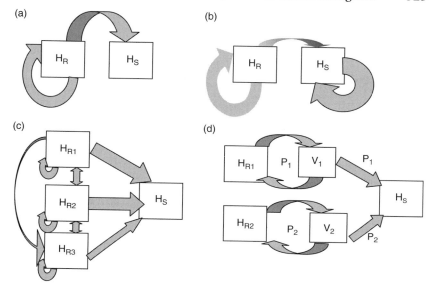

Fig. 8.1. Emerging infectious disease systems. (a) Simple spillover from a reservoir host (H_R) to a novel susceptible host species (H_S). Arrows indicate transmission routes. (b) Evolutionary jump from ancestral host (H_R) to novel host (H_S); the novel disease is maintained by intraspecific transmission in the novel host; the ancestral pathogen strain may continue to propagate in the ancestral host (pale arrow). (c) Multiple reservoir hosts H_R with interspecific transmission of a shared pathogen and spillover to a novel host. (d) Multiple parasite strains (P_1 and P_2) with strain-specific reservoir hosts (H_R) and strain-specific vectors (V) that also transmit the infection to a novel host.

emergence in humans, animals and plants (Section 8.4). We then go on to consider the role of environmental change and anthropogenic impact on the process of emergence (Section 8.5), concluding with a brief examination of control strategies for EIDs from a community ecology perspective (Section 8.6). We have selected a limited set of recent human and wildlife EIDs as representative examples. Our aim in this chapter is to examine the community ecology of emergence, and provide links between examples that may foster greater cross-disciplinary understanding of the similarity of causes and control for human and wildlife EIDs.

8.2 The process of disease emergence

We are constrained to viewing an EID relationship as a snapshot; in reality it reflects a long-term evolving relationship that can potentially

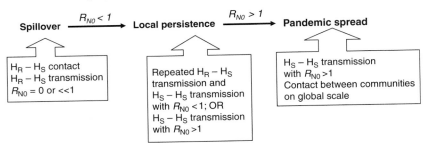

Fig. 8.2. Stages in disease emergence. Spillover of disease from a reservoir host (H_R) to a novel susceptible host (H_S) may lead to persistence, and may go on to spread globally (pandemic spread) if intraspecific transmission in the novel host is sufficiently high. Arrowed boxes indicate the conditions required for the disease to reach each of the stages; R_{N0} is the basic reproductive number in the novel host; this must exceed 1 for maintenance and spread of the EID without additional spillover. The stages of disease emergence are equivalent to those for biological invasion (see Fig. 7.1).

pass through a number of stages (Fig. 8.2). Many more pathogens are involved in the initial stages of emergence than reach the final stage (recall a similar scenario for biological invasions (Fig. 6.1)), and at each stage a number of factors alter the likelihood of transition to the next stage. Several models of disease emergence have been proposed (Childs *et al.*, 2007b; Wolfe *et al.*, 2007; Lloyd-Smith *et al.*, 2009; WHO, http://www.who.int/csr/disease/avian_influenza/phase/en/index.html), and all take a similar step form, although the terminology and number of stages varies somewhat. Ours, as depicted in Fig. 8.2, follows Childs *et al.* (2007b), in which the similarities between emergence and biological invasion are emphasised.

8.2.1 Spillover

Emergence starts with a series of initial spillover events from a reservoir host to a novel host species (Fig. 8.2). Spillover requires contact between the novel (susceptible) host and the original (reservoir) host, or contact with infected vectors or free-living stages of the pathogen from a contaminated environment. The spillover phase is equivalent to the transport and introduction phases for biological invasion (Fig. 6.1). For viral diseases, which arguably comprise the majority of human EIDs (Section 8.4.1), spillover has been variously termed 'viral traffic', 'viral noise' or 'viral chatter' (Daszak *et al.*, 2007). These terms evoke the transient on–off nature of spillover events: they may be very frequent, but the majority, occurring in isolated populations and failing to transition further towards emergence, will remain unnoticed. Many human EIDs cluster in the early spillover stage of emergence; human infection results from independent spillover incidents from wild or domestic animal hosts or vectors, and human–human transmission does not occur. Despite this, such diseases can have considerable public health and economic implications, and are of concern not least because of the risk that they may go on to evolve the capacity for fully pandemic emergence.

Infectious disease is recognised increasingly as a threat to wildlife, and can be particularly problematic for the conservation of small, isolated populations. Basic one-host–one-parasite theory tells us that parasites with density-dependent transmission cannot drive a host population extinct even if they are highly pathogenic. This is because maintenance of the disease requires a threshold host population size (N_T; Box 1.2), below which the parasite is driven extinct, allowing the host population to recover. However, in the real world, other factors (interactions with other species, habitat loss or degradation, for example (Section 8.5)) can combine with infectious disease to reduce populations still further when they reach a critical size. In addition, generalist parasites that are maintained in multi-host reservoir populations are not reliant on a single host population for persistence. Consequently, the dynamics of spillover and spillback often dominate wildlife EIDs and control measures must focus on disease control in the reservoir host or control of contact between reservoir and susceptible host (Section 8.6). Disease may also spillover from livestock or crops to infect wild species. For example, pathogen (*Crithidia bombi*) spillover from commercially reared bumblebees (*Bombus* spp.; used for greenhouse crop pollination) has been implicated in the ongoing decline of wild bumble bees in North America (Otterstatter & Thomson, 2008). Spillover of infectious diseases from invasive to indigenous species can be an important determinant of the impact of an invasion (Chapter 6).

8.2.1.1 Spillover and spillback

The term 'spillover' is used to refer to cross-species transmission of a parasite from a competent host species that acts as a reservoir for the infection into a new host (Fenton & Pedersen, 2005). The term spillback is also used; in invasion biology it refers to cases where introduced species act as a reservoir for a native parasite species (Kelly *et al.*, 2009). In the context of emerging diseases, spillback can occur between wild and domesticated animals (Daszak *et al.*, 2000). For example, domestic birds act as a reservoir for the influenza virus, which originated in wild birds (Box 8.3). Bovine tuberculosis (*Mycobacterium bovis*) provides another example: in the United Kingdom, it was prevalent in cattle in the nineteenth and early twentieth centuries, and (presumably) at some point in this period spilled over into wildlife populations, including the European badger (*Meles meles*). Eradication of the disease from cattle seemed likely in the 1970s following vaccination programmes; however, disease incidence has since increased, with badgers acting as the main reservoir enabling spillback into cattle (MacDonald *et al.*, 2006), with occasional spillover from cows to humans (Smith *et al.*, 2004). Spillover and spillback thus refer to the same process: in each case the parasite is maintained in a reservoir host population, but can cause outbreaks in another host species via cross-species transmission; the terms reflect different histories of emergence; in spillover, the reservoir host is the original host; in spillback, the reservoir is a more recently acquired host.

8.2.2 Persistence

Spillover events may result in *persistence*, whereby the parasite becomes locally established in a novel host population, but has yet to spread to many other populations (equivalent to the establishment phase in biological invasions; see Fig. 6.1). In most cases this will require adaptation of the parasite (Section 8.2.3) for increased probability of infection and replication within the host combining to increase intraspecific transmission efficiency. Theory for one-host–one-parasite population dynamics informs us that, for large populations of susceptible hosts, when the basic reproductive number (R_0; the number of secondary cases produced by each primary case – Box 1.2) exceeds 1, the pathogen can spread. In the context of an emerging disease, this means that any spread in the novel host requires an $R_{N0} > 1$ (R_{N0} is R_0 in the novel host), a characteristic it is unlikely to possess *a priori*, but one that will be subject to natural selection acting on both pathogen and host (Woolhouse *et al.*, 2005). For vector-borne diseases, persistence may require changes in vector behaviour and location to complete the transmission route. Indeed, for some infections there may be barriers

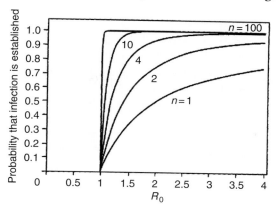

Fig. 8.3. The importance of initial propagule size and $R_0 \gg 1$ in achieving sustained emergence. From May *et al.* (2001).

to the evolution of intraspecific transmission in the novel host (an infection principally transmitted through biting, for instance; consider rabies in humans). During the persistence phase, R_{N0} may be close to, but less than 1, and hence insufficient to maintain the parasite by intraspecific transmission alone. However, repeated spillover from the reservoir host can make up this shortfall, resulting in local persistence in populations in contact with the reservoir host. During this phase, evolutionary adaptation to the novel host can occur, with the reservoir host acting as a source and the novel host a sink for the disease (Dennehy *et al.*, 2006).

In reality, populations are finite and as an infection spreads the pool of susceptible hosts becomes depleted. For this reason, and because transmission is a probabilistic event, even diseases with an $R_{N0} > 1$ will often stutter to extinction as by-chance successive transmission events fail, or the local pool of susceptible hosts is exhausted. Even for quite a high R_{N0}, if the initial propagule of infected individuals is small (less than four or so), the chances of extinction are high (May *et al.*, 2001; Fig. 8.3).

However, probabilistic transmission also means that parasites with $R_{N0} < 1$ can achieve appreciable chains of successful transmission before they eventually fail. The existence of these 'stuttering chains' is important for the evolution of transmission efficiency in the novel host; the longer the chain, the more generations of selection can occur (Box 8.1).

8.2.3 Pandemic emergence

Finally, with $R_{N0} > 1$, the situation is set for a *spread phase*, or *pandemic emergence*. This can occur when the pathogen has spread throughout relatively isolated local populations of the novel host, and local chains of transmission begin to link into larger populations with stronger transmission routes. The pandemic spread phase for an emerging disease bears

---

Box 8.1 Stuttering chains and the evolution of emergence

In the early stages of emergence, a disease is introduced repeatedly into a novel susceptible species through contacts with the reservoir host. At this stage, it is unlikely that parasite transmission between members of the novel species is particularly efficient, but some transmission may occur. Antia *et al.* (2003) examined theoretically the consequences of such inefficient transmission for the evolution of emergence in humans (although the arguments apply to emergence in any novel species). Initial outbreaks are short-lived and result from the combined forces of repeated cross-species introductions and occasional within-host transmission (Fig. 8.4a). As a result of chance failures and successes in transmission, individuals contracting the disease from cross-species introduction (primary transmission) generate chains of secondary cases of varying length. Even though the average number of secondary cases per primary case is less than one (basic reproductive number in the novel host, $R_{N0} < 1$), some chains can be several 'generations' deep. Imagine that the pathogen undergoes mutation affecting R_{N0}; each round of within-host replication and secondary transmission provides a selective round for natural selection to act on R_{N0}. What is the likelihood of the pathogen evolving the ability to self-sustain in the new host ($R_{N0} > 1$) given its mutation rate and current transmission properties? This question was analysed by Antia *et al.* using branching process mathematics (a branch of probability theory designed for dealing with stochastic generative processes). For a variety of models they found that the chances of evolving $R_{N0} > 1$ increased with mutation rate, current R_{N0} and spillover rate (Fig. 8.4b). Hence, even when $R_{N0} < 1$, there is an advantage to having a slightly higher R_{N0}, as it will result in longer chains of secondary cases (and hence more chance of evolving a higher R_{N0}) before the chain stutters to extinction.

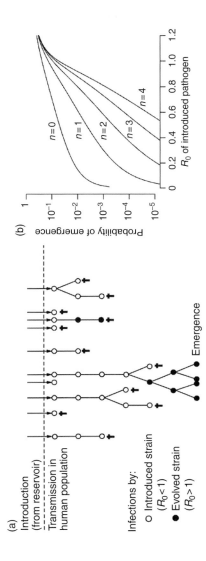

Fig. 8.4. The consequences of probabilistic transmission in early stages of emergence. (a) Schematic of stuttering chains of transmission in the novel host following spillover events from a reservoir host. Circles indicate successive infected individuals; if the infection is not passed on the chain is terminated (daggers). Occasionally, the strain evolves an $R_{N0} > 1$ in an individual (black circles); if transmission of this strain is initially successful, emergence on a larger scale can occur. (b) The probability of emergence given multiple steps in the evolution of R_{N0}, given n mutations are required to bring $R_{N0} > 1$. In this model, mutations occur at a rate of 0.1 per transmission cycle, and the parasite's R_{N0} remains at its introductory value until all n mutations have occurred. Emergence is much less likely as more mutations are required, and for lower initial values of R_{N0}. From Antia *et al.* (2003).

Box 8.2 The emergence of HIV and AIDS

Human immunodeficiency viruses (HIVs) are lentiviruses derived from simian immunodeficiency viruses (SIVs), which circulate in at least 40 species of African primate. There have been at least four separate spillovers from primates to humans resulting in several distinct virus lineages; one of these (HIV-1 group M) has reached pandemic status, while the others form more localised epidemics largely confined to Central Africa. By the end of 2009 nearly 60 million people had been infected with HIV globally, with 25 million HIV-related deaths. There are approximately 33 million people living with HIV. Of current ongoing cases, 67% are in sub-Saharan Africa, where the epidemic has already orphaned 14 million children; 91% of new infections among children occur in this region (UNAIDS: http://www.unaids.org/en/KnowledgeCentre/HIV-Data/EpiUpdate).

Multiple origins of HIV. The most likely route for HIV emergence is through human consumption or handling of SIV-infected primates, which traditionally form part of the diet in sub-Saharan Africa. Sequence analysis suggests that the reservoir host from which HIV-1 is derived is our closest relative, the chimpanzee (*Pan troglodytes*); HIV-2 probably came from sooty mangabeys (*Cercocebus atys*) (Hahn *et al.*, 2000). The M, N and O groups of HIV-1 represent separate spillover events of SIVcpz from chimpanzees (Keele *et al.*, 2006). The recent finding of a new HIV variant (group P), from a Cameroonian woman, with greater sequence similarity to the gorilla SIV (Plantier *et al.*, 2009) may be evidence of a further spillover event from a third primate species (Fig. 8.5). The chimpanzee SIV also appears to have arisen as a spillover event, most likely via consumption of two monkey species and recombination of their respective SIVs within chimpanzees (Bailes *et al.*, 2003; Fig. 8.5). The restricted prevalence of SIVcpz and its moderate pathogenicity suggests that SIVcpz is of relatively recent evolutionary origin compared to other SIVs, which are generally not pathogenic (Keele *et al.*, 2009).

From village to city to pandemic. Recent molecular studies have produced a detailed picture of the emergence of pandemic HIV-1 (reviewed in Sharp & Hahn, 2008). Sequence evolution data and comparisons from chimpanzees and humans place the likely spillover event for the group M virus within a forested region in southeastern

Box 8.2 (*continued*)

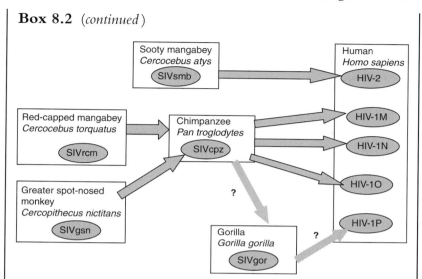

Fig. 8.5. HIV/SIV cross-species transmission and origination of HIV subtypes. From the sequence analysis, the relationship between SIVcpz, SIVgor and HIV P is unclear; SIVgor may have originated by spillover from chimpanzees, or both could have deeper ancestral roots.

Cameroon, around 1910. From there the infection spread along the river trade routes of the Congo basin, 700km south to the recently founded city of Kinshasa, in the Democratic Republic of Congo (then Leopoldville, Belgian Congo), where the group M pandemic was spawned. By 1960 HIV-1M variants were already diversifying in Kinshasa (Worobey *et al.*, 2008), more than 20 years before AIDS was given a name in 1982. At the time HIV reached Kinshasa, no later than 1933, it was the largest of the Central African cities and a major trading destination. As the virus amplified in this high-density population, it continued to diversify; Kinshasa today harbours the greatest diversity of group M subtypes anywhere in the world (Vidal *et al.*, 2000). From there, the virus was spread worldwide. North American and European infections initially originated from a single strain (now subtype B) that was carried to Haiti (around 1966), then to the United States (around 1970), from where it spread to Europe (Gilbert *et al.*, 2007).

Why did it take so long to recognise HIV? If cases grew exponentially from 1910, there would probably have been a few thousand cases in 1960, all in Central Africa (Sharp & Hahn, 2008). Given the frequently long latent period of infection before symptoms, and likely

Box 8.2 (*continued*)

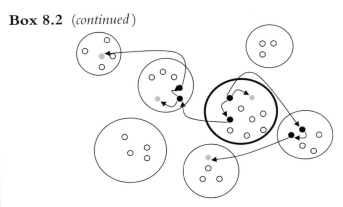

Fig. 8.6. May and Anderson's (1990) interlinked villages model. Infected individuals (shaded dots) infect individuals within the village and in other villages (circles, see also May *et al.*, 2001). Black dots indicate individuals whose transmission cycle is complete; grey dots have yet to transmit, white dots are susceptible. In this example, components of R_0 can be calculated from the six infected individuals for whom the transmission cycle has completed: $R_0(within villages)=5/6$, $R_0(between villages)=4/6$; so total $R_0=1.5$, sufficient for deterministic spread of the infection. Infection in the focal village (bold circle) will be extinguished if the single recently infected case (grey dot) does not pass on the infection, but it has spread to four other villages.

initial misdiagnosis of the diverse symptoms, it is not surprising that HIV remained unrecognised until it was already established in the United States and beyond. May and Anderson (1990) provide theoretical insight into the initial spread of HIV. They model the emergence of the disease in a network of 'loosely interlinked villages'; their model is deterministic, so the influence of chance events is ignored. After initial spillover in a single village, infection propagates within the village in the standard way for a frequency-dependent (sexually transmitted) disease (Box 1.4). However, individuals may also choose sexual partners from other villages, so the total reproductive number for HIV is the sum of two terms: an R_0 term for spread within the 'seed' village, and another for spread to new villages (Fig. 8.6). The disease can spread provided total $R_0 > 1$; hence (barring chance events), it would spread even if within-village $R_0 < 1$, provided there was sufficient diffusion to other villages. This pattern would lead to an initial drop in prevalence in the seed village whilst prevalence in others slowly increased, and could explain some of the delay in the recognition of HIV. This concept can be applied at a larger scale; only when cases become sufficiently frequent within a particular community is an EID likely to be recognised as such.

similarity to the invasive stage reached by some introduced species that spread rapidly throughout the introduced range (Fig. 6.1). For many human EIDs, scale-dependent changes in population structure, host movement and contact rates can result in increased transmission once infection reaches sufficiently large population centres. For instance, in the case of HIV, initial spillover and persistence phases occurred in isolated villages with relatively high contact rates with the simian reservoir species, and it was probably several decades before cases reached a city. However, once HIV became established in a centre of international trade, it was rapidly disseminated across the globe (Box 8.2).

8.2.4 Heterogeneity in R_0: superspreaders and their effect on disease dynamics

One of the lessons from HIV is that a disease can have different rates of spread at different stages in its emergence; R_0 was probably low initially in village communities, but would likely have been much higher once it reached the first city, and changed again once it began to spread between continents. Transmission and contact rates will also differ between subpopulations (inhabitants of village or city; heterosexual or homosexual, for example); this heterogeneity, unless it is recognised, can seriously hamper estimation of disease prevalence, rates of spread and eventual severity of the epidemic.

Superspreading events can be particularly important in determining disease spread. Superspreaders are individuals that are responsible for many more than the average number of transmission events. For sexually transmitted diseases like HIV, individual variation in contact rates and transmission are well documented and conform to a rule of thumb, the '20/80 rule' (20% of cases cause 80% of transmission; Woolhouse et al., 1997). In the more recent emergence of SARS (Box 8.9), contact-tracing data demonstrated the importance of superspreading events in fuelling epidemics in Hong Kong and Singapore of this directly transmitted disease (Section 8.6.2).

The existence of superspreaders requires another approach to modelling R_0, which represents the average number of secondary cases generated by each infectious individual in a large susceptible population. Conventionally, we assume that the population is comprised of individuals with identical 'individual reproductive numbers' (the number of secondary cases each will produce), equal to R_0, or there is some variation around a mean of R_0, to describe variation due to chance events. In mathematical models of sexually transmitted diseases, high-risk 'core groups' of sexually active individuals are included as a separate group with a higher R_0 (May & Anderson, 1987). Lloyd-Smith et al. (2005) examined how to include superspreading for directly transmitted diseases, where it is not usually possible to identify such core groups. Instead, they used different probability distributions to allow for individual differences in transmission, reflecting underlying differences in behaviour, contact rates and pathogen load, for example. This approach allows us to include superspreaders as natural variants along the continuum of transmission, rather than having to identify and model separate classes of individual for set levels of transmission. This enables an elegant assessment of the role of superspreaders in determining disease dynamics. Lloyd-Smith et al. found that their heterogeneity models produced quite different predictions to conventional average-based models (see also Lipsitch et al., 2003). When the transmission distribution was skewed to reflect superspreading, disease extinction was more likely for a given R_0, but disease outbreaks, when they occurred, were much more explosive. Both outcomes result from the skewed distribution of individual reproductive numbers: extinction becomes more likely because most individuals have a low reproductive number so the

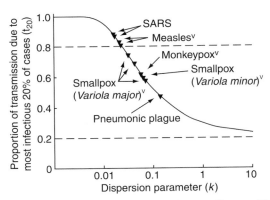

Fig. 8.7. Transmission heterogeneity for selected human diseases. The proportion of secondary cases arising from the most infectious 20% of cases is above 80% for diseases with highly skewed transmission, such as SARS. On the horizontal axis, a low value of the dispersion parameter k indicates a high degree of skew in the distribution of pathogen reproductive numbers from individual hosts, assumed to come from a negative binomial distribution. Dashed lines indicate expectation from 20/80 rule (upper) and homogeneous transmission (lower). From Lloyd-Smith et al. (2005).

infection stutters to extinction along each chain; outbreaks are rare but more severe because occasionally a superspreader is encountered, boosting subsequent cases disproportionately. Lloyd–Smith *et al.* used data from different human diseases to quantify the skew in transmission and its influence on disease spread. The diseases differed in their distributions, with SARS in particular showing a high degree of skew, from which we can estimate that a very high proportion of secondary cases are caused by a minority of superspreaders (Fig. 8.7).

The theoretical studies discussed in this section highlight a number of important lessons with relevance for control of the initial stages of emergence. Attaining $R_{N0} > 1$, required for self-sustained persistence in the novel host and eventual pandemic emergence, depends on a number of factors:

(1) the frequency of cross-species introduction events, and the propagule size at introduction (the number of concurrent cross-species infections); limiting contact or reducing propagule size may preclude emergence (Section 8.6.2);
(2) the structure of the population into which spillover occurs: the dynamics will differ in isolated or loosely interlinked villages compared to a larger population, and transmission probability may vary with scale (air travel becomes important once infection reaches large population centres but is irrelevant in isolated villages); control through reduction of particular contact routes may be effective at some stages, less so at others (Section 8.6.2);
(3) distribution of transmission: highly skewed distributions with superspreaders will have different disease dynamics than homogeneous populations; control focusing on superspreaders will be more effective (Lipsitch *et al.*, 2003). Transmission distribution may have both genetic and environmental underpinnings;
(4) the propensity of the pathogen to mutate (pathogens with error-prone replication may be more likely to emerge; Sections 8.3 and 8.4.1);
(5) current R_{N0} for the pathogen; EIDs may be more likely between phylogenetically close host species or for pathogens with a broad host range.

The first three factors will have ecological/environmental determinants, which we discuss in Sections 8.5 and 8.6. The latter three factors underline the importance of evolution in determining EIDs, which we explore in the following section.

8.3 The evolution of emergence

A number of evolutionary factors can influence transitions between stages of emergence. These appear to be particularly important in determining effectiveness of spillover from reservoir to novel hosts, and transition to competent intraspecific transmission within the novel host. Ecological factors, discussed in Section 8.5, influence these stages to a lesser extent, but are key in determining outcomes of the other stages of contact and transition to pandemic emergence (Fig. 8.1).

The likelihood that spillover into a new host will go on to form an EID depends primarily on the pathogenicity of the disease in the new host, and on the efficiency of host–host transmission in the new host (Section 8.2.2). In the spillover phase, as the pathogen has no history of adaptation to the new host, transmission within the new species may be low and new cases of the EID will result from spillover from the reservoir host. In contrast, diseases that are adapted to or acquire the necessary adaptations to sustain transmission in the new host can persist and spread. Compare the strains of swine flu and avian flu that are currently circulating. The recent (2009) swine flu pandemic was caused by a strain of influenza A that jumped from pigs to humans and has efficient human–human transmission. In contrast, ongoing outbreaks of avian flu have not become epidemic as human–human transmission is rare (Box 8.3).

The degree of specialisation and the transmission route of a parasite are key factors for the risk of emergence (Woolhouse, 2002). Transmission route influences the likelihood of encountering new host species, and hence the selective advantages of having a wide host range. For example, a review of human EIDs (Woolhouse & Gaunt, 2007) found that microparasites that were transmitted by indirect contact, and were hence likely to encounter new host species, were more likely to be zoonotic than were directly transmitted species. Vector-borne parasites were also more likely to be zoonotic. The route of transmission is linked to the degree of host specialisation. For example, a parasite that utilises a blood-feeding, generalist vector will be under selective pressure to use a range of host species so as to maximise opportunities for transmission. At the molecular level, parasites that use cell receptors which are conserved across host species are more likely to be generalists. For example, the rabies virus enters host cells by binding to the conserved nicotinic acetylcholine receptor, and is hence able to infect a wide host range (Baranowski et al., 2001). Parasites may also acquire adaptations that enable them to use new receptors (Parrish et al., 2008). For example, recombination between introduced and native

Box 8.3 Influenza A

Influenza A is an RNA virus, the genome of which is made up of eight segments, each containing one or two genes. The different influenza subtypes are named according to two surface proteins that induce the host immune response; haemaglutinin (HA), which is involved in binding to the target cell, and neuraminidase (NA), which is involved in virus release from the cell. To date, there are 16 known HA serotypes and nine NA serotypes (Gatherer, 2009). For example, the current swine flu outbreak is caused by a strain of H1N1.

Influenza evolution. The potential for novel strains of influenza to emerge is high because the virus undergoes rapid evolution as a result of both mutation and reassortment (Gatherer, 2009; Wertheim, 2009). High levels of mutation occur in RNA viruses. In the influenza literature, 'antigenic drift' refers to adaptive evolution as a result of mutations (this contrasts with the use of the term in evolutionary biology to refer to neutral evolution). The high rate of mutation of HA allows escape from the host immune system. Reassortment is a source of rapid evolution that may occur if an individual animal is host to two different influenza subtypes. Segments of different virus subtypes may be exchanged, giving rise to new flu subtypes. Reassortment of NA and HA segments may lead to the emergence of strains with a novel antigenic profile and is termed 'antigenic shift'. A risk of such reassortment is the emergence of novel subtypes to which the host has no immunity. For example, the current H1N1 outbreak has resulted from a reassortment of two types of swine flu, one of which was itself a triple-reassorted strain comprising segments that originated in avian, swine and human seasonal influenza (Smith *et al.*, 2009a).

Bird reservoir hosts. Flu is endemic in waterfowl and shorebirds, which are asymptomatic and act as the primary reservoir hosts. The virus replicates in the intestinal epithelia and, to a lesser extent, the respiratory epithelia. The route of transmission is oral–faecal (Van Reeth, 2007). The virus can survive in water for several days, and bird movement and migration provide opportunities for virus dispersal. Phylogenetic analysis reveals that all of the flu strains recorded in mammals and birds originated in water birds (Webster *et al.*, 1992). Influenza can spill over into domestic birds, with transmission assisted by high densities and commercial movement of domestic flocks. Most viruses that affect domestic poultry are of low pathogenicity, but two subtypes (H5 and H7) cause

Box 8.3 (*continued*)

high levels of mortality. The subtype H5N1 caused severe outbreaks in domestic flocks in Southeast Asia in 2003–2004. Interestingly, in China, spillback into migratory wild birds has caused high mortality, indicating an increase in virulence in the virus (Olsen *et al.*, 2006).

The role of pigs in flu evolution. In contrast with birds, influenza targets the respiratory tract of mammals, and mortality is a result of severe respiratory distress. The relative inefficiency of H5N1 to infect humans may be because bird flu and human flu target different receptors on the host cell. However, pigs possess respiratory receptors that are similar to both the human respiratory and bird enteric epithelial cells targeted by human and bird flu, respectively. Hence, dual infections in pigs may provide an opportunity for emergence of a reassorted virus able to bind to human-type receptors but having bird-like antigenic activity. Such a virus could avoid pre-existing human immunity (Greger, 2007), resulting in great pandemic potential. The H1N1 virus responsible for the 2009 swine flu pandemic is thought to have undergone reassortment in pigs (see below).

History of recent pandemics. Since 1900 there have been five flu pandemics. The most devastating was Spanish flu (caused by an H1N1 subtype), which occurred in 1918–1919. Recreation of the 1918–1919 virus (Tumpey *et al.*, 2005) revealed a virus of avian origin, but with an HA that can bind to human receptors. The virus affected the respiratory tract and was unusual in that mortality was high in younger (age 20–40) people. High-density populations in the trenches during the First World War may have provided opportunities for virulence to evolve (Table 8.2), whilst demobilisation of troops at the end of the war and food-stressed populations provided an opportunity for the virus to spread around the world (Gatherer, 2009). Spanish flu persisted in human populations, where it caused seasonal flu until the 1950s. It re-emerged in 1978, causing the Russian flu outbreak (Table 8.2). However, the outbreak was less dramatic than previous pandemics, as older members of the population had immunity from the previous pandemic. Two other subtypes that derive from human–avian reassortments were responsible for the Asian flu and Hong Kong flu outbreaks in the 1950s and 1960s, respectively. H2 died out of human populations, but H3 persists in seasonal flu.

Box 8.3 (*continued*)

The 2009 outbreak of swine flu in humans is related to a triple-reassorted flu strain that had been circulating in pigs for a number of years. This strain appears to have undergone a further reassortment to generate a virus that incorporates genome segments from bird flu, human flu and two lineages of swine flu. This is the first documented outbreak of swine flu in humans for which human–human transmission is sufficient to sustain the virus in the new host. At the time of writing, it is difficult to predict the outcome. The virus does not currently cause high mortality (in comparison with the 1918 outbreak, where mortality was 2.5%; mortality for the current H1N1 is estimated by the WHO at 1%: http://www.who.int/csr/don/2010_03_19/en/index.html), although this is an overestimate as many (non-fatal) cases go unreported. However, further evolution could increase the virulence either through mutation (antigenic drift) as occurred in 1918–1920, or through reassortment occurring in dual infections with seasonal flu (Gatherer, 2009).

Potential future pandemics. Flu pandemics have all involved subtypes with avian-origin HA, to which humans have no immunity (Table 8.1). An avian subtype which is being closely monitored by the WHO is the highly pathogenic H5N1 avian flu. Prior to 1997 bird–human cases were very rare (17 cases) and benign. However, during severe H5N1 outbreaks in domestic ducks in Asia in 2003 and 2004, several (97) cases of transmission from birds to humans occurred and outbreaks have since occurred across a wide geographical range. To date, there have been 492 confirmed human cases (291 deaths); with ongoing activity in Egypt, Indonesia and Viet Nam (*WHO Global Alerts and Response Tables*: http://www.who.int/csr/en). Mortality is very high (~60%) in comparison to the 2.5% mortality for 1918–1919 Spanish flu (Table 8.2). Human–human transmission is currently very rare. However, should H5N1 evolve efficient human–human transmission whilst retaining its virulence, there is the potential for a devastating pandemic.

---

fungi can lead to the emergence of new strains with an expanded host plant range (Brasier, 2000), and mutations and reassortment have allowed the influenza virus to infect new species (Box 8.3). CPV emergence resulted from multiple mutations in feline parvovirus (FPV) that allowed it to bind to the canine transferring receptor (Parrish *et al.*, 2008).

The host genotype is also of key importance in determining susceptibility or resistance to a new disease. For example, about 1% of Europeans are resistant to HIV as they carry a mutation in a cell surface receptor for the virus (Morens *et al.*, 2008). The origin of the mutation predates HIV, and its increase in frequency and geographic spread over past centuries is thought to have resulted from the selective advantage it imparted in resistance to other diseases, such as smallpox (Novembre *et al.*, 2005). Genetic diversity is necessary for the host to evolve in response to a novel pathogen (Woolhouse *et al.*, 2002). However, farming practices can reduce the genetic diversity of livestock and crops, increasing their vulnerability to EIDs. For example, bananas (the cultivated variety of which is descended from a single clone, the Cavendish cultivar) are vulnerable to outbreaks of Panama disease (Ploetz, 1994), caused by strains of the fungus *Fusarium oxysporum* f. sp. *cubense*, prompting the investigation of new resistant clones (Ploetz, 1994).

The EID that presents the greatest threat to humans is influenza (Table 8.2; Greger, 2007; Gatherer, 2009). The influenza virus is endemic in aquatic birds and, as a result of intensive domestication of birds and pigs, it is frequently (in evolutionary terms) transmitted to humans. Flu may show rapid and effective transmission between humans, but, in comparison with other human EIDS such as HIV (Box 8.2) or SARS (Box 8.9), the mortality rate of flu pandemics to date has been low. The 2009 H1N1 swine flu pandemic infected a large proportion of the global population within months, but mortality was low, and even the 1918 flu pandemic killed only 2.5% of infected individuals (Table 8.2). However, models predict dramatic impacts on human populations should the virulent avian flu H5N1 (case fatality 60% – Greger, 2007) acquire efficient human–human transmission while retaining high human pathogenicity (Parrish *et al.*, 2008). The potential for the emergence of new strains of flu results from anthropogenic and evolutionary factors. Intensive bird farming and animal and human transportation provide multiple opportunities for transmission between birds and from birds to mammal hosts, including humans. Furthermore, the flu virus has the potential for rapid evolution to generate strains to which humans have little immunity (Box 8.3).

8.3.1 Virulence evolution of emerging diseases

Influenza A remains a problem for public health, in part because of its high evolutionary rate, in common with other RNA viruses such as

Table 8.2. *Human flu pandemics since 1900*

Subtype	Date	Origins	Virulence and transmission
H1N1 Spanish flu	1918–1920	Virus of avian origin; mutations resulted in waves of renewed virulence	25–30% of world's population infected, more than 40 million deaths
H2N2 Asian flu	1957–1963	Segments from avian (HA, NA, PB1) and human flu	1.5–4 million deaths
H3N2 Hong Kong flu	1968–1970	Avian (HA and PB1) and human flu	1–2 million deaths
H1N1 Russian flu	1977–1979	Avian origin, identical to 1918 Spanish flu	0.7 million deaths
H1N1 Swine flu	2009–	Reassortant virus containing segments derived from North American swine flu, Eurasian swine flu, North American avian flu and H3N2	Less than 100 000 deaths (at July 2010)

Data from Horimoto and Kawaoka (2005) and Michaelis *et al.* (2009).

HIV, which can evolve up to one million times faster than animal DNA (Sharp & Hahn, 2008). Subtypes of these diseases are also highly pathogenic to man (Boxes 8.2 and 8.3). The commonly held supposition that virulent parasites evolve to become less virulent as the association with their host species continues does not necessarily apply in the case of emerging diseases (Bull & Ebert, 2008).

It is generally assumed that a tradeoff exists between virulence (pathogenicity) and parasite transmission, such that higher transmission efficiency can only be achieved by increasing the negative impact on the host. Such a tradeoff is likely because virulence probably reflects parasite replication processes within the host, with higher parasite burdens resulting in higher transmission. For endemic parasites (those that have reached a dynamic equilibrium with their host population and are no longer spreading through it), theory suggests that parasites will be selected to optimise virulence (Anderson & May, 1982), maximising their effective reproductive number R by optimising the tradeoff between transmission efficiency and the duration of infection (reflected by the length of the infectious state). However, parasites introduced into a novel host can spread provided their basic reproductive rate R_0 in this novel population, comprising only susceptible individuals, is greater than 1; this may be achieved for a range of virulence levels (including higher or lower values than the optimum for an endemic pathogen). In general, higher virulence is optimal in these initial stages; as the invasion proceeds and the pool of susceptibles decreases, selection favours reduced virulence (Lenski & May, 1994; Day & Gandon, 2007). In addition, even if a pathogen reaches equilibrium in the host population, variants that can escape the host's immune response can invade over a relaxed range of virulence, depending on the extent to which the whole host population is susceptible to them. These 'escape mutants' can potentially spread, with much higher virulence, and may be especially problematic if there is covariance between resistance to host defences and virulence (Bull & Ebert, 2008).

Furthermore, selection between competing pathogen strains within hosts can lead to higher, or maintained, virulence. For instance, in the case of HIV-1M, selective processes operate within individual hosts that harbour a great diversity of viral genomes introduced at transmission or formed by mutation within the host. Within–host selection between viral strains may maintain pathogenicity as HIV virus production is linked to chronic hyperactivation of the host immune system (Grossman *et al.*, 2002). If viral strains have structured populations within the host, localised pathogenicity may be advantageous because it disproportionately benefits the pathogenic strain; systemic pathogenicity may be a by-product of this process (Bartha *et al.*, 2008). However, a sufficiently long evolutionary history of association may decouple these processes; most primate species do not suffer pathogenicity as a result of infection with their species-specific SIVs; understanding this process may provide avenues for vaccine development (Sodora *et al.*, 2009).

8.4 Phylogenetic and temporal patterns of emergence

8.4.1 Which diseases emerge, and in which hosts?

8.4.1.1 Which diseases?

The type of parasite most likely to emerge differs with host taxonomic group. For humans, reviews by Taylor *et al.* (2001) and Woolhouse and Gowtage-Sequeria (2005) indicate that zoonotic diseases (infections which are naturally transmitted between vertebrate hosts and humans;

Chapter 1) are strongly over-represented among emerging diseases of humans compared to non-zoonotic diseases. The most common reservoirs are ungulates, carnivores and rodents, although primates, bats and birds are also important (Woolhouse & Gaunt, 2007). Viruses and bacteria are more likely to emerge than fungi and helminths, with viruses accounting for the majority of human EIDs (Cleaveland et al., 2007). For plants, fungal pathogens show a strong tendency to emerge, often as a result of anthropogenic influences leading to contact with new host populations or species (Slippers et al., 2005). For mammals (Pedersen et al., 2007), artiodactyls are most likely to be infected by bacteria (47% of diseases) followed by viruses (29%), whereas carnivores are more affected by viruses (56%) then helminths (19%). Relationships for other vertebrates have yet to be fully elucidated with data dominated by host groups deemed under particular threat: viral infections (morbillivirus, distemper) for marine mammals (Box 8.4), fungal infections (chytridio-mycosis; Box 8.6) for frogs, and avian malaria for many species of Hawaiian birds (Smith et al., 2009a). In general, the majority of wildlife EIDs are caused by viruses, bacteria and protists, with very few reports of emergence of macroparasitic diseases (Dobson & Foufopoulos, 2001). However, emerging (or re-emerging) macroparasitic diseases are an ongoing public health problem (Chomel, 2008).

8.4.1.2 Which hosts?

A survey of EIDs in mammals highlights the importance of spillover and spillback in determining which diseases emerge, and in which animals. Using the IUCN Red List for threatened and endangered animals, Pedersen et al. (2007) examined which species under threat were listed with disease as a contributing factor to their demise. Of the mammals identified as threatened by parasites, 88% came from two taxonomic orders: artiodactyls and carnivores. Among carnivores, almost all threatened species were either canids or felids, and among artiodactyls, the clades containing our most common domesticated animals (cows, pigs, sheep, goats and horses) were most strongly represented. Hence, close phylogenetic relatedness to a common domestic animal is a strong risk factor for acquiring a novel infectious disease; it is noteworthy that more than 80% of known domestic animal pathogens are capable of infecting wildlife species (MacDonald & Laurenson, 2006). These patterns suggest that the majority of diseases have emerged as a result of spillover rather than cospeciation with the host.

8.4.2 Are EIDs increasing?

For humans, a recent examination of global trends suggests that diseases have emerged more frequently in recent years. Jones *et al.* (2008) examined first reports of new diseases between 1940 and 2004; controlling for increases in reporting over the period, EID events rose significantly over the time period, with a peak in the late 1980s coinciding with the reporting of opportunitistic diseases associated with immune-compromised AIDS patients. Distinct categories of human EIDs were evident: drug-resistant variants of known pathogens; novel diseases such as SARS and HIV, often with a zoonotic origin; and vector-borne diseases. Increases occurred across all categories, although the strongest temporal trends were evident for drug-resistant pathogens and zoonotic diseases originating from domestic animals.

There have been some attempts to quantify the impact of EIDs on wildlife populations. However, despite a growing feeling that infectious disease is an increasingly important factor (Wilcove *et al.*, 1998; MacDonald & Laurenson, 2006), the available data do not make it easy to verify this conjecture. Smith *et al.* (2006) surveyed IUCN Red List data for cases where infectious disease was reported as a contributing factor to species extinction or endangerment. Less than 4% of extinctions and 8% of critical endangerments were attributed to disease, and in some cases it was unclear from searches of the scientific literature to what diseases the reports might refer. Thirty-one species extinctions since 1500 were facilitated by disease, the majority either birds (especially Hawaiian endemics, in which avian malaria contributed to their demise) or amphibians (for which chytridiomycosis may have been a contributing factor – Box 8.6). Lack of detailed records, including uncertainty over disease identity and the level of threat, makes it difficult to assess the real extent of the problem (Smith *et al.*, 2006; 2009b). The evidence suggests, however, that infectious disease may have its strongest impact via interactions with other major drivers of extinction such as habitat loss, invasive species and climate change (Smith *et al.*, 2009b; Tompkins *et al.*, 2010). This is apparent when we look at the distribution of extinctions and endangerments with regard to the number of contributing factors listed in the Red List; when disease is listed, it occurs more frequently in concert with other driving factors than would be expected if it acted independently (Fig. 8.8). To the extent that these other driving factors, many of which result from anthropogenic influence (Section 8.5), are increasing in severity, we can anticipate that wildlife EIDs, or their impact, will also be on the increase. An example of the interactions

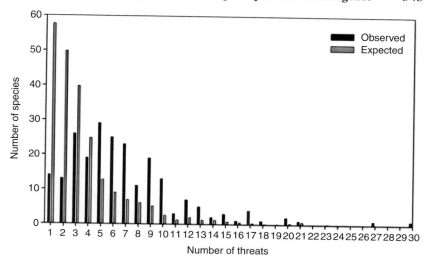

Fig. 8.8. Frequency of species for which infectious disease is listed among factors contributing to critical endangerment. Disease is more often observed (black bars) acting with multiple other threats (horizontal axis) than would be expected from random associations between independent factors (grey bars). From Smith *et al.* (2006).

between disease and environmental drivers is provided in marine ecosystems (Box 8.4), where widespread patterns of disease increase in a broad range of hosts have been documented.

As illustrated by the examples in Box 8.4, environmental stress may be an important factor in emerging disease in the oceans, affecting diverse taxa. The complex interactions between climate change, host and parasite dynamics are an area of increasing importance for marine and other systems (Section 8.5). Elucidating the mechanisms that underpin these outbreaks presents an ongoing challenge for marine management and conservation.

Box 8.4 Are marine diseases increasing?

Over the last 30 years, diseases have increased in a range of marine organisms. In a key review, Ward and Lafferty (2004) found evidence for increased disease in five (echinoderms, corals, molluscs, turtles and mammals) of the nine taxonomic groups studied, whilst a decrease in disease reports was found for fish. The mechanisms underpinning disease emergence are unknown for many systems.

Box 8.4 (*continued*)

However, for systems that have been studied, disease emergence is linked with environmental stresses, in particular climate change and changes in host density as a result of fishing and aquaculture (Harvell *et al.*, 2002).

Corals. Climate change has been linked to an exponential increase in coral diseases over recent decades (Sokolow, 2009). Ocean warming, changes in rainfall and sea level and ocean acidification are all likely to affect disease emergence in corals. The interaction between temperature and disease has recently been studied for two corals. The red gorgonian *Paramuricea clavata*, a key species in coral communities, has suffered bacterial-induced mass mortalities in years of elevated temperature anomalies in the Mediterranean (Bally & Garrabou, 2007). The interaction between infection and temperature is thought to result from temperature-dependent expression by the bacterium of an extra-cellular protease that lyses coral tissues. The role of temperature in disease emergence was supported by laboratory studies in which exposure to *Vibrio coralliilyticus* led to rapid disease development and colony mortality at 25°C, whereas colonies maintained at 16°C did not develop the disease. Ward *et al.* (2007) measured the impact of temperature on host resistance and pathogen (*Aspergillus*) growth on another species of coral, the sea fan *Gorgonia ventalina*. Surprisingly, elevated temperatures led to an increase in the host's response to infection, with higher production of anti-fungal compounds in response to the parasite. However, the growth rate of the pathogen also increased more rapidly in response to temperature than did host resistance, allowing the infection to become established (Ward *et al.*, 2007). This fungus has caused massive mortality in sea fans in Florida (Kim & Harvell, 2004), a problem likely to be exacerbated by climate change.

Amphipods. Climate change has also been predicted to lead to parasite outbreaks and population crashes in amphipods. *Corophium volutator* is a keystone species whose tube-building behaviour maintains the sediment stability of intertidal mudflats (Poulin & Mouritsen, 2006). The development of trematode infections in this host is dependent on temperature. Poulin and Mouritsen (2006) modelled the impact of an increase in temperature on the dynamics of the amphipod/trematode system in the Wadden Sea, Denmark. The resulting increase in cercarial production and associated host

Box 8.4 (*continued*)

mortality is predicted to cause a decline in amphipod populations, with extinctions predicted if temperature increases by 4°C (a rise of 6°C over 70 years is in fact predicted for this region). The tube-building behaviour of this amphipod is key for sediment stability in this habitat; a previous parasite-induced crash in amphipod populations was found to result in an increase in sediment particle size and increased primary production (Chapter 7). Hence, the increase in parasite-induced mortality predicted under global warming is likely to have dramatic effects on the intertidal ecosystem.

Molluscs. In addition to climate change, changes in population density may be important factors underlying the documented increases in marine disease. Aquaculture of marine organisms provides increased opportunities for the introduction of parasites, for example, through movement of stocks and for parasite transmission between organisms kept at high densities (Berthe *et al.*, 2004; Lafferty *et al.*, 2004). For example, oyster diseases (resulting from protist, bacterial and viral parasites) are a major economic cost, and have increased in frequency in recent years (Lafferty *et al.*, 2004).

Fish. Aquaculture may increase opportunities for disease emergence by providing reservoir populations for disease maintenance; for instance, in the last two decades, sea lice have become an increasing problem for farmed and wild salmon in the Atlantic and Pacific oceans. The prevalence of sea louse (*Lepeophtheirus salmonis*) in wild pink (*Oncorhychus gorbuscha*) and chum (*Oncorhychus keta*) salmon has increased in line with the rise in salmon farming (Krkosek *et al.*, 2006). In addition, the location of fish farms can result in disease spreading to more susceptible life stages, resulting in higher pathogenicity. For instance, the migratory route of wild salmon passes extensive fish farms off the west coast of Canada, bringing juveniles from the wild population into contact with sea lice from the farms. Juveniles suffer high mortality as a result of sea lice infestation; this is normally avoided by migration, which separates juveniles from adults (which suffer little morbidity or mortality from sea lice infestation in the wild). Hence, fish farms act as reservoirs for infection and also disrupt one of the functions of wild stock migration (Krkosek *et al.*, 2006). Fishing of wild stocks may have the reverse effect; as the host population declines below the threshold size for parasite maintenance, parasites can become 'fished out' (Section 3.6.3). For example, the decline in

Box 8.4 (*continued*)

fish diseases reported by Ward and Lafferty (2004) reflects mostly species that have suffered population declines due to overfishing. Fishing can also alter the dynamics of parasites that affect species at lower trophic levels. For example, increased bacterial disease in sea urchins in California has been linked to increased population density, itself a result of overfishing of their predators (lobsters and sea otters) (Ward & Lafferty, 2004; Hayes *et al.*, 2005).

Seals. Epidemics of phocine distemper virus (PDV) caused massive mortality of harbour seals (*Phoca vitulina*) in the North Sea in 1988 and 2002, and of Caspian seals (*Pusa caspica*) in 2000. PDV killed up to 60% of seals, with strong age- and sex-dependent mortality and evidence of acquired immunity (Harkonen *et al.*, 2007). A number of factors have been hypothesised to contribute to these complex outbreak patterns. Lavigne and Schmidtz (1990) found that distemper outbreaks appear to coincide with warm temperatures and high seal densities, providing opportunities for disease spread. Migration of harp seals (*Pagophilus groenlandicus*) outside of their normal range (possibly in response to overfishing or climate change) suggested a route of infection into harbour seals during the 1998 outbreak. However, Lavigne and Schmitz (1990) found that the pattern of migration did not fit the geographical spread of distemper. It has also been suggested that a build-up of environmental contaminants might reduce the immune defences of seals, although toxicology of Caspian seals that died during the 2000 outbreak revealed no relationship between levels of organochlorines and infection (Kuiken *et al.*, 2006).

8.5 Environmental change and emergence

There has been much speculation as to the role of environmental change in the apparent increase in rates of emergence for both wildlife and human diseases in recent decades. Environmental change results from many different processes but can be roughly classed into three broad categories: changes in land use as a result of farming, urbanisation or habitat destruction; transportation of animals and plants and human travel; and climate change. These environmental changes can affect various stages of emergence as summarised in Fig. 8.9 and outlined in the sections below; more detailed discussion can be found in Daszak *et al.* (2000; 2007) and Childs *et al.* (2007b).

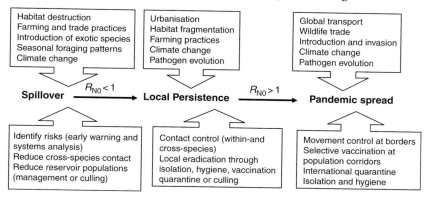

Fig. 8.9. Stages of emergence: influence of environmental factors and control options. Boxes above the transition line indicate factors influencing transition between stages; boxes below indicate possible control options for limiting outbreaks to the current stage and/or reducing them to the previous stage. As with biological invasions (Fig. 6.1), effective control options diminish with each transition.

8.5.1 Land use changes

The destruction of natural habitats for forestry, farming or settlement appears to be a key factor in the emergence of several EIDs, in particular influencing the contact and spillover phases of emergence.

8.5.1.1 *Destruction of natural habitats*

Reduction of natural habitats influences disease emergence in two ways. Firstly, reduction in habitat size or quality may increase nutritional (or territorial) stress on resident species, forcing them into human-occupied zones. Secondly, habitat destruction through deforestation and settlement often brings humans and their domestic animals into direct contact with species seldom previously encountered. Both these processes result in increased contact rates between wildlife and humans or their domestic animals. This can provide opportunities for spillover of wildlife infections to cause novel diseases in humans or domestic animals. For example, range expansion by fruit bats as a result of habitat destruction may have contributed to the emergence of Nipah and Hendra viruses (Box 8.5) and SARS (Box 8.9; Daszak *et al.*, 2007). Deforestation in the tropics for settlement or logging not only results in habitat loss, but also increases connectivity (affecting the contact and spread phases of emergence). Logging routes traffic wildlife, diseases and people to or from remote locations; associated trade in wildlife for the pet or trophy, bushmeat and 'medicinal' markets is also increased (Daszak *et al.*, 2007).

Box 8.5 Nipah and Hendra viruses

Nipah and Hendra viruses are closely related paramyxoviruses that emerged in the southern hemisphere in the 1990s, causing neurological or respiratory disease in humans. Hendra virus was first reported in Australia in 1994, and Nipah virus was first reported in Malaysia in 1998. Several species of fruit bat (*Pteropus* spp.) are the reservoir hosts for these viruses; in both cases, domestic animals (pigs for Nipah virus and horses for Hendra virus) are implicated in virus amplification and subsequent spillover to man (reviewed in Field *et al.*, 2007). Both EIDs appear to be in the early spillover stage of emergence, with no human–human transmission reported for Hendra virus or the Malaysian outbreak of Nipah virus. However, more worrying are the recent outbreaks of Nipah virus in Bangladesh and India, in which human–human transmission did occur (Chadha *et al.*, 2006; Field *et al.*, 2007).

Nipah virus: from bats to pigs to humans. Both pigs and humans are novel, susceptible hosts for Nipah virus, with fever, respiratory distress or neurological symptoms in pigs and fever with rapid onset of encephalitis in humans. The 1998–1999 outbreak in Malaysia led to substantial economic loss through mortality and culling of over one million pigs and 105 human fatalities from 265 cases (Chua *et al.*, 2000). The human cases had direct contact with pigs, so human–human transmission seemed unlikely. An outbreak in Singapore (1999, 11 cases) also involved transmission from pigs to humans. The highly contagious nature of the infection in pigs (many of which had only mild symptoms) made this a concerning new EID, as without adequate monitoring and control of pig movements it could rapidly escalate.

Human–human transmission of Nipah virus. A further 11 outbreaks have occurred since 1999, in Bangladesh and India (case totals: 210; deaths: 146 (WHO, http://www.who.int/mediacentre/factsheets/fs262/en)). These outbreaks show clear evidence of spillover transmission from bats to humans, and direct human–human transmission in the larger outbreaks. The most likely route of transmission from bats to humans is from consumption of fruit and raw date palm juice contaminated with bat urine or faeces; case distribution has even been matched to the routes taken by local palm juice sellers (Dobson *et al.*, 2006). Roughly half of the cases in the Bangladesh and India outbreaks were the result of human–human transmission, which occurs via close contact with human secretions or

Box 8.5 (*continued*)

excretions (WHO, http://www.who.int/mediacentre/factsheets/ fs262/en/index.html). In Siliguri, India, human–human transmission chains developed in hospitals in association with patient care, demonstrating the need for vigilance and appropriate hygiene precautions against this disease (Chadha *et al.*, 2006).

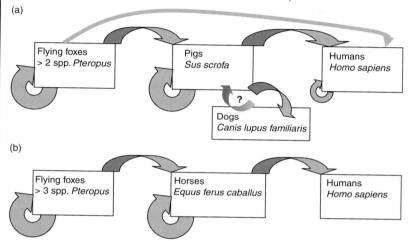

Fig. 8.10. Community ecology of (a) Nipah and (b) Hendra viruses. Arrows show transmission routes; for Nipah virus, there is good evidence for bat–human and human–human transmission, reasonable evidence of pig–dog transmission at a pig farm and possibly also dog–pig transmission (Field *et al.*, 2007).

Hendra virus: from bats to horse to humans. Hendra virus has caused 11 outbreaks since its discovery, all in Australia (human cases: 6; deaths: 3; equine cases: 38 (WHO: http://www.who.int/mediacentre/factsheets/images/hendra_virus_table.gif)). All the human cases were closely associated with the infected horses. Although there have been very few cases, mortality in horses and humans is high, and outbreaks continue to occur (two in 2008). Recent outbreaks have shown a shift from respiratory to mainly neurological symptoms in horses, which might indicate viral adaptation towards increased transmission (WHO: http://www.who.int/mediacentre/factsheets/fs329/en). Outbreaks occur during the fruiting season when fruit bats are attracted to orchards surrounding stables, resulting in contamination of pasture from bat excreta (Field *et al.*, 2007).

8.5.1.2 Urbanisation

In developed countries, pockets of wildlife habitat exist interspersed with urban development; the expansion of urban and suburban zones leads to further reduction in habitat patch abundance and size. Reduced habitat patch size, number and connectivity often leads to reduced biodiversity within habitat patches. Biodiversity has been implicated as a buffer against the spread of infectious disease in at least three human/wildlife EIDs: tick-borne encephalitis (discussed in Dobson *et al.*, 2006), Lyme disease (Box 7.1) and West Nile virus (Box 8.8). In the latter two cases, the more competent reservoir species are relatively robust to habitat loss, persisting in smaller patches at the expense of other species that contribute to dilution effects. In addition, habitat fragmentation can result in differential loss of predators, reducing their regulatory impact on the reservoir host populations (Section 7.3.3).

8.5.1.3 Urban sprawl

In less-developed countries, urban sprawl can take the form of so-called shanty towns, where humans coexist at high densities with domestic animals and disease vectors in conditions of poverty and poor sanitation. Childs *et al.* (2007b) caution that such environments could be potential breeding grounds for zoonotic EIDs, with high human–animal or vector contact rates, allowing many potential spillover events. The demographics of urban sprawl have been implicated in the emergence or re-emergence of several diseases, including cholera, dengue fever, Japanese encephalitis and human tuberculosis (Childs *et al.*, 2007b).

When spillover occurs, high local densities of people provide ideal conditions for human–human transmission and potentially long transmission chains through which pathogen adaptation to its novel host may occur (Section 8.2.2). Hence pathogens spilling over in human population centres run a higher risk of evolving to a status capable of pandemic emergence, compared to those spilling over in small, isolated village communities. In addition, larger host populations are vulnerable to persistence of a greater range of pathogens, because they will meet the criterion of threshold population size for parasite establishment (N_T; Box 1.2) in more cases. Several modern human diseases such as measles and rubella have characteristics requiring large threshold host populations for establishment ($N_T > 100\,000$); their emergence was only possible once human populations reached these thresholds following the agricultural revolution (Dobson & Carper, 1996; Diamond, 1997).

8.5.1.4 Farming practices

Farming practices can influence disease emergence in a variety of ways. For plant diseases in particular, the modern agricultural practice of monoculture results in high-density host populations with reduced genetic variance that may have greater susceptibility to pathogens (Section 8.3) and are above the critical size for the establishment of density-dependent pathogens (Section 5.1.2). In addition, monocultures do not benefit from the buffering effect of diversity on disease spread (i.e. there is little dilution effect: see Section 7.3.2). Modern farm feed practices (the use of animal parts in animal feeds) were key to the emergence of bovine spongiform encephalitis (BSE) in the United Kingdom (Pattison, 1998). Irrigation systems put in place for crop maintenance may increase arthropod vector breeding grounds; on the other hand, draining of wetlands may reduce vector habitat. Both patterns have been implicated in the changing distribution of human malaria (Box 8.7). Domestic stock raised in large, high-density populations provide good opportunities for disease amplification; these populations may act as reservoirs of infection to wildlife (as in the case of rinderpest in ungulates, discussed in Section 7.2.1) or as amplifiers of zoonoses to man (as is the case for novel influenza strains (Box 8.3) and early outbreaks of Nipah virus (Box 8.5)).

In developing countries, movement of farming operations into forest areas influences contact rates with exotic species through its effects on habitat destruction (see above). In some cases, forest animals may be attracted into farmlands. In Malaysia, for example, there has been a large increase in pig farming over the last decade. Fruit orchards are often planted next to pig enclosures, with damaged fruit used as a dietary supplement for the pigs. During the fruiting season, fruit bats (*Pteropus* spp.) are attracted from neighbouring forest to the orchards, bringing them into closer contact with pigs and man. This process led to the emergence of Nipah virus, an EID of pigs and man; Hendra virus, an EID of horses and man, emerged through a similar process (Box 8.5). Control for Hendra virus, for which there has been no human–human transmission, focuses on reducing the contact between the bat reservoir species and horses, for instance, by siting stables away from orchards or grazing horses away from trees in flower or fruit (Field *et al.*, 2007). Similar strategies may help to minimise spillover risk of Nipah virus to pigs, but highly efficient pig–pig transmission, together with bat–human and human–human transmission require additional contact control, including disinfectant cleaning of pig enclosures, public education about the risk associated with raw fruit where bats have fed and rigorous hospital hygiene measures (WHO, http://www.who.int/mediacentre/factsheets/fs262/en).

Nipah and Hendra viruses, together with SARS (Box 8.9) and possibly Ebola and Marburg viruses (Leroy *et al.*, 2005), all utilise bats as reservoir hosts. This has led some to suggest that bats pose a particular threat as reservoirs of EIDs, although Dobson (2005) points out that since 20% of all mammals are bats, they are, in fact, under-represented as reservoir hosts of EIDs. However, some unique features of bat ecology, physiology and immunology may predispose them towards propagation and spillover of viruses (Dobson, 2005; Calisher *et al.*, 2006). For instance, bats appear to harbour a broad range of viruses but suffer little pathogenicity; the habit of chewing and discarding partially eaten food enhances opportunities for cross-species oral transmission; many species congregate in very dense feeding and breeding colonies, providing ample opportunities for intraspecific and interspecific transmission; and many species range large distances, the subpopulations mixing on a seasonal basis, providing a mechanism for widespread dissemination of viruses.

8.5.2 Trade and transport changes

Increased connectivity of populations through transport and trade can increase the rate of spread of diseases through increased contact rates between and within species. This can influence all phases of emergence from spillover to pandemic spread for wildlife, domestic animal and human EIDs. Increased trade in wildlife species enhances opportunities for introductions of exotic pests and diseases, and is implicated in the worldwide spread of chytridiomycosis, an often fatal disease of amphibians with global conservation implications (Box 8.6). Disease vectors have also been disseminated by human–assisted transport, as evidenced by the introduction of the mosquito *Anopheles gambiae* to Brazil in the 1930s, which led to an increase in the incidence of malaria, resulting in a estimated loss of 16 000 lives. Combined transport and climatic mapping reveals a close association between the worldwide distribution of the mosquito *Aedes albopictus* (a vector of dengue, yellow fever and WNV) and traffic volumes to ports with favourable conditions for vector establishment (Tatem *et al.*, 2006). Timber trading is thought to have brought Dutch elm disease *Ophiostoma ulmi* and *O. novo-ulmi* and its beetle vector to the United Kingdom in the

Box 8.6 Chytridiomycosis in amphibians

Chytridiomycosis is an EID of amphibians, caused by the fungus *Batra-chochytrium dendrobatidis (Bd)*, first reported in 1998 (Berger *et al.*, 1998). The fungus infects the epidermis of frogs and toads, causing thickening and lesions which interfere with the skin's physiological function in water balance and respiration; it is also linked to reproductive anomalies. Infection with *Bd* has been reported for hundreds of species, and there appears to be wide variation in pathogenicity, with up to 100% fatality in many species but little impact on others; there is also variation in pathogenicity for different *Bd* isolates (reviewed in Rosenblum *et al.*, 2010). Over the last decade, reports have amassed of *Bd* in natural populations across the globe, including Europe, the Americas, Australia and Indonesia (http://www.spatialepidemiology.net/bd/maps).

Over the last 30 years, there has been a dramatic worldwide decline in anuran biodiversity, with 34 known extinctions and 43% of all amphib-ian species experiencing severe declines (Stuart *et al.*, 2004). Chytridio-mycosis appears to be one of several factors responsible for the declines; habitat loss, overhunting for the food and pet trades, pollution and climate change are other principle drivers (Brook *et al.*, 2008). Recently, the extinction of the sharp-snouted day frog (*Taudactylus acutirostris*) has been attributed solely to *Bd* (Schloegel *et al.*, 2006); this is quite excep-tional, as whilst other cases of disease-induced extinction are known (Section 8.4), there are usually other factors involved.

Global warming versus invasive spread. Causes for the rise of chytridio-mycosis remain the subject of controversy (Rohr *et al.*, 2008). One hypothesis is that climate change has facilitated the spread of *Bd* or has resulted in greater pathogenesis owing to its combined impact on the distribution and life history parameters for frogs and/or fungus. Under this scenario, *Bd* may exist in a benign form and becomes pathogenic under particular environmental conditions. The climate-linked hypoth-esis is supported by strong correlations between temperature increase and declines in harlequin frog (genus *Atelopus*) diversity in Central and South America (Pounds *et al.*, 2006). However, the data are also consistent with the spread of *Bd* in these regions as an invasive pathogen. By mapping reports of *Bd* and species decline dates, Lips *et al.* (2008) make a convincing case for a wave-like spread of *Bd* consistent with a novel pathogen moving into naïve populations, following a small number of introduction events. Long-term sampling of yellow-legged frogs (*Rana muscosa* and *Rana sierrae*) in California suggests a wave-like

Box 8.6 (*continued*)

pattern of spread within lakes, with most populations becoming infected within three years of *Bd* arrival (Vredenburg *et al.*, 2010). This research also reveals that frogs do not die from the disease until a threshold intensity of infection (about 10 000 fungal zoospores per frog) is reached. This could explain the rapid spread and sudden collapse of some populations (Fig. 8.11; Briggs *et al.*, 2010): *Bd* does not 'burn out' like other virulent pathogens through lack of infectious individuals, because infected frogs carry on transmitting the infection until spore production reaches lethal levels. High-density populations will build up lethal doses more quickly, suggesting that one avenue for conservation may be to reduce the population density of susceptible frogs (by capture or antifungal treatment (Lubick, 2010)) before they succumb to the wave of infection.

Global trade. The rapid global spread of *Bd* since its presumed emergence between 1960 and 1980 leads one to ask how it has been distributed. International trade in the African clawed frog (*Xenopus laevis*) for human pregnancy tests in the 1950s was originally implicated (Ouellet *et al.*, 2005), but genetic analyses of *Bd* strains now suggest that *Xenopus* from Africa were not the source of worldwide introduction. Anthropogenic introduction is nevertheless implicated in the spread of *Bd*. Mapping estimates suggest that *Bd* can spread at anything from 1.1 km to 282 km per year, depending on scale, with the slowest rates reported for local studies, intermediate rates at regional scales and the fastest rates at intercontinental scales (Lips *et al.*, 2008). This pattern strongly suggests that different processes operate at different scales, with human-facilitated transport implicated between continents. A staggering 10 000 tonnes of frog parts are traded globally each year, mostly to the United States and European food industries (Warkentin *et al.*, 2009), with substantial trade in live specimens for the exotic pets industry (more than one million animals per annum to the United States, for instance (Smith *et al.*, 2009c)).

Reservoir frogs. Among the species now implicated in the spread of *Bd* is the American bullfrog, *Rana catesbeiana*, which is farmed commercially for the global food industry. This species suffers little pathogenic effect from chytridiomycosis, but can transmit the disease; frog farming is widespread in Ecuador and Brazil and *Bd* infection on farms has been reported (Hanselmann *et al.*, 2004). These farms export principally to North America; in addition to their potential role in long-distance dispersal, infected farms may act as a local source of disease to threatened indigenous species.

Box 8.6 (*continued*)

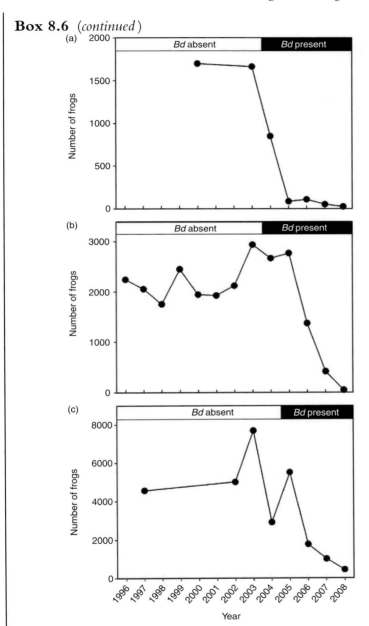

Fig. 8.11. Estimated numbers of adult and subadult frogs in three metapopulations in the United States before and after the detection of *Batrachochytrium dendrobatidis*. Dramatic population declines occurred at all three locations, but the pattern of decline varies. From Vredenburg *et al.* (2010).

1950s, and also introduced chestnut blight *Cryphonectria parasitica* to the United States (Loo, 2009). Modern farming practices in the United Kingdom, including rapid long-distance trading of cattle, pigs and sheep between farms, facilitated the early spread of foot and mouth disease in the 2001 UK outbreak (http://archive.cabinetoffice.gov.uk/fmd/fmd_report). For human EIDs, air travel was involved in the rapid pandemic spread of HIV and SARS; river trade routes also facilitated the early persistence and spread of HIV (Box 8.2). The practice of trading many different wildlife and farmed species at live animal markets (wetmarkets) is implicated in the spillover of SARS from bats to palm civets and humans (Box 8.9). Several authors highlight the increasing trade in bushmeat as a high-risk activity likely to lead to further human EIDs (Wolfe *et al.*, 2005; Daszak *et al.*, 2007). The trade in bushmeat continues to grow, despite diminishing stocks. Many indigenous people in Africa and Asia prefer wild animal to domestic animal meat, but in Asia it has become a luxury product mostly eaten by the urban affluent. Up to 4.5 million tonnes of bushmeat (mostly primates) are extracted from the Congo basin each year (Fa *et al.*, 2002). Spillover of HIV types 1 and 2 were almost certainly the result of (traditional, rural) bushmeat consumption or handling (Box 8.2). Outbreaks of Ebola in humans have also been attributed to consumption or handling of great apes (Leroy *et al.*, 2004).

8.5.3 Climate change and emerging diseases

Biodiversity is higher in the tropics and this pattern also holds for infectious diseases of humans (Guernier *et al.*, 2004). Vector-borne diseases in particular, such as malaria, shistosomiasis and dengue fever are more prevalent in tropical and subtropical zones (Ostfeld, 2009b). Climate change can, by altering host distributions or parasite transmission dynamics, allow diseases to emerge without the need for evolutionary changes in the ability of a parasite to use the host (Brooks & Hoberg, 2007). Since the 1990s there has been a surge of interest in the likely impact of climate change on the distribution and prevalence of human diseases, with many authors predicting that global warming will lead to an expansion in the range of tropical diseases, particularly vector-borne diseases (Epstein, 2000; Harvell *et al.*, 2002).

However, more recently a series of articles in *Ecology* (volume 90) have considered the ecological impacts of climate change on disease distribution. These find that neither long-term datasets nor mathematical models show an expansion in the range or incidence of vector-borne diseases. Lafferty and colleagues discuss three main explanations. Although disease transmission is affected by climate, other factors such as host immunity, competition and predation will affect the suitability of a habitat. In addition, anthropogenic activities, in particular land use and vector and disease control, may alter disease distribution independently of climate. This is illustrated by Hay *et al.* (2004), who map malaria distribution over the twentieth century, showing a strong correlation between disease control and distribution (Box 8.7). Finally, vector and disease distributions may be affected by the maximum and minimum temperatures at which an organism can survive. Hence, more recent models predict that, rather than leading to a net expansion in infectious diseases, climate change is likely to lead to a range shift in disease (Lafferty, 2009).

Although earlier predictions of a massive increase in infectious diseases under global warming now appear to be unrealistic, changes in distribution may still have dramatic health and economic effects. Indeed, the current human distribution in tropical areas results in part from disease, with higher population densities in areas of less disease. Hence the emergence of novel diseases in these populous areas would affect high numbers of people who, lacking previous exposure, are likely to have little immunity to these diseases (Box 8.7).

8.5.3.1 *The relative importance of different environmental drivers*

From the review in Box 8.7, it seems apparent that changes in land use are involved in the emergence of many diseases, often in the early spillover phase or because of pathogen amplification in a farmed species. Transport and trade changes are frequently implicated in the spread of EIDs during the persistence or pandemic phases, but may also play a part in the dynamics of spillover. Climate change can potentially have an impact on all of the stages of emergence, although its relative importance in comparison with these other factors remains to be elucidated.

We certainly cannot rule out climate change as a driving force of disease emergence, as seasonal weather patterns are known to influence infection prevalence in many one-host–one-parasite systems (reviewed in Altizer *et al.*, 2006). Such 'seasonal forcing' may be particularly important in multi-host systems at the spillover stage of emergence, where changes in the weather affect vector abundance or resource availability, influencing contact rates between reservoir and susceptible host species. For example, Ebola outbreaks in primates occur at the end

Box 8.7 Malaria and climate change

Malaria in humans is caused by five species of the protist parasite *Plasmodium*. It is a vector-borne parasite, transmitted by mosquitoes of the genus *Anopheles*. The disease currently affects tropical and sub-tropical regions and is a serious problem in Africa. The WHO estimates that 50% of the world's population live in areas where they are at risk of malaria and that there are 250 million cases of malaria each year, leading to about one million fatalities. In Africa, 20% of childhood deaths are a result of malaria. Malaria is linked with economic development, with feed back between disease and the economic growth. One perspective holds that the economic growth of poorer countries is kept in check by the disease (Gallup & Sachs, 2001; Sachs & Malaney, 2002); an alternative view is that poorer countries have agricultural practices that encourage malaria (or its vectors) and less resources to control it (e.g. Sachs & Malaney, 2002).

Temperature and malaria emergence. The maintenance of malaria in a geographical region is strongly influenced by temperature, which affects the life cycle of both *Plasmodium* and its vector. Sustained malaria outbreaks vectored by *Anopheles* mosquitoes occur when temperatures are regularly greater than 15°C (Epstein, 2000). Temperature affects the development time of the parasite in the vector. Diurnal fluctuations are also important and may mask or exacerbate the effect of temperature trends on predictions for epidemiology (Paaijmans *et al.*, 2009). Host distribution and development times are also affected by humidity and temperature, leading to complex tradeoffs between *Plasmodium* and host development, which make it difficult to predict the relationship between transmission and temperature. Finally, higher temperatures result in increased biting persistence by the mosquito vector, increasing opportunities for malaria transmission.

Climate change. Since the 1990s there has been widespread interest in the impact of climate change on infectious disease. Harvell *et al.* (2002) predicted that climate warming should have a particularly strong effect on parasites, such as malaria, that affect poikilothermic hosts as warming should increase the habitat and activity of these hosts. Many studies have warned that warming will result in an expansion in the range of tropical disease such as malaria, so that they spread into more temperate areas (e.g. Martens *et al.*, 1997; 1999; Epstein, 2000). However, the effects of global warming on malaria distribution are not simple to predict. Randolph (2009)

Box 8.7 (*continued*)

points out that mosquitoes (and other vectors of disease) are as vulnerable to the damaging effects of climate change as those arthropods that deliver ecosystem services such as pollination. For example, increased flooding under climate change might increase malaria if stagnant pools are created where mosquitoes breed, but flooding might wash away breeding pools (Epstein, 2000).

Range shifts. Lafferty (2009) reviewed recent models predicting the range of infectious diseases, including malaria, and found that rather than a net increase in the range of disease, most models predicted a shift in the range of malaria, with an expansion into more temperate zones, but a loss of malaria in regions that are predicted to become hotter and drier. For example, Rogers and Randolph (2000) modelled current and future malaria distribution in response to rainfall and saturation vapour pressure as well as temperature, based on climate change scenarios predicted by the UK Hadley Centre for Climate Prediction and Research. In contrast to models that considered only temperature, they did not predict a large increase in malaria, but small shifts in the distribution. Small net changes were predicted; with less than 1% increase in the number of people at risk under the medium–high climate change scenario, and less than 1% decrease under high climate change.

Altitudinal patterns. Malaria is more prevalent in lowland areas of Africa, as lower minimum temperatures disrupt transmission at higher altitudes. During the 1980s, a resurgence of malaria occurred in the highlands of East Africa with an increase in prevalence and mortality. Several authors have attributed this increase to global warming over the twentieth century (e.g. Epstein *et al.*, 1998). Pascual *et al.* (2006) found that the temperature records for these regions showed significant warming of 0.5°C in the 1980s and 1990s. They went on to model the biological impact of such a temperature change, predicting that the effect of temperature was magnified ten-fold by changes in the mosquito population dynamics, which will in turn increase malaria abundance in these areas.

Consequences of range shift. Range shifts will result in an increase in malaria at higher altitudes as global warming expands the habitat and/ or season for malaria, and reduction of malaria from more arid areas (Lafferty, 2009). However, any shift in distribution to a novel area is unlikely to be simply cancelled out by a decrease elsewhere, because newly exposed populations may be larger or smaller, and mortality is

Box 8.7 (*continued*)

likely to be higher in naïve populations. In Africa and South America, population densities are higher at high altitudes, in part because humans have migrated to less malarious areas, with population growth and economic growth higher in such areas. Pascual and Bouma (2009) caution that a shift in the distribution of malaria to affect more highland areas could therefore have dramatic consequences, with large numbers of people affected, and malaria transmission facilitated by high population densities. There is evidence that, in India, malaria has become prevalent in regions 1500–2000 m above sea level which were previously malaria free (McMichael *et al.*, 2000). Furthermore, Pascual and Bouma (2009) point out that the predicted decrease in malaria in more arid areas (Lafferty, 2009) may not occur if agricultural development in these areas results in dams and irrigation that provide mosquito breeding sites. In addition, emergence of malaria into new geographical areas is likely to affect populations that have had little prior exposure to malaria, leading to higher mortality (Dobson, 2009). This is because individuals exposed to malaria exhibit temporary immunity with frequent re-exposure required for the host to mount an effective immune response. In addition, populations exposed over a longer timescale have evolved adaptations to the disease, such as the sickle cell trait.

Other factors affecting malaria distribution. Examination of historical patterns of malaria distribution reveals that the population dynamics do not depend solely on climate, but are strongly affected by other anthropogenic factors, in particular land use and strategies for vector and disease control (Lafferty, 2009; Ostfeld, 2009b). For example, malaria used to be endemic in England in medieval times (Lafferty, 2009) and historical reports indicate that the disease did not disappear as a result of cold temperatures during the little ice age of the sixteenth century, but declined as a result of agriculture and the draining of wetlands during the nineteenth century. Hay *et al.* (2004) mapped changes in the global distribution of malaria over the twentieth century. They concluded that reduction in malaria resulted from elimination of mosquito breeding sites in the early twentieth century, followed by the introduction of insecticide control in the 1940s which was combined with chemoprophylaxis during the WHO malaria eradication programme in the 1950s and 1960s. These control strategies led to a dramatic decrease (from 53% to 27% of the earth's

Box 8.7 (*continued*)

terrestrial surface) during the twentieth century in the geographical areas where malaria is endemic. However, it is interesting to note that the decreases were most marked at the latitudinal extremes of the malaria range, rather than in regions of holoendemic malaria (although countries at lower latitudes tend also to be the more developed). Furthermore, Hay *et al.* (2004) found little reduction in the geographic distribution of malaria between 1994 to the present day, whilst demographic changes have in fact increased the total numbers of people living in areas of malaria transmission.

Conclusion. Patterns of malaria distribution are a result of a combination of interacting factors, including climate change, control strategies and socioeconomic factors. Although earlier predictions of a massive increase in the range of malaria appear to be unfounded, even small shifts in the range of this disease (or its vectors) could have health and economic implications for large numbers of people and present a challenge for control.

of the dry season when resources are scarce and susceptible species feed together with fruit bats (the likely reservoir for Ebola) on the limited fruiting trees (Pinzon *et al.*, 2004). In the United States, outbreaks of hantavirus infection in humans are triggered by El-Nino-associated prior increases in precipitation that enhance resource availability for the rodent reservoir hosts (Yates *et al.*, 2002). A similar 'trophic cascade' model (Fig. 8.12) linking weather to increases in the population densities of small rodents and fleas (the vector) has been used to explain outbreaks of plague (*Yersinia pestis*) in humans (Enscore *et al.*, 2002) and prairie dogs (*Cynomys ludovicianus*) in the United States (Collinge *et al.*, 2005).

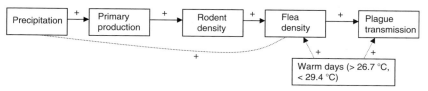

Fig. 8.12. Enscore *et al.*'s 'trophic cascade' model for plague transmission. Statistical analysis of climate and human plague incidence patterns for the southwestern United States suggest that rainfall directly affects primary production, but negatively impacts flea populations; warm weather increases flea populations and biting activity (but extremes of hot weather are associated with reduced flea populations). Adapted from Ray and Collinge (2006).

8.6 Conservation and control

Recent initiatives such as Conservation Medicine and One Health (Table 8.1) recognise the threats posed by emerging diseases to human and animal health. In this section we explore how conservation and control measures can be used to limit the impact of EIDs through monitoring and surveillance (Section 8.6.1) to identify outbreaks, contact control (Section 8.6.2) to reduce spillover or spread, and responsive vaccination once an outbreak has occurred (Section 8.6.3).

Recent reviews and meta-analyses suggest that EIDs in general and zoonoses from wildlife in particular are a growing threat to human health and the global economy (Taylor *et al.*, 2001; Childs *et al.*, 2007a; Jones *et al.*, 2008). The risk of emergence is likely to differ depending on geographic location and the types of disease involved. Jones *et al.* (2008) mapped the relative risk of emergence events for human diseases and identified several 'hotspots' at high risk of disease emergence. Emergence of drug-resistant variants, most likely centred around hospitals, clustered in Europe and the United States; zoonotic diseases from wildlife and vector-borne diseases were more likely to emerge in lower latitude countries, with strong activity in Southeast Asia and sub-Saharan Africa, but also surprisingly strong activity in northern Europe for the zoonoses; zoonoses from domestic animals were also more likely to emerge from northern Europe.

Risks to wildlife from EIDs are also likely to increase, for several reasons (Section 8.4). Firstly, environmental factors interact with disease to drive population declines and extinctions (Smith *et al.*, 2009b); as anthropogenically influenced environmental change continues or increases, the negative impacts of these interactions will probably become increasingly apparent (see chytridiomycosis in amphibians for a possible example – Box 8.6). Secondly, threatened species with small populations are at increasing risk of generalist parasite spillover from larger reservoir populations; many specialist parasites cannot be supported below a minimum host population size, but this threshold will not apply for multi-host parasites (see diseases in wild canids and felids, for example – Box 8.10). Thirdly, declining populations are at risk of losing genetic variation for resistance, making them more susceptible to novel diseases for which immunity is limited. This may explain the dramatic spread of devil facial tumour, an emerging infectious cancer of Tasmanian devils (*Sarcophilus harrisii*), which potentially threatens this species with extinction (McCallum *et al.*, 2007). Conversely, diseases may drive loss of genetic diversity: the outbreak of canine parvovirus (CPV) in grey wolves, Isle Royal (United States) appears to have caused a genetic bottleneck, resetting wolf population density to a lower equilibrium, despite the absence of CPV for over 20 years (Box 3.6).

8.6.1 Monitoring

Ideally, we would like to be able to identify the diseases that present the greatest threat before they emerge. Whilst meta-analyses give some indication of the types of disease that present the greatest threat for particular taxonomic groups (Section 8.4.1), a more targeted risk analysis would be desirable. For instance, can we use the outbreak history of a newly emerging disease (for instance, a zoonosis with poor intraspecific transmission in humans) to predict its likelihood of pandemic emergence? This question was addressed by Arinaminpathy and McLean (2009), who used branching process models to examine whether signs of ongoing adaptation can be signalled by previous outbreaks. They found that the frequency of outbreak reports gives some indication of the potential for a disease to become persistent or pandemic in the novel host (that is, whether it evolves an $R_{N0} > 1$ in the novel host – see Section 8.2.2), but cannot be used to rule out this scenario (Fig. 8.13). Hence, if a pathogen has caused hundreds of cases through separate spillover events (highly pathogenic H1N5 avian flu, for example), this indicates that it is unlikely to cross the species barrier; however, it *may* still do so. The likelihood of emergence will also depend on how mutations combine to influence R_{N0}; if mutations act additively, each will influence R_{N0} to some degree, making emergence more likely and easier to detect. However, if mutations interact so that most do not influence R_{N0} until a threshold number of mutations have occurred (as in Fig. 8.4b and 8.13), time to emergence will tend to be longer and imminent emergence harder to detect.

The realisation that we cannot necessarily predict which diseases will emerge, or when, highlights the need to develop monitoring approaches that can detect disease emergence in its early stages. Examples include web-based forums such as the Emerging Health Threats forum (http://www.eht-forum.org), and event-based biosurveillance, in which different sources of data, including media reports and internet interactions, are monitored (in some cases, with automated protocols) for evidence of emerging health threats. Developments in spatial epidemiology (Ostfeld et al., 2005) have also contributed to state-of-the-art mapping procedures enabling prediction of disease ranges, tracking of disease spread and test of

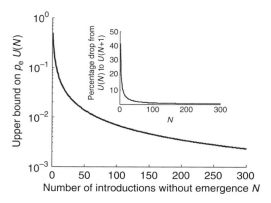

Fig. 8.13. Predicting large-scale emergence ($R_{N0} > 1$) from previous outbreaks. The main graph plots the upper bound $U(N)$ on the probability of pandemic emergence (P_e), given that N spillovers have occurred without emergence. The probability of emergence drops rapidly with the first 30 or so spillovers, but is still non-zero even after hundreds of events. Inset plots the relative improvement in information that each new spillover provides (measured as the percentage drop in the upper bound between $U(N)$ and $U(N+1)$). Initially, each new event improves predictive power, but after many introductions, little extra information is gained. From Arinaminpathy and McLean (2009).

hypotheses for disease range change (Tatem *et al.*, 2006; see also malaria (Box 8.7) and chytridiomycosis in amphibians (Box 8.6) for examples).

The arrival of West Nile virus (WNV) in the United States in 1999 occurred in an era of improving data capture and presentation. Its increasing incidence was recorded on one of the earliest events-based surveillance tools, ProMED (Program for Monitoring Emerging Diseases; http://www.promedmail.org). This open-access programme posts daily reports of human, animal and plant disease outbreaks from media outlets, office and personal subscribers, which are then screened and researched by expert staff. The US Centers for Disease Control and Prevention collate annual figures and publish maps of many human diseases, including WNV incidence in mosquitoes, birds and humans, capturing the dramatic spread of this pathogen (Box 8.8). The US Geological Survey (USGS: http://diseasemaps.usgs.gov) publish weekly case maps for vector-borne EIDs in the United States at the county level. Coordinating data collection on this scale clearly requires a strong health monitoring and control infrastructure already in place; in the United States this is the remit of several offices within the CDC. This system includes the reporting of listed notifiable diseases which, in the United

States, was established from 1878 when the forerunner of the public health service began collating reports of cholera, yellow fever, smallpox and plague mortalities from consuls overseas. In the case of WNV, the rapid availability of data facilitated design of control and public awareness campaigns, and contributed to their success. These and other databases have proved valuable for investigating relationships between disease spread and biodiversity, and for examining the conservation implications of WNV (Box 8.8).

Community ecological-motivated studies, such as those described for WNV (Box 8.8) and Lyme disease (Box 7.1), provide evidence of a link between biodiversity and disease risk. This has led to renewed calls for inclusion of disease buffering in the list of ecosystem services provided by

---

Box 8.8 West Nile virus

WNV is a flavivirus transmitted by mosquitoes feeding on infected birds. A variety of species (mainly *Culex* genus) transmit the virus to a broad range of bird species. Human infection by WNV was first reported in 1937 from Uganda (formerly West Nile district). The native range of WNV includes Africa and the Middle East, with occasional outbreaks in Europe. WNV received little media attention until 1999 when there was an outbreak in the city of New York; in the following few years, WNV spread dramatically throughout the United States. In humans, the disease is often asymptomatic, but can cause flu-like symptoms and fever, or more serious neurological symptoms (encephalitis and meningitis). In the United States, roughly 4% of reported cases are fatal, but many mild or asymptomatic cases are probably not reported.

Vectors and amplifiers. Many species of mosquito can transmit WNV, but they vary in their competence as vectors; generally, a small number of species are the principal vectors in any given region (Hayes *et al.*, 2005; Reiter, 2010). Birds are the amplifying hosts of WNV; many species can become infected, but vary in their competence as amplifying hosts. In the United States, all of the 25 bird species tested from 17 families were demonstrated as susceptible in laboratory infection trials (Komar *et al.*, 2003). The strength and duration of viraemia varied: song birds (Passeriformes), shorebirds (Charadriiformes), owls (Strigiformes) and hawks (Falconiformes) were the most competent orders; within these, certain passerines

Box 8.8 (*continued*)

(including various crows, sparrows and finches) were particularly viraemic. Pathogenicity also varied between bird species; those of Old World origin were generally asymptomatic, whereas many New World species suffered considerable morbidity and mortality; up to 100% for the most susceptible species, the American crow (*Corvus brachyrhynchos*) (Komar *et al.*, 2003).

Other hosts and transmission. Humans are generally a 'dead-end' host as the viraemia produced is insufficient to infect biting mosquitoes. However, the role of alternative transmission routes requires further evaluation. Vertical transmission has been documented in humans and mosquitoes, and direct transmission between birds and alligators has been reported or suspected in farm and laboratory settings (Hayes *et al.*, 2005). Oral transmission may be important in natural populations, particularly for species feeding on the carcasses of infected birds that then go on to transmit the infection to offspring via meal regurgitation (Hartemink *et al.*, 2007; Reiter, 2010). Infection has been recorded in many other vertebrates, including cats, dogs, bats, rodents, reptiles and amphibians (Marra *et al.*, 2004; Hayes *et al.*, 2005). In horses, the infection can cause serious neurological disease with up to 40% mortality of those with symptomatic encephalitis; a WNV vaccine is available for horses in the United States (CDC: http://www.cdc.gov.ncidod/dvbid/westnile/birds& mammals.htm).

Emergence in the United States. The first indications of WNV in the United States came from reports of large numbers of dead and dying birds, together with a cluster of human cases in the vicinity of the Bronx Zoo, New York in the summer of 1999. It is unclear how WNV was transported to the United States; accidental shipment of an infected mosquito or bird is most likely. In subsequent years WNV spread west, south and north in a wave-like pattern, with recorded cases peaking in succession in neighbouring states as each was invaded (Fig. 8.14). Peak national incidence occurred in 2003, when 9862 cases were reported, with foci in the midwest (264 fatalities); by 2004 human WNV cases peaked in many east coast states; by the end of 2006 every state except Hawaii and Alaska had reported human cases. To date there have been nearly 30 000 reported cases, with 3.9% of these fatal (1999–2009: 29 624 cases, 1161 deaths; CDC: http://www.cdc.gov.ncidod/dvbid/westnile/surv&control.htm#maps).

Box 8.8 (continued)

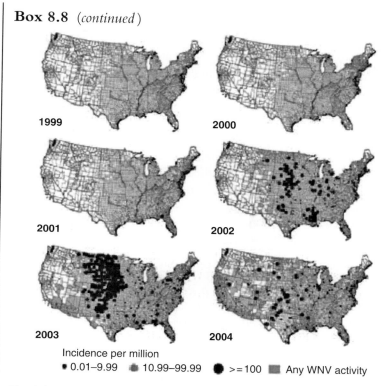

Fig. 8.14. WNV incidence 1999–2004. Incidence of neuroinvasive WNV reported to US Centers for Disease Control and Prevention. From Hayes *et al.* (2005).

Monitoring and public awareness. As the wave of infection spread, a control strategy was rapidly implemented, focusing on public awareness, personal protection against mosquito bites ('Fight the Bite' campaign; CDC: http://www.cdc.gov.ncidod/dvbid/westnile/index.htm), roadside insecticidal fogging and reduction of larval habitat through drain and sewer maintenance. Incidence in the United States has declined since the 2003 peak, although WNV remains a serious healthcare issue (it is the second most prevalent vector-borne disease in the United States, after Lyme disease). Incidence may also have reduced owing to the increase in the immune class of birds and humans (vertebrates attain life-long immunity after infection (Hayes *et al.*, 2005)).

Box 8.8 (*continued*)

Conservation implications. In the early stages of the US epidemic, arrival of WNV in a new county was often heralded by reports of dead and dying birds, and the high mortality sustained by some species has led to concerns for their conservation; e.g. greater sage grouse (*Centrocercus urophasiamu*) (Naugle *et al.*, 2004), and for the effects of WNV on wildlife biodiversity (Marra *et al.*, 2004). LaDeau *et al.* (2007) examined data from the North American Breeding Bird Survey (BBS; an annual roadside bird count running since 1966; http://www.mbr-pwrc.usgs.gov/bbs) for evidence of WNV-induced population declines. Seven of the 20 species examined had suffered significant population declines in the years following the introduction of WNV; all were highly susceptible species including the American crow (*Corvus brachyrhynchos*), American robin (*Turdus migratorius*) and eastern bluebird (*Sialia sialis*). Although the populations of some have since recovered, the long-term impact on bird populations and feed back on WNV transmission is unknown. Several recent studies identify biodiversity as a buffer against WNV spread at local and national scales (Ezenwa *et al.*, 2006; 2007; Allan *et al.*, 2009). Allan *et al.* (2009) integrated data from four US surveys (the BBS for bird biodiversity; CDC for mosquito populations and WNV infection prevalence estimates; USGS for county-level human WNV incidence (http://diseasemaps.usgs.gov/wnv); and the US Census Bureau for human population data (http://www.census.gov)) to investigate prevalence patterns during the epidemic peak across the United States. Prevalence of WNV in mosquitoes and humans was negatively correlated with bird biodiversity, following a similar pattern to that found for Lyme disease (Box 7.1); similar processes, involving loss of dilution hosts and increased abundance of competent hosts in urban and suburban areas appear to operate in both cases.

WNV as a threat to Galapagos communities. The unique vertebrate communities of the Galapagos Islands may also be under threat from WNV, which has spread into Central and South America and is predicted to reach mainland Ecuador in the next few years. The wide host range of WNV means that many species of birds and reptiles unique to these islands may be at risk. Risk assessment models suggest the most likely route of WNV introduction is through accidental transfer of infected mosquitoes on airplanes (Kilpatrick *et al.*, 2006);

Box 8.8 (*continued*)

dissemination to smaller islands in the archipelago would be likely via transport of an infected mosquito on one of the many small tourist cruise boats (Bataille *et al.*, 2009a). Three mosquito species are found on the islands (two introduced and one indigenous). If WNV were to spread into the native mosquito species *Aedes taeniorhynchus*, which is found throughout the islands and feeds on a variety of bird and reptile species, there would be few if any refugia from the disease for these small island populations (Bataille *et al.*, 2009b).

biodiversity (Ostfeld, 2009a), and lends further support to conservation-based approaches (Table 8.1) to EID management (Aguirre *et al.*, 2002; MacDonald & Laurenson, 2006).

8.6.2 Contact reduction

When novel diseases emerge, often there is no vaccine or prophylaxis available. Under these circumstances, disease control relies on measures to reduce disease spread through reduction in contact rates between infectious and susceptible individuals. The initial rate of spread of an EID is determined by R_{N0} (Box 1.2); the disease is likely to spread barring stochastic extinction provided $R_{N0} > 1$ (Section 8.2.2). Control of the disease is achieved if the realised reproductive rate R_{int} (for R under intervention) is brought down to values less than 1. The realised rate can be described by various formulations; one highlighting the possible points of control being:

$$R_{int} = k\beta D, \tag{8.1}$$

where k is the contact rate per infectious individual; β is the probability that the infection is passed on when an infectious and a susceptible individual meet; and D is the duration of the infectious period. This model (from Lipsitch *et al.*, 2003) again assumes that all contacts are susceptible, as is probable for a novel disease. To reduce R_{int} below 1, we have the following options (all could be applied to human EIDs, most could be applied in some form to domestic animals; less are feasible for wildlife populations):

- *Reduce contact rate (reduce k)*: restrict contacts through isolation of symptomatic individuals and quarantine of their recent contacts;

restrict movement of individuals in general (e.g. by restricting air travel or travel between districts); voluntary reduction in contacts (staying at home, avoiding restaurants, for example; this behaviour became apparent during the 2003 SARS epidemic); reduce high-density contacts (for instance, close schools, public transport systems or workplaces; a policy of school closure was adopted by many countries in the early stages of the 2009 swine flu outbreak). Isolation of infected patients is documented in the Old Testament; the word 'quarantine' derives from legislation introduced to control the spread of plague in Europe in the fourteenth century when newly arrived ships were isolated for 40 days (*quaranta giorni*) before coming to port.

- *Reduce transmission (reduce β)*: identify the transmission route and take appropriate hygiene precautions. Isolation wards with negative pressure can prevent spread by fine-droplet-transmitted respiratory diseases, but these facilities are not available in all countries. Stringent use of sterile masks and gowns with strict hand hygiene was key to reducing transmission of SARS on hospital wards. Use of antiviral medicines may reduce pathogen load, reducing effectiveness of transmission. During the 2009 flu pandemic, the UK government used television and newspaper advertising to encourage the use and disposal of tissues to reduce droplet transmission through sneezing.

- *Reduce duration of infectiousness (reduce D)*: by removing infectious individuals from the general population (isolation), the effective period of infectiousness is ended (provided hygiene measures are sufficient to prevent transmission to carers whilst the subject is isolated); antiviral medicines can also limit the infectious period by reducing the severity of infection.

Most of these strategies are only as effective as the monitoring scheme to identify infected individuals; for human EIDs, effective public awareness campaigns of the symptoms and what to do if symptoms arise are key, in addition to an effective public health infrastructure. For animal EIDs, monitoring of populations at risk is crucial for taking prompt action (see the examples in Box 8.10). We illustrate the use of contact control measures with reference to SARS, the first human EID to reach pandemic status in the twenty-first century (Box 8.9).

Box 8.9 Community ecology of SARS

Severe acute respiratory syndrome (SARS), a previously unknown human disease, was first reported from Guangdong province, southern China in November 2002. The subsequent epidemic across five continents resulted in 774 known deaths and 8098 reported cases by the time the outbreak ceased at the end of July 2003. SARS was caused by a previously unrecognised coronavirus (now named SARS-CoV), the genome of which had been fully sequenced well before the cessation of the 2003 outbreak. Despite these rapid advances on the molecular front, the community ecology of this virus took longer to clarify.

Emergence from animal reservoirs. Early indications were of an animal origin for SARS, which was first reported in workers associated with the food trade or wetmarkets in Guangdong. These markets sell live animals for food, the pet industry and traditional medicine; traders stock a variety of species, often housed in close proximity to each other for variable periods of time until sold. Molecular studies initially identified the palm civet (*Paguma larvata*) as the most likely reservoir species for spillover to man, with viral sequences closely related to SARS-CoV isolated from palm civets at the markets associated with the first index cases. However, subsequent molecular studies suggest that three species of horseshoe bats (*Rhinolophus* spp.) are the primary reservoirs; wild-caught horseshoe bats in Hong Kong and mainland China have a high SARS-CoV antibody prevalence (Lau et al., 2005; Li et al., 2005), whereas palm civets on farms have not tested seropositive. This strongly suggests a chain of transmission events occurred in wetmarkets: spillover from bats to palm civets, which amplified the virus with symptomatic infections; then spillover from civets to man. Live bats are frequently traded at wetmarkets and the main route of transmission from bats (which remain asymptomatic) is via faecal shedding. The practice of housing and handling multiple species in wetmarkets provides many opportunities for cross-species transmission. Individuals from several other species have also been infected (Wang & Eaton, 2007), including other species associated with wetmarkets, as well as cats and pigs (probably infected from contact with humans) with the further possibility that some human cases may have become infected from domestic cats (Fig. 8.15).

Box 8.9 (*continued*)

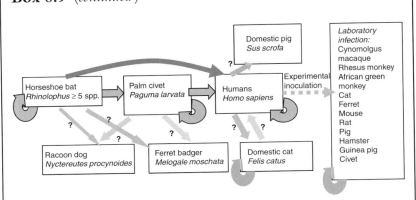

Fig. 8.15. Multi-species transmission of SARS. Arrows indicate transmission routes; it is unclear whether species at wetmarkets caught the virus from bats or palm civets (pale arrows); many species have been shown by experimental inoculation to be susceptible (dashed arrow).

The 2003 pandemic. During winter and spring 2003, a human SARS epidemic developed in mainland China, with Beijing badly hit (5327 probable cases in mainland China), causing new epidemics in Hong Kong (1755 cases), Singapore (206 cases), Toronto (246 cases), Taiwan (346) and Viet Nam (63), with sporadic outbreaks (< 30 cases) seeded through long-range travel of infected individuals to many other countries (WHO: http://who.int/csr/sars/table2004_04_21/en/index.html). Case fatality rate was recorded as 9.6%, but may have been closer to the 14–17% estimates from Singapore, Hong Kong and Toronto, which all had significant outbreaks and rapid reporting and tracing mechanisms.

Virus adaptation in civets and man. Genome sequence analysis of virus samples from early, middle and late phase human infections in the 2002–2003 epidemic show that the viral genotype evolved as the epidemic progressed. Sequences from early phase infections showed strong sequence homology to those from civets sampled at the onset of the outbreak. As the human epidemic progressed, the neutral mutation rate remained constant, but the rate of mutations in coding regions reduced in later stages, indicating rapid evolution. Similar results have been found in samples from civets; in both cases sequences associated with receptor binding and resistance to antibodies show the strongest signal of evolution, suggesting that the

Box 8.9 (*continued*)

virus was adapting differently to each of its novel hosts (reviewed in Wang & Eaton, 2007).

Further spillover risk. Epidemiological and molecular studies suggest that the 2002–2003 outbreak was the result of several independent spillover events from bats to civets to man. There was a further outbreak in Guangdong province, confined to four human cases with no secondary transmission and no mortality, in December 2003–January 2004. Civet trading in wetmarkets had been banned during the 2002–2003 outbreak, but the ban was lifted in August 2003, suggesting that further sporadic spillovers from bats to civets to man occurred (WHO: http://who.int/csr/don/archive/disease/severe_acute_respiratory_syndrome/en). Because of the practices inherent in wetmarkets, they remain a potential hotspot for the emergence of novel zoonotic EIDs, or for the re-emergence of SARS variants.

8.6.2.1 *Control strategies for SARS*

A global strategy to contain SARS was instigated by the WHO, emphasising rapid identification of SARS cases and their contacts. The chief mode of transmission in humans occurred during close contact via large-droplet respiratory secretions. Suspected SARS cases were isolated and treated in hospitals; contacts of SARS were quarantined at home and monitored for symptoms. Initially, sufficiently strict hygiene regulations were not always observed, leading to a large proportion of the secondary infections centred around hospitals (Lloyd-Smith *et al.*, 2003; Gumel *et al.*, 2004). The resources available for isolation, treatment and monitoring vary from country to country and ideals of personal freedom versus public health also vary, complicating health policy formulation and implementation.

Rapid reporting networks such as that provided by the WHO and proMED, together with contact tracing data, enabled the rapid analysis of mathematical models for disease spread and efficacy of control measures following the epidemics in Hong Kong (Riley *et al.*, 2003), Singapore (Lipsitch *et al.*, 2003) and Toronto (Gumel *et al.*, 2004). The models, based on a SEIR template (individuals transition through four states: susceptible, exposed, infectious, recovered; Box 1.4 and Fig. 8.16)

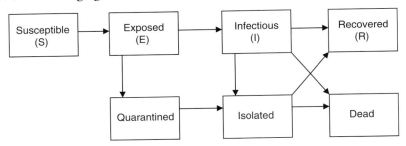

Fig. 8.16. Flowchart of SEIR model, with isolation and quarantine. Variants of this format were used for many of the mathematical studies of SARS. Quarantine refers to removal of exposed individuals from contact with the population; symptomatic individuals are isolated.

produced similar estimates for R_0 for SARS CoV in the range of 2.2–3.7. These moderate values suggested that control through contact reduction was feasible, but in the absence of control, a SARS pandemic of 'catastrophic' proportions (Riley *et al.*, 2003) would develop.

The models demonstrate that a combination of strategies could be used effectively to bring SARS under control (Lipsitch *et al.*, 2003; Riley *et al.*, 2003). For instance, reduction by two days in the hospital admission time for new cases alone was insufficient to control spread in Hong Kong, but this combined with halving the population contact rate and hospital infection rate would bring an R_0 of 2.7 to an R_{int} below 1, rapidly curtailing the epidemic (Riley *et al.*, 2003). However, combined strategies can have non-linear payoffs, such that it is more efficient to focus on one control strategy rather than dividing resources between two. This pattern is found in the relationship between isolation and quarantine; efficient isolation (or quarantine) was found to be more effective at reducing spread than a partial combination of both (Gumel *et al.*, 2004).

The timing of action is also crucial for control; the general rule here being one of prompt action (Anderson & May, 1991). If cases can be detected and isolated early in an outbreak, there will be less secondary cases to control. Treatment, isolation and quarantine facilities become increasingly stretched as an epidemic grows, with lapses in hygiene and contact control more likely. In addition, keeping an outbreak very small maximises the chances that it will stutter to extinction through stochastic (chance) events, and minimises the likelihood that a superspreading event will occur (Section 8.2.3). Efficient and early quarantine of the contacts of an infected case actually reduces the cost of management (in

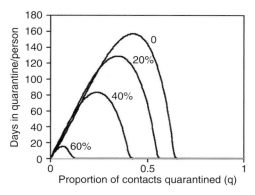

Fig. 8.17. Cost-effectiveness of quarantine scenarios. The total number of days spent in quarantine per person is plotted for given levels of effective quarantine. For more effective quarantine, the number of days spent in quarantine is reduced. From Lipsitch *et al.* (2003).

terms of total days in quarantine per person) because it eliminates the contribution that these individuals would otherwise make to R_{int}; either as susceptibles that may catch the infection or as infectious individuals if they are already incubating the disease (Lipsitch *et al.*, 2003; Fig. 8.17). Screening at points of entry such as airports and district or national borders can be effective in the early stages of disease spread, preventing the initiation of new foci of disease. However, once new centres of infection become seeded, such screening contributes little to overall reduction in caseload (Gumel *et al.*, 2004).

Vigilance must also be maintained during the tail end of an epidemic, however. For SARS, voluntary behavioural changes (reduced mixing and contact at restaurants and sports venues, for instance) were reported in the affected cities, but their relative influence can be hard to measure. As an epidemic wanes, prior patterns of behaviour may be resumed, making observance of other measures (such as hospital hygiene and contact tracing) even more critical. In addition, because disease progression to recovery or death takes time, most individuals in a decaying epidemic are in hospital, making continued hygiene observance essential (Riley *et al.*, 2003).

Lloyd-Smith *et al.* (2003) developed a compartment model (based on matrix approaches – Box 7.5) to examine the dynamics of infection at the local scale of a hospital and its surrounding community. They modelled the different components of transmission (with elements for within-hospital, within-community and hospital–community transmission, for instance). This approach enabled examination of the efficacy of different healthcare regimes on the various components and dynamics overall. Within–community transmission is effectively reduced by isolation and quarantine, provided isolation occurs early in an individual's infectious period. As the cases become concentrated in hospital, healthcare workers are exposed to infection risk and within-hospital transmission increases. Modern isolation wards and hygiene measures are key to reducing spread to front-line healthcare workers, but just as important is strict observance of basic hygiene regulations within hospitals to prevent spread between healthcare workers. Another important component is hospital–community transmission, which occurs when off-duty healthcare workers incubating the disease re-enter the community. Voluntary quarantine (irrespective of infection status) can effectively eliminate this component; if community cases are brought quickly into hospital this approach can rapidly bring the epidemic to an end. Such an approach was adopted successfully in some hospitals in Toronto (Dwosh et al., 2003), and also in Hanoi, Viet Nam, where self-imposed containment of hospital staff contributed to the fastest curtailment of all the major SARS outbreaks (Twu et al., 2003). Because this approach effectively reduces disease leakage back into the community, it results in a high proportion of cases being healthcare workers (unless hospital hygiene is absolutely rigorous), which could erroneously be interpreted as an indication of failure to curtail the epidemic. In Hanoi, a sobering 63% of the cases were healthcare workers; however, had the strategy not been undertaken, much higher absolute figures (both in the hospital and the community at large) would have been likely.

Decisions about management can also be complicated by difficulties interpreting case data to estimate R_{int} and the shape of the epidemic curve. For instance, independent communities or subpopulations may have fundamentally different transmission parameters; the rate at which these populations contribute to new reports will differ and can conflate interpretation of epidemic curves (Dye & Gay, 2003). In addition, whilst clear examples of superspreading events were identified early on in the outbreaks (Lipsitch et al., 2003; Riley et al., 2003), only one of the contemporary analyses (Lipsitch et al., 2003) included an explicit formulation for superspreading. The existence of superspreading can lead to underestimation of reproductive numbers (Lipsitch et al., 2003) and has important implications for disease spread (Section 8.2.3).

An important point emerging from these studies is that control of an epidemic depends crucially on the length of the presymptomatic (latent) period of infection (Fraser et al., 2004). Diseases with long latent (but infectious) periods, like HIV (and unlike SARS), are very difficult to control by contact reduction because the infection has many opportunities to spread before the extent of the problem is detected. Such diseases can have devastating consequences for populations before effective control measures can be put in place. The recent finding of a threshold burden of infection for chytridiomycosis in frogs (Vredenburg et al., 2010) means that the infection is effectively latent, but infectious, before the threshold is reached and could explain why this disease causes such catastrophic declines in frog populations (Box 8.6).

8.6.3 Vaccination

When immunisation programmes are used to eradicate a disease, in the absence of other measures a proportion P_c of the population must be vaccinated (Anderson & May, 1991):

$$P_c = 1 - (1/R_0). \qquad (8.2)$$

This is the level at which the proportion of susceptibles remaining is too small to support the parasite (i.e. it is below the threshold population size for establishment, N_T). Hence, not all individuals need to be vaccinated for the disease to be eradicated; the remaining susceptibles benefit from the 'herd immunity' provided by the immune class. However, the value of P_c rises sharply with increasing R_0; this means, whilst it may be feasible to eradicate diseases with a low R_0 such as smallpox ($R_0 \approx 4$), those with a high R_0 represent an unassailable challenge (for instance, measles, with $R_0 \approx 13$–18, would require a vaccination coverage of 90–95% (Anderson & May, 1991)).

Vaccination can, however, have an alternative use in conservation as a protective measure for threatened populations. Instead of attempting to eradicate a disease, we can attempt to reduce the size of an outbreak. This may be achieved through targeted vaccination of boundary populations to prevent disease spreading further to other subpopulations. Alternatively, vaccination may be used to protect a subset of the population sufficient to prevent its extinction as the disease runs its course. For many species, there is a non-linear relationship between fitness and population size, with breeding success or offspring survival to reproduction falling disproportionately in small populations. Such Allee effects create a minimum viable population (MVP), below which the population will go deterministically extinct. Provided vaccination can cover at

Box 8.10 Canid and felid diseases

A number of infectious diseases including rabies, canine parvovirus (CPV; Box 3.6) and canine distemper virus (CDV) have caused population declines of wild canids, including Ethiopian wolves, the grey wolf (Box 3.6) and African wild dogs. Although felids have generally been less affected than canids (possibly because of differences in ranging and social behaviour), there have also been recent outbreaks of rabies in Serengeti lions (RoelkeParker *et al.*, 1996) and feline leukaemia (FLV) in the Iberian lynx. These diseases can be classed as EIDs because the frequency or severity of outbreaks has increased over the last 30 years, in parallel with increasing populations of reservoir hosts.

Reservoir cats and dogs. Wild felids and canids are at particular risk from disease because of their close phylogenetic relatedness to domestic cats and dogs (Pedersen *et al.*, 2007), which often exist in large feral populations in contact with threatened wild carnivore populations. These unmanaged reservoir populations act as continual sources of spillover infection. The Serengeti population of African wild dogs (*Lycaon pictus*) had gone extinct by 1992 following rabies outbreaks (however, a small population has recently re-established in the eastern plains); an outbreak of CDV in 1994 dramatically reduced Serengeti lion (*Panthera leo*) population numbers. Unlike farmed cattle, which have previously acted as reservoirs for rinderpest (Section 8.2), there is little economic incentive for vaccination of feral dogs and cats, although in the Serengeti area there is an ongoing public health and conservation programme to vaccinate feral dogs to reduce outbreaks of rabies in humans, canids and lions (Cleaveland *et al.*, 2003; http://www.tusk.org). Feral populations have increased in recent years. Sadly, in parts of South Africa, increases in dog populations are partly the result of another EID: HIV in man. The high mortality caused by HIV in some villages has led to the dissolution of many family units and consequent abandonment of the domestic dogs that are kept by most families for hunting or herding (Nel & Rupprecht, 2007).

Rabies in Ethiopian wolves. The Ethiopian wolf (*Canis simensis*) is one of the world's rarest canid, existing in several small subpopulations connected by narrow habitat corridors. Rabies outbreaks in the 1990s and 2003 caused severe population declines. In August 2003 a rabies outbreak began in one of the larger subpopulations, eventually

Box 8.10 (*continued*)

killing 72 (76%) of the one-year-old wolves. Three months later, a vaccination campaign commenced, targeting the packs occupying the corridor connecting the infected subpopulation (where the outbreak was deemed too far advanced to benefit from vaccination). Eventually, the other main subpopulations were also subject to vaccination (achieving 37.5% coverage). Despite the low coverage, this targeted campaign prevented the epidemic from spreading to other subpopulations (Haydon *et al.*, 2006). Simulation models suggest that such reactive vaccination programmes can be highly effective, even if only a small percentage are targeted (Fig. 8.18). The models predict the vaccination coverage required to sufficiently reduce the risk that the population falls below an MVP size as a result of an outbreak, rather than eliminating the disease entirely. Accurate predictions are contingent on the quality of data used to parameterise these often complex models; see Vial *et al.* (2006) for an example applied to rabies outbreaks in African wild dogs.

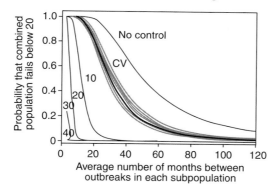

Fig. 8.18. Predicting catastrophic population declines for the Iberian lynx. The graph plots the probability that the population falls below its minimum viable size in relation to frequency of disease outbreaks for different control strategies. Extinction probability is highest assuming no control; corridor vaccination scenarios (CV) reduce extinction probability to intermediate values; most effective of all are reactive vaccination programmes (with coverages of 10–40%, as labelled). From Haydon *et al.* (2006).

FLV in the Iberian lynx. The Iberian lynx (*Lynx pardinus*) is the most critically endangered felid in the world, existing in a meta-population comprising two breeding populations and a number of sink populations in the Iberian peninsula, southern Spain. Populations fell dramatically in the twentieth century, mainly because this predator

Box 8.10 (*continued*)

specialises on wild rabbits (*Oryctolagus cuniculus*), which have declined because of two EIDs: myxomatosis in the 1950s and rabbit haemorrhagic disease in the 1980s. Although occasional spillover to Iberian lynx of FLV from domestic cats had been recorded before, an outbreak in one of the main breeding subpopulations in 2007 was particularly alarming. In this outbreak, lynx–lynx transmission was apparently highly efficient, and pathogenicity was also high, resulting in the deaths of 7 of the 12 infected adults (infection prevalence in adults was 27%). A targeted control and vaccination programme prevented the spread of FLV to the isolated subpopulations and other main breeding colony (Lopez *et al.*, 2009). In the infected population, 80% of adults were trapped and their infection status determined before removal (if infected) or vaccination (if uninfected). Viraemic individuals were monitored; of the three surviving infection, two reverted to latency and were released into the population. Although this presents a future threat, the benefits (of releasing two females which went on to breed that year) were deemed to outweigh the risks (but will necessitate high-level monitoring in the future). The disease may have spread rapidly during this outbreak because it occurred just before the mating season, when interspecific contacts and fights are at their highest (the disease is transmitted through biting). The initial spillover event may have resulted from the use of artificial feeding stations put out as part of conservation measures to boost the lynx's food supply. Feeding stations present two potential risks: they attract domestic and feral cats to the same vicinity as the lynx, and they also increase the local lynx density and hence their rate of contact, thus contributing both to disease spillover and spread. The use of feeding stations has been curtailed at this site since the outbreak and subsequent research (Lopez *et al.*, 2009).

least the number of adults needed for an MVP, threatened populations can be protected from disease-induced extinction. These approaches have been used in recent outbreaks of rabies in the Ethiopian wolf and FLV in the Iberian lynx, as discussed in Box 8.10.

The experience with canid and felid diseases (Box 8.10) raises some important points for conservation:

- Vaccination need not cover a large proportion of the population, provided it is timely, targeted and its outcome monitored.
- Monitoring for novel disease outbreaks is an essential part of control measures: reactive vaccination campaigns are more effective (or will be effective with lower coverage) if outbreaks are detected early, before the infection spreads to many individuals or subpopulations.
- Contact control can be an effective counterpart to vaccination; reducing susceptible-infectious contact rates by monitoring and removal of infected individuals, or reduction of reservoir populations.
- Contact control should include identifying and removing or reducing risks at disease hotspots; for instance, feeding stations that encourage high-density aggregation and increased interspecific contact exacerbate transmission.

A corollary of the points above is that some diseases will be easier to control than others. Diseases with a long period of infectiousness before symptoms become apparent will be particularly difficult to control through reactive campaigns (Section 8.6.2). For these diseases in particular, management must focus on risk reduction, with monitoring and assessment of spillover risks, reduction of spillover risk through pre-emptive reduction in contact between populations at risk and potential reservoir hosts, and reduction in intraspecific contacts or risky behaviour (Childs *et al.*, 2007b; Daszak *et al.*, 2007). Two human EIDs echo this problem. SARS, for which the onset of infectiousness is fairly much concurrent with the onset of symptoms, was brought under control during its 2003 pandemic after initiation of reactive contact and isolation control measures. In contrast, HIV can be infectious for many years prior to the onset of symptoms; reactive control of the symptomatic alone would clearly have been insufficient to prevent its spread.

8.7 Conclusions

This chapter, as with the previous chapter, ends by raising more questions than it answers. Firstly, we need to address the apparent mismatch between the general conclusion readers will draw from this chapter (namely that newly emerging parasites can have damaging effects on populations) and our tentative conclusion from the previous chapter (that 'healthy ecosystems' are associated with a diversity of parasites). This reflects differences in the state of the ecosystems under consideration. The conclusions drawn in the previous chapter stemmed from studies of relatively undisturbed ecosystems in which host and parasite communities have had a long history of co-adaptation. In contrast, the systems considered in this chapter are, in essence, already perturbed, with the majority of novel diseases arising as a result of unusual contact between species resulting from anthropogenic influence on habitat, climate or movement patterns. It is quite possible that both parasite diversity loss *and* emergence of novel diseases could be driven by ecosystem change, and both could feed back, resulting in further change. We know of no empirical or theoretical studies that explicitly investigate links between these processes.

Emerging diseases present many other areas requiring active research. In particular, there are few mathematical models examining the early stages of emergence (spillover in multi-host systems), and few incorporating interspecific and intraspecific transmission in the novel host. It could be that these processes operate on such different timeframes and scales that the inclusion of spillover is irrelevant once the disease can spread by intraspecific transmission in the novel host, but we do not have an adequate model base on which to make this judgement.

Another area of mismatch is that between models examining the evolution of virulence (Section 8.3), which come from a one-host–one-parasite perspective, with those examining the evolution of transmission in a novel host (Section 8.2.2), which use a multi-host perspective. The former set tends to assume virulence and transmissibility are initially high, with virulence and per capita transmission tending to evolve downwards; the latter assumes that transmission in the novel host is initially low, and evolves upward. To what extent would reconciliation of these two approaches provide new insights into disease emergence? A consensus approach might pay dividends in the field of virulence management, in which the influence of social practices on the evolution of virulence is examined with the aim of informing public healthcare strategies (Read & Taylor, 2001; Koella *et al.*, 2009).

There are many areas in which we require further empirical and statistical research. To what extent do the modern events-based surveillance systems provide an advantage over traditional medical and veterinary notification systems? Can the relative influence of multiple environmental drivers, including climate change, be quantified for disease systems such as malaria? Basic research into the community ecology of many emerging diseases requires further work; for instance, the bat reservoirs of the Ebola Zaire strain have only recently been tentatively identified (Leroy *et al.*, 2005), and little is known about the reservoirs for other Ebola strains, or for the closely related Marburg virus. The many viruses causing hantavirus pulmonary syndrome in man, and their apparently species–specific rodent reservoirs, clearly represent a complex community about which much remains to be discovered. Interactions with climatic factors appear to be important in this system, and are probably relevant to many others (both wildlife and human). Several recent studies attempt to identify the parasite groups most likely to cause EIDs in different hosts, and the locations where diseases are most likely to emerge. Perhaps this approach could be extended to identify 'risky reservoirs', those species or taxonomic groups most likely to harbour potential EIDs. The role of diverse bat species as reservoirs of many recent human EIDs requires further investigation in this respect.

These activities require multidisciplinary approaches. Throughout this book we have highlighted the importance of integrating expertise from parasitology and community ecology. To develop our understanding of EIDs and to predict and manage their spread and impact requires an unprecedented level of collaboration. Only by integrating the expertise of ecologists, parasitologists, evolutionary biologists, geographers, social scientists and healthcare professionals can we elucidate the complex ecological, evolutionary and anthropogenic processes that underlie emerging disease.

9 · *Where do we go from here?*

A survey of the average date of references cited in this book serves to illustrate how our understanding of the importance of parasites in ecology has developed. We have come from a realisation that parasites can influence competitive interactions between other species (starting with Park's laboratory study of flour beetles in 1948), through the development of ideas for how parasites interact with predation (perhaps exemplified by manipulative field studies on red grouse (Hudson *et al.*, 1992a; 1992b), to recognition of their effects in more complex modules such as intra-guild predation (MacNeil *et al.*, 2003c; Borer *et al.*, 2007a; Hatcher *et al.*, 2008). As this 'pure' science has evolved, so has its application to real-world scenarios. Along with the application of Hudson *et al.*'s work to gamebird management, we now recognise the importance of parasites in biological invasions (red squirrels: Tompkins *et al.*, 2003; amphipods: Dunn, 2009; enemy release: Torchin *et al.*, 2003), and in influencing ecosystem function (ecosystem engineering: Thomas *et al.*, 1999; food webs: Lafferty *et al.*, 2008a; bioenergetics: Kuris *et al.*, 2008; carbon balance: Holdo *et al.*, 2009). As this process of exporting basic model systems (theoretical and empirical) to inform our understanding of natural systems has developed, so has the urgency with which we need to tackle an increasing problem: that of emerging infectious diseases. This problem perhaps exemplifies, above all other subjects in this book, the need for integration between a range of disciplines and services (basic ecological research; medical, veterinary, humanitarian research and practice) in order to understand and control how parasites interact in ecological communities.

This process of exporting and applying ideas (Fig. 9.1) is ongoing. This is why, in this book, we have tried to balance the treatment of current research at the frontiers of conservation and public health with sufficient background in the theory and empirical research that preceded it. We believe there is still a great deal to learn (and develop and apply) from the basic research on interactions between species, as

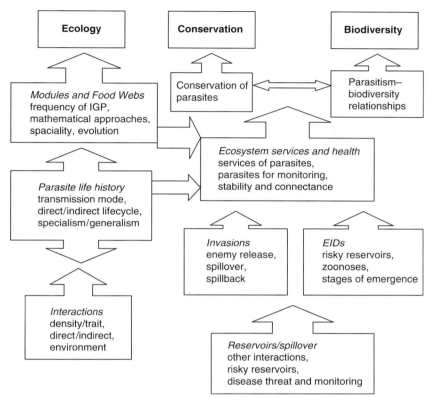

Fig. 9.1. Parasites in ecological communities: current problems in ecology, biodiversity and conservation. For simplicity, not all links between areas are shown.

well as discoveries at the 'cutting edge'. So what have we learned, and where do we need to extend this knowledge, from the perspective of practice in ecology, biodiversity and conservation? We examine some of the more current areas of research below.

Biological interactions

An aspect of basic science that has implications throughout the ecological arena is the study of biological interactions. Traditionally envisaged as two-component processes (how A interacts with B and B with A), these are the building blocks of the structures we perceive as ecological communities. However, it is becoming increasingly clear that some processes involve three or more components, and cannot be easily reduced. For instance, trophically transmitted parasites link species at different trophic levels, but are not merely mediators of predation by one on another. Similarly, parasites can mediate 'apparent competition' between shared host species; importantly, in the absence of the parasite, there is no interaction between these species at all. Is it meaningful to talk about 'interactions between interactions' (e.g. parasitism and predation; Moller, 2008), or should the component species (parasite, predator, host/prey) each be considered linked by consumer–resource interactions and indirect interactions? In this regard, the relative importance of direct and indirect interactions in determining ecological dynamics is under-studied. Interactions with parasites bear many similarities to better-studied forms such as predation, but there are also important differences (Hatcher et al., 2006). One that we highlight throughout this book is the propensity of parasites to produce trait-mediated indirect effects (TMIEs). Whilst the potential importance of TMIEs in general has been recognised, their impact in real communities requires exploration. A further problem highlighted in this book is how environmental factors alter biotic interactions. This may be particularly important in host–parasite systems as environmental variables such as temperature may influence many aspects of reproductive and transmission ecology, and can interact in complex ways with host reproductive biology (see, for example, the case of malaria and climate change; Box 8.7). As we become increasingly aware of the need to predict the complex outcomes of environmental change, systems involving parasites, many of which are amenable to experimental culture and manipulation, can provide a valuable tool for investigating these questions.

Life history traits

Impacting on, and influenced by, basic ecological interactions are the life history traits of parasites. In this book we have seen how different transmission modes can produce fundamentally different predictions for population and community outcomes. For instance, density- and frequency-dependent transmission modes produce opposing predictions for the relationship between parasite and host biodiversity (Section 7.4). When we attempt to integrate parasitology into community ecology, important questions arise. For instance, it is unclear how to represent the links between parasites with indirect life cycles in food web analyses, because they utilise multiple hosts but are not ecological generalists. What aspects of life history and parasite–host interactions determine the degree of specialism/generalism for a parasite (see, e.g., Barrett *et al.*, 2009)? Issues of host range are of fundamental applied concern, affecting how parasites respond to and influence biodiversity, and their propensity to become EIDs.

Modules and food webs

In this book we have advocated use of a community module approach to represent groups of strongly interacting species, with the aim that modules can then be put together to build food webs. We are perhaps still lacking an appropriate mathematical framework for this endeavour. Even simple modules involving three species, one of which is a parasite, can become analytically intractable because representing a (micro)parasite requires that we represent infected and uninfected population classes separately (or include functions for the distribution of parasites between hosts, in the case of macroparasites). We also need to incorporate biological realism in the form of spatial structure and evolutionary dynamics into our models (something that applies beyond parasite ecology). Spatial studies are largely confined to systems of conservation concern, such as red squirrels (Box 6.3) and Iberian lynx (8.10); broader implications should also be considered. The empirical studies showing evidence for rapid host–parasite evolution (Section 3.3) suggest that new approaches in both population ecology and evolution may be necessary. In addition, the ramifications of rapid evolution for virulence management (Section 8.3) require urgent examination. Questions about food webs and their constituent modules remain: how frequent is intraguild predation (IGP), and how often do parasites engage in it? From the evidence so far, IGP appears to be surprisingly common, and interactions involving parasites appear to be especially prone to it. This is relevant to our understanding of the stability of ecological communities, as contrary to earlier views, it now seems that IGP modules may enhance the robustness and resilience of communities (Section 7.5).

Reservoirs and spillover

Fundamental to our understanding of the dynamics of parasites in multi-host communities are the concepts of reservoir hosts and parasite spillover. Reservoir hosts can enable parasite persistence or repeated outbreak in host species that would not on their own support them, and are key to the success of some invasive species (Section 6.5) and to the emergence of infectious diseases of wildlife and humans (Section 8.2). The chapters on competition, predation and IGP reveal how parasite prevalence and abundance in a focal host are influenced by interactions with other species; we now need to examine how these mechanisms influence spillover to other potential hosts. For instance, the risk of spillover into other species is related to parasite abundance in the reservoir (i.e. for microparasites, the density of the infected subclass). Predation and harvesting have both been shown theoretically to influence parasite abundance (Chapter 3), and hence could influence spillover. However, predicting outcomes for real communities requires more detailed knowledge of the biology of the systems involved; factors including acquired immunity of the reservoir host (Section 3.6), the relationship between predation risk and dilution hosts (Section 7.3), and transmission modes (Section 7.4) each alter spillover risk. Knowledge of the basic biology of infection is therefore needed for a comprehensive approach to assessment of disease threat and monitoring.

Emerging infectious diseases

The flu pandemic of 2009 proved fortunately to be of low virulence, relative to previous flu pandemics. Nonetheless, it highlighted the potential for EIDs to spread rapidly with dramatic health and economic costs (Box 8.3), and the need for accurate and timely approaches to monitoring. One aspect of a biologically based monitoring programme would be to identify 'risky reservoirs'; species in particular communities that possess the characteristics most likely to result in disease spillover. This would complement existing work on the parasite types most likely to lead to EIDs (Section 8.4). Characteristics of parasites that make them liable to spillover include possession of a broad host range and high prevalence in the host. Zoonoses are strongly over-represented amongst emerging diseases of humans (Section 8.4.1), and also present ongoing problems for the conservation of wildlife (Mathews, 2009). We also need a more integrated approach to the theory of EIDs. Currently, most models focus either on the spillover process, or on spread through the novel host (usually man) once the new disease has become established (i.e. it has evolved sufficient intraspecific transmission capability in the new host). Few models combine spillover and intraspecific transmission; even fewer examine the timeframes for evolution of transmission in the novel host (which we know from recent epidemics of SARS and avian flu can occur quite rapidly; certainly within an ecological timescale).

Invasions

Biological invasions come second only to habitat destruction as a threat to biodiversity (Section 6.6). There are many examples of parasite loss in invasive species, as illustrated by meta-analyses of plants and animals (Section 6.3). However, the mechanisms underpinning enemy release are poorly understood and may be a complex mix of population dynamic and evolutionary factors that call for further exploration. Invasions may also lead to parasite introduction with the potential for emergent disease. The impact of introduced parasites can be seen in the extirpation of the native squirrel in regions of the United Kingdom (Box 6.3) and in global amphibian declines (Box 8.6). Invaders can also act as reservoirs for native parasites (e.g. transmission of porcelain disease to native crayfish – Box 6.2), and recent studies suggest that the spillback may be of underestimated significance both for invasion biology and managed species. In addition, trait-mediated indirect effects (TMIEs) of parasites have also been shown to have dramatic effects on native–invader interactions (Dick *et al.*, 2010) and there is a need to develop our theoretical understanding of these processes.

Ecosystem services and health

Whilst parasites can facilitate biological invasions, they can also provide a variety of ecosystem services that are only now being recognised. In the few food webs where parasites have been evaluated, parasites contribute substantially to food web connectance and thereby may enhance community resilience and robustness (Section 7.5). They can be responsible for a sizeable portion of ecosystem productivity, providing a substantial fraction of resources available to mid and higher trophic levels (Section 7.6), and their regulatory impact on hosts at higher trophic levels can ramify throughout the community (Section 7.2). Parasites can act as ecosystem engineers, changing the physical structure of ecosystems, and in the process provide habitats for exploitation by other species (Section 7.7). Some parasites alter the chemical composition of ecosystems through nutrient cycling; others play a role in detoxification of ecosystems by sequestering pollutants. Parasites may provide services to ecosystem research; in situations where complete ecosystem monitoring is infeasible, parasites might be used as proxy measures of ecosystem health (Section 7.8). Many of these studies concern parasites of aquatic ecosystems (salt marshes, in particular). In view of these identified roles, we urge further examination of the role of parasites as beneficial players in a variety of ecosystems.

Conservation of parasites and parasite–biodiversity relationships

The points raised in the preceding sections are clearly relevant to conservation and biodiversity. Here we highlight two issues that arise from the above discussion. Firstly, if parasites do provide ecosystem services, then for this reason alone, they too need conserving. This is no easy task. Parasites may be particularly vulnerable to extinction, because in many cases they require a minimum threshold host population for establishment, so they will go extinct well before their hosts. Many parasites have indirect life cycles, relying on more than one host species for their own propagation; whichever of these species is exposed to population decline will be a weak link. Secondly, if we are to use parasites in ecosystem monitoring, or we are to predict disease threats to man or other species, we need to understand the relationship between parasite diversity and biodiversity as a whole. Parasitism is regarded by many as the most common consumer strategy of all, and parasites account for a sizeable portion of the world's biodiversity. Parasite diversity may increase or decrease with community diversity, depending on their underlying biology (such as transmission mode – Section 7.4). Depending on how parasite and host diversity are connected, parasite diversity has the potential to feed back on biological diversity as a whole (Section 7.8). We need to understand these processes if we are going to formulate meaningful community-level conservation action plans in the future.

References

Abrams, P. A. (1992). Predators that benefit prey and prey that harm predators: unusual effects of interacting foraging adaptations. *American Naturalist*, **140**, 573–600.

Abrams, P. A. & Matsuda, H. (1996). Positive indirect effects between prey species that share predators. *Ecology*, **77**, 610–616.

Abrams, P. A. & Roth, J. D. (1994). The effects of enrichment of three-species food-chains with nonlinear functional-responses. *Ecology*, **75**, 1118–1130.

Agrawal, A. A. & Kotanen, P. M. (2003). Herbivores and the success of exotic plants: a phylogenetically controlled experiment. *Ecology Letters*, **6**, 712–715.

Agrawal, A. A., Kotanen, P. M., Mitchell, C. E., Power, A. G., Godsoe, W. & Klironomos, J. (2005). Enemy release? An experiment with congeneric plant pairs and diverse above- and belowground enemies. *Ecology*, **86**, 2979–2989.

Aguirre, A. A., Ostfeld, R. S., Tabor, G. M., House, C. A. & Pearl, M. C. (eds) (2002). *Conservation Medicine: Ecological Health in Practice*. New York: Oxford University Press.

Alexander, H. M. & Holt, R. D. (1998). The interaction between plant competition and disease. *Perspectives in Plant Ecology Evolution and Systematics*, **1**, 206–220.

Aliabadi, B. W. & Juliano, S. A. (2002). Escape from gregarine parasites affects the competitive interactions of an invasive mosquito. *Biological Invasions*, **4**, 283–297.

Allan, B. F., Keesing, F. & Ostfeld, R. S. (2003). Effect of forest fragmentation on Lyme disease risk. *Conservation Biology*, **17**, 267–272.

Allan, B. F., Langerhans, R. B., Ryberg, W. A., Landesman, W. J., Griffin, N. W., Katz, R. S., Oberle, B. J., Schutzenhofer, M. R., Smyth, K. N., de St Maurice, A., Clark, L., Crooks, K. R., Hernandez, D. E., McLean, R. G., Ostfeld, R. S. & Chase, J. M. (2009). Ecological correlates of risk and incidence of West Nile virus in the United States. *Oecologia*, **158**, 699–708.

Altizer, S., Dobson, A., Hosseini, P., Hudson, P., Pascual, M. & Rohani, P. (2006). Seasonality and the dynamics of infectious diseases. *Ecology Letters*, **9**, 467–484.

Ameloot, E., Verlinden, G., Boeckx, P., Verheyen, K. & Hermy, M. (2008). Impact of hemiparasitic *Rhinanthus angustifolius* and *R. minor* on nitrogen availability in grasslands. *Plant and Soil*, **311**, 255–268.

Amundsen, P. A., Lafferty, K. D., Knudsen, R., Primicerio, R., Klemetsen, A. & Kuris, A. M. (2009). Food web topology and parasites in the pelagic zone of a subarctic lake. *Journal of Animal Ecology*, **78**, 563–572.

Anderson, A., McCormack, S., Helden, A. J., Sheridan, H., Kinsella, A. & Purvis, G. (2011). The potential of parasitoid Hymenoptera as bioindicators of arthropod diversity in agricultural grasslands. *Journal of Applied Ecology*. In press.

Anderson, R. A., Koella, J. C. & Hurd, H. (1999). The effect of *Plasmodium yoelii nigeriensis* infection on the feeding persistence of *Anopheles stephensi* Liston throughout the sporogonic cycle. *Proceedings of the Royal Society of London Series B – Biological Sciences*, **266**, 1729–1733.

Anderson, R. C. (1972). The ecological relationships of meningeal worm and native cervids in North America. *Journal of Wildlife Diseases*, **8**, 304–310.

Anderson, R. M. & May, R. M. (1981). The population dynamics of micro-parasites and their invertebrate hosts. *Philosophical Transactions of the Royal Society of London Series B – Biological Sciences*, **291**, 451–524.

Anderson, R. M. & May, R. M. (1982). Coevolution of hosts and parasites. *Parasitology*, **85**, 411–426.

Anderson, R. M. & May, R. M. (1985). Age-related changes in the rate of disease transmissions: implications for the design of vaccination programs. *Journal of Hygiene*, **94**, 365–436.

Anderson, R. M. & May, R. M. (1986). The invasion, persistence and spread of infectious-diseases within animal and plant communities. *Philosophical Transactions of the Royal Society of London Series B – Biological Sciences*, **314**, 533–570.

Anderson, R. M. & May, R. M. (1991). *Infectious Diseases of Humans: Dynamics and Control*. Oxford: Oxford University Press.

Andreadis, T. G. (1980). *Nosema pyrausta* infection in *Macrocentrus grandii*, a braconid parasite of the European corn-borer, *Ostrinia nubilalis*. *Journal of Invertebrate Pathology*, **35**, 229–233.

Antia, R., Regoes, R. R., Koella, J. C. & Bergstrom, C. T. (2003). The role of evolution in the emergence of infectious diseases. *Nature*, **426**, 658–661.

Arim, M. & Marquet, P. A. (2004). Intraguild predation: a widespread interaction related to species biology. *Ecology Letters*, **7**, 557–564.

Arinaminpathy, N. & McLean, A. R. (2009). Evolution and emergence of novel human infections. *Proceedings of the Royal Society of London Series B – Biological Sciences*, **276**, 3937–3943.

Arnold, A. E., Maynard, Z., Gilbert, G. S., Coley, P. D. & Kursar, T. A. (2000). Are tropical fungal endophytes hyperdiverse? *Ecology Letters*, **3**, 267–274.

Aron, J. L. & May, R. M. (1982). The population dynamics of malaria. In Anderson, R. M. (ed.), *The Population Dynamics of Infectious Diseases: Theory and Applications*. London: Chapman & Hall.

Askary, H. & Brodeur, J. (1999). Susceptibility of larval stages of the aphid parasitoid *Aphidius nigripes* to the entomopathogenic fungus *Verticillium lecanii*. *Journal of Invertebrate Pathology*, **73**, 129–132.

Auger, P., Mchich, R., Chowdhury, T., Sallet, G., Tchuente, M. & Chattopadhyay, J. (2009). Effects of a disease affecting a predator on the dynamics of a predator–prey system. *Journal of Theoretical Biology*, **258**, 344–351.

Augspurger, C. K. (1983). Seed dispersal of the tropical tree, *Platypodium elegans*, and the escape of its seedlings from fungal pathogens. *Journal of Ecology*, **71**, 759–771.

Augspurger, C. K. (1984). Seedling survival of tropical tree species – interactions of dispersal distance, light-gaps, and pathogens. *Ecology*, **65**, 1705–1712.

Bailes, E., Gao, F., Bibollet-Ruche, F., Courgnaud, V., Peeters, M., Marx, P. A., Hahn, B. H. & Sharp, P. M. (2003). Hybrid origin of SIV in chimpanzees. *Science*, **300**, 1713.

Bally, M. & Garrabou, J. (2007). Thermodependent bacterial pathogens and mass mortalities in temperate benthic communities: a new case of emerging disease linked to climate change. *Global Change Biology*, **13**, 2078–2088.

Bandi, C., Dunn, A. M., Hurst, G. D. D. & Rigaud, T. (2001). Inherited micro-organisms, sex-specific virulence and reproductive parasitism. *Trends in Parasitology*, **17**, 88–94.

Baranowski, E., Ruiz-Jarabo, C. M. & Domingo, E. (2001). Evolution of cell recognition by viruses. *Science*, **292**, 1102–1105.

Bardgett, R. D., Smith, R. S., Shiel, R. S., Peacock, S., Simkin, J. M., Quirk, H. & Hobbs, P. J. (2006). Parasitic plants indirectly regulate below-ground properties in grassland ecosystems. *Nature*, **439**, 969–972.

Barrett, L. G., Kniskern, J. M., Bodenhausen, N., Zhang, W. & Bergelson, J. (2009). Continua of specificity and virulence in plant host–pathogen interactions: causes and consequences. *New Phytologist*, **183**, 513–529.

Bartha, I., Simon, P. & Muller, V. (2008). Has HIV evolved to induce immune pathogenesis? *Trends in Immunology*, **29**, 322–328.

Bascompte, J., Jordano, P., Melián, C. J. & Olesen, J. M. (2003). The nested assembly of plant–animal mutualistic networks. *Proceedings of the National Academy of Sciences of the United States of America*, **100**, 9383–9387.

Bascompte, J. & Melian, C. J. (2005). Simple trophic modules for complex food webs. *Ecology*, **86**, 2868–2873.

Bataille, A., Cunningham, A. A., Cedeno, V., Cruz, M., Eastwood, G., Fonseca, D. M., Causton, C. E., Azuero, R., Loayza, J., Martinez, J. D. C. & Goodman, S. J. (2009a). Evidence for regular ongoing introductions of mosquito disease vectors into the Galapagos Islands. *Proceedings of the Royal Society of London Series B – Biological Sciences*, **276**, 3769–3775.

Bataille, A., Cunningham, A. A., Cedeno, V., Patino, L., Constantinou, A., Kramer, L. D. & Goodman, S. J. (2009b). Natural colonization and adaptation of a mosquito species in Galapagos and its implications for disease threats to endemic wildlife. *Proceedings of the National Academy of Sciences of the United States of America*, **106**, 10230–10235.

Bauer, A., Trouve, S., Gregoire, A., Bollache, L. & Cezilly, F. (2000). Differential influence of *Pomphorhynchus laevis* (Acanthocephala) on the behaviour of native and invader gammarid species. *International Journal for Parasitology*, **30**, 1453–1457.

Bauer, L. S., Miller, D. L., Maddox, J. V. & McManus, M. L. (1998). Interactions between a *Nosema* sp. (Microspora: Nosematidae) and nuclear polyhedrosis virus infecting the gypsy moth, *Lymantria dispar* (Lepidoptera: Lymantriidae). *Journal of Invertebrate Pathology*, **72**, 147–153.

Beckstead, J. & Parker, I. M. (2003). Invasiveness of *Ammophila arenaria*: release from soil-borne pathogens? *Ecology*, **84**, 2824–2831.

Bedhomme, S., Agnew, P., Vital, Y., Sidobre, C. & Michalakis, Y. (2005). Prevalence-dependent costs of parasite virulence. *PLoS Biology*, **3**, 1403–1408.

Begon, M. & Bowers, R. G. (1994). Host–host–pathogen models and microbial pest-control: the effect of host self-regulation. *Journal of Theoretical Biology*, **169**, 275–287.

Begon, M. & Bowers, R. G. (1995). Beyond host–pathogen dynamics. In Grenfell, B. T. & Dobson, A. P. (eds), *Ecology of Infectious Diseases in Natural Populations*. Cambridge: Cambridge University Press, pp. 478–509.

Begon, M., Bowers, R. G., Kadianakis, N. & Hodgkinson, D. E. (1992). Disease and community structure: the importance of host self-regulation in a host–host–pathogen model. *American Naturalist*, **139**, 1131–1150.

Begon, M., Sait, S. M. & Thompson, D. J. (1999). Host–parasite–parasitoid systems. In Hawkins, B. A. & Cornell, H. V. (eds), *Theoretical Approaches to Biological Control*. Cambridge: Cambridge University Press, pp. 327–348.

Bell, S. S., White, A., Sherratt, J. A. & Boots, M. (2009). Invading with biological weapons: the role of shared disease in ecological invasion. *Theoretical Ecology*, **2**, 53–66.

Belliure, B., Janssen, A., Maris, P. C., Peters, D. & Sabelis, M. W. (2005). Herbivore arthropods benefit from vectoring plant viruses. *Ecology Letters*, **8**, 70–79.

Belliure, B., Janssen, A. & Sabelis, M. W. (2008). Herbivore benefits from vectoring plant virus through reduction of period of vulnerability to predation. *Oecologia*, **156**, 797–806.

Ben-Ami, F., Mouton, L. & Ebert, D. (2008). The effects of multiple infections on the expression and evolution of virulence in a Daphnia–endoparasite system. *Evolution*, **62**, 1700–1711.

Bennetts, R. E., White, G. C., Hawksworth, F. G. & Severs, S. E. (1996). The influence of dwarf mistletoe on bird communities in Colorado ponderosa pine forests. *Ecological Applications*, **6**, 899–909.

Benson, J., Van Driesche, R. G., Pasquale, A. & Elkinton, J. (2003). Introduced braconid parasitoids and range reduction of a native butterfly in New England. *Biological Control*, **28**, 197–213.

Berger, L., Speare, R., Daszak, P., Green, D. E., Cunningham, A. A., Goggin, C. L., Slocombe, R., Ragan, M. A., Hyatt, A. D., McDonald, K. R., Hines, H. B., Lips, K. R., Marantelli, G. & Parkes, H. (1998). Chytridiomycosis causes amphibian mortality associated with population declines in the rain forests of Australia and Central America. *Proceedings of the National Academy of Sciences of the United States of America*, **95**, 9031–9036.

Bernot, R. J. & Lamberti, G. A. (2008). Indirect effects of a parasite on a benthic community: an experiment with trematodes, snails and periphyton. *Freshwater Biology*, **53**, 322–329.

Berthe, F. C. J., Le Roux, F., Adlard, R. D. & Figueras, A. (2004). Marteiliosis in molluscs: a review. *Aquatic Living Resources*, **17**, 433–448.

Bertolino, S., Lurz, P. W. W., Sanderson, R. & Rushton, S. P. (2008). Predicting the spread of the American grey squirrel *(Sciurus carolinensis)* in Europe: a call for a co-ordinated European approach. *Biological Conservation*, **141**, 2564–2575.

Bever, J. D., Westover, K. M. & Antonovics, J. (1997). Incorporating the soil community into plant population dynamics: the utility of the feedback approach. *Journal of Ecology*, **85**, 561–573.

Bezemer, T. M. & van Dam, N. M. (2005). Linking aboveground and belowground interactions via induced plant defenses. *Trends in Ecology & Evolution*, **20**, 617–624.

Bilu, E. & Coll, M. (2007). The importance of intraguild interactions to the combined effect of a parasitoid and a predator on aphid population suppression. *Biocontrol*, **52**, 753–763.

Blackmore, M. S., Scoles, G. A. & Craig, G. B. (1995). Parasitism of *Aedes aegypti* and *Ae albopictus* (Diptera, Culicidae) by *Ascogregarina* spp. (Apicomplexa, Lecudinidae) in Florida. *Journal of Medical Entomology*, **32**, 847–852.

Blossey, B. & Notzold, R. (1995). Evolution of increased competitive ability in invasive nonindigenous plants: a hypothesis. *Journal of Ecology*, **83**, 887–889.

Bolker, B., Holyoak, M., Krivan, V., Rowe, L. & Schmitz, O. (2003). Connecting theoretical and empirical studies of trait-mediated interactions. *Ecology*, **84**, 1101–1114.

Bolzoni, L., De Leo, G. A., Gatto, M. & Dobson, A. P. (2008). Body-size scaling in an SEI model of wildlife diseases. *Theoretical Population Biology*, **73**, 374–382.

Bonsall, M. B. (2004). The impact of diseases and pathogens on insect population dynamics. *Physiological Entomology*, **29**, 223–236.

Bonsall, M. B. & Hassell, M. P. (1997). Apparent competition structures ecological assemblages. *Nature*, **388**, 371–373.

Bonsall, M. B. & Hassell, M. P. (2000). The effects of metapopulation structure on indirect interactions in host–parasitoid assemblages. *Proceedings of the Royal Society of London Series B – Biological Sciences*, **267**, 2207–2212.

Bordes, F. & Morand, S. (2008). Helminth species diversity of mammals: parasite species richness is a host species attribute. *Parasitology*, **135**, 1701–1705.

Borer, E. T., Briggs, C. J. & Holt, R. D. (2007a). Predators, parasitoids, and pathogens: a cross-cutting examination of intraguild predation theory. *Ecology*, **88**, 2681–2688.

Borer, E. T., Briggs, C. J., Murdoch, W. W. & Swarbrick, S. L. (2003). Testing intraguild predation theory in a field system: does numerical dominance shift along a gradient of productivity? *Ecology Letters*, **6**, 929–935.

Borer, E. T., Hosseini, P. R., Seabloom, E. W. & Dobson, A. P. (2007b). Pathogen-induced reversal of native dominance in a grassland community. *Proceedings of the National Academy of Sciences of the United States of America*, **104**, 5473–5478.

Bosch, J. & Rincon, P. A. (2008). Chytridiomycosis-mediated expansion of *Bufo bufo* in a montane area of Central Spain: an indirect effect of the disease. *Diversity and Distributions*, **14**, 637–643.

Bostock, R. M. (2005). Signal crosstalk and induced resistance: straddling the line between cost and benefit. *Annual Review of Phytopathology*, **43**, 545–580.

Bouwma, A. M., Ahrens, M. E., DeHeer, C. J. & DeWayne Shoemaker, D. (2006). Distribution and prevalence of *Wolbachia* in introduced populations of the fire ant *Solenopsis invicta*. *Insect Molecular Biology*, **15**, 89–93.

Bowers, R. G. & Begon, M. (1991). A host–host–pathogen model with free-living infective stages, applicable to microbial pest-control. *Journal of Theoretical Biology*, **148**, 305–329.

Bowers, R. G. & Turner, J. (1997). Community structure and the interplay between interspecific infection and competition. *Journal of Theoretical Biology*, **187**, 95–109.

Bradley, D. J., Gilbert, G. S. & Martiny, J. B. H. (2008). Pathogens promote plant diversity through a compensatory response. *Ecology Letters*, **11**, 461–469.

398 · **References**

Brasier, C. (2000). Plant pathology: the rise of the hybrid fungi. *Nature*, **405**, 134–135.

Briggs, C.J. (1993). Competition among parasitoid species on a stage-structured host and its effect on host suppression. *American Naturalist*, **141**, 372–397.

Briggs, C.J., Knapp, R.A. & Vredenburg, V.T. (2010). Enzootic and epizootic dynamics of the chytrid fungal pathogen of amphibians. *Proceedings of the National Academy of Sciences of the United States of America*, **107**, 9695–9700.

Brisson, D., Dykhuizen, D.E. & Ostfeld, R.S. (2008). Conspicuous impacts of inconspicuous hosts on the Lyme disease epidemic. *Proceedings of the Royal Society of London Series B – Biological Sciences*, **275**, 227–235.

Brodeur, J. & Rosenheim, J.A. (2000). Intraguild interactions in aphid parasitoids. *Entomologia experimentalis et Applicata*, **97**, 93–108.

Brook, B.W., Sodhi, N.S. & Bradshaw, C.J.A. (2008). Synergies among extinction drivers under global change. *Trends in Ecology & Evolution*, **23**, 453–460.

Brooks, D.R. & Hoberg, E.P. (2007). How will global climate change affect parasite–host assemblages? *Trends in Parasitology*, **23**, 571–574.

Brown, D.H. & Hastings, A. (2003). Resistance may be futile: dispersal scales and selection for disease resistance in competing plants. *Journal of Theoretical Biology*, **222**, 373–388.

Brown, S.P., Inglis, R.F. & Taddei, F. (2009). Evolutionary ecology of microbial wars: within-host competition and (incidental) virulence. *Evolutionary Applications*, **2**, 32–39.

Bruce, T.J. & Pickett, J.A. (2007). Plant defence signalling induced by biotic attacks. *Current Opinion in Plant Biology*, **10**, 387–392.

Bull, E.L., Heater, T.W. & Youngblood, A. (2004). Arboreal squirrel response to silvicultural treatments for dwarf mistletoe control in northeastern Oregon. *Western Journal of Applied Forestry*, **19**, 133–141.

Bull, J.J. & Ebert, D. (2008). Invasion thresholds and the evolution of nonequilibrium virulence. *Evolutionary Applications*, **1**, 172–182.

Bullock, J.M. & Pywell, R.F. (2005). *Rhinanthus*: a tool for restoring diverse grassland? *Folia Geobotanica*, **40**, 273–288.

Burden, J.P., Nixon, C.P., Hodgkinson, A.E., Possee, R.D., Sait, S.M., King, L.A. & Hails, R.S. (2003). Covert infections as a mechanism for long-term persistence of baculoviruses. *Ecology Letters*, **6**, 524–531.

Bush, S.E. & Malenke, J.R. (2008). Host defence mediates interspecific competition in ectoparasites. *Journal of Animal Ecology*, **77**, 558–564.

Byers, J.E. (2009). Including parasites in food webs. *Trends in Parasitology*, **25**, 55–57.

Byrne, C.J., Holland, C.V., Kennedy, C.R. & Poole, W.R. (2003). Interspecific interactions between Acanthocephala in the intestine of brown trout: are they more frequent in Ireland? *Parasitology*, **127**, 399–409.

Caesar, A.J. & Caesar-Ton That, T. (2008). Rhizosphere bacterial communities associated with insect root herbivory of an invasive plant, *Euphorbia esula/virgata*. In Julien, M.H., Sforza, R., Bon, M.C., Evans, H.C., Hatcher, P.E., Hinz, H.L. & Rector, B.G. (eds), *Proceedings of the XII Symposium on Biological Control of Weeds*. Wallingford: CAB International, pp. 13–19.

Calisher, C. H., Childs, J. E., Field, H. E., Holmes, K. V. & Schountz, T. (2006). Bats: important reservoir hosts of emerging viruses. *Clinical Microbiology Reviews*, **19**, 531–545.

Callaway, R. M. & Pennings, S. C. (1998). Impact of a parasitic plant on the zonation of two salt marsh perennials. *Oecologia*, **114**, 100–105.

Callaway, R. M. & Ridenour, W. M. (2004). Novel weapons: invasive success and the evolution of increased competitive ability. *Frontiers in Ecology and the Environment*, **2**, 436–443.

Callaway, R. M., Thelen, G. C., Rodriguez, A. & Holben, W. E. (2004). Soil biota and exotic plant invasion. *Nature*, **427**, 731–733.

Cameron, T. C., Wearing, H. J., Rohani, P. & Sait, S. M. (2005). A koinobiont parasitoid mediates competition and generates additive mortality in healthy host populations. *Oikos*, **110**, 620–628.

Cardoza, Y. J., Lait, C. G., Schmelz, E. A., Huang, J. & Tumlinson, J. H. (2003). Fungus-induced biochemical changes in peanut plants and their effect on development of beet armyworm, *Spodoptera exigua* Hubner (Lepidoptera: Noctuidae) larvae. *Environmental Entomology*, **32**, 220–228.

Carroll, B., Russell, P., Gurnell, J., Nettleton, P. & Sainsbury, A. W. (2009). Epidemics of squirrelpox virus disease in red squirrels (*Sciurus vulgaris*): temporal and serological findings. *Epidemiology and Infection*, **137**, 257–265.

Chadha, M. S., Comer, J. A., Lowe, L., Rota, P. A., Rollin, P. E., Bellini, W. J., Ksiazek, T. G. & Mishra, A. C. (2006). Nipah virus-associated encephalitis outbreak, Siliguri, India. *Emerging Infectious Diseases*, **12**, 235–240.

Chapuis, J. L., Bousses, P. & Barnaud, G. (1994). Alien mammals, impact and management in the French sub-Antarctic islands. *Biological Conservation*, **67**, 97–104.

Chattopadhyay, J. & Bairagi, N. (2001). Pelicans at risk in Salton Sea: an eco-epidemiological model. *Ecological Modelling*, **136**, 103–112.

Chattopadhyay, J., Srinivasu, P. D. N. & Bairagi, N. (2003). Pelicans at risk in Salton Sea: an eco-epidemiological model-II. *Ecological Modelling*, **167**, 199–211.

Chen, H. W., Liu, W. C., Davis, A. J., Jordan, F., Hwang, M. J. & Shao, K. T. (2008). Network position of hosts in food webs and their parasite diversity. *Oikos*, **117**, 1847–1855.

Chilcutt, C. F. & Tabashnik, B. E. (1997). Host-mediated competition between the pathogen *Bacillus thuringiensis* and the parasitoid *Cotesia plutellae* of the diamondback moth (Lepidoptera: Plutellidae). *Environmental Entomology*, **26**, 38–45.

Childs, J. E., Mackenzie, J. S. & Richt, J. A. (eds) (2007a). *Wildlife and Emerging Zoonotic Diseases: The Biology, Circumstances and Consequence of Cross-species Transmission*. Berlin: Springer.

Childs, J. E., Richt, J. A. & Mackenzie, J. S. (2007b). Introduction: conceptualizing and partitioning the emergence process of zoonotic viruses from wildlife to humans. In Childs, J. E., Mackenzie, J. S. & Richt, J. A. (eds) *Wildlife and Emerging Zoonotic Diseases: The Biology, Circumstances and Consequences of Cross-species Transmission*. Berlin: Springer, pp. 1–31.

Choisy, M. & Rohani, P. (2006). Harvesting can increase severity of wildlife disease epidemics. *Proceedings of the Royal Society of London Series B – Biological Sciences*, **273**, 2025–2034.

Chomel, B. B. (2008). Control and prevention of emerging parasitic zoonoses. *International Journal for Parasitology*, **38**, 1211–1217.

Chong, J. H. & Oetting, R. D. (2007). Intraguild predation and interference by the mealybug predator *Cryptolaemus montrouzieri* on the parasitoid *Leptomastix dactylopii*. *Biocontrol Science and Technology*, **17**, 933–944.

Choo, K., Williams, P. D. & Day, T. (2003). Host mortality, predation and the evolution of parasite virulence. *Ecology Letters*, **6**, 310–315.

Chua, K. B., Bellini, W. J., Rota, P. A., Harcourt, B. H., Tamin, A., Lam, S. K., Ksiazek, T. G., Rollin, P. E., Zaki, S. R., Shieh, W. J., Goldsmith, C. S., Gubler, D. J., Roehrig, J. T., Eaton, B., Gould, A. R., Olson, J., Field, H., Daniels, P., Ling, A. E., Peters, C. J., Anderson, L. J. & Mahy, B. W. J. (2000). Nipah virus: a recently emergent deadly paramyxovirus. *Science*, **288**, 1432–1435.

Clay, K. (1996). Interactions among fungal endophytes, grasses and herbivores. *Researches on Population Ecology*, **38**, 191–201.

Clay, K. & Holah, J. (1999). Fungal endophyte symbiosis and plant diversity in successional fields. *Science*, **285**, 1742–1744.

Clay, K., Holah, J. & Rudgers, J. A. (2005). Herbivores cause a rapid increase in hereditary symbiosis and alter plant community composition. *Proceedings of the National Academy of Sciences of the United States of America*, **102**, 12465–12470.

Clay, K., Marks, S. & Cheplick, G. P. (1993). Effects of insect herbivory and fungal endophyte infection on competitive interactions among grasses. *Ecology*, **74**, 1767–1777.

Clay, K. & Schardl, C. (2002). Evolutionary origins and ecological consequences of endophyte symbiosis with grasses. *American Naturalist*, **160**, S99–S127.

Cleaveland, S., Haydon, D. T. & Taylor, L. (2007). Overviews of pathogen emergence: which pathogens emerge, when and why? In Childs, J. E., Mackenzie, J. S. & Richt, J. A. (eds) *Wildlife and Emerging Zoonotic Diseases: The Biology, Circumstances and Consequences of Cross-species Transmission*. Berlin: Springer, pp. 85–111.

Cleaveland, S., Kaare, M., Tiringa, P., Mlengeya, T. & Barrat, J. (2003). A dog rabies vaccination campaign in rural Africa: impact on the incidence of dog rabies and human dog-bite injuries. *Vaccine*, **21**, 1965–1973.

Cohen, J. E., Jonsson, T. & Carpenter, S. R. (2003). Ecological community description using the food web, species abundance, and body size. *Proceedings of the National Academy of Sciences of the United States of America*, **100**, 1781–1786.

Colautti, R. I., Muirhead, J. R., Biswas, R. N. & MacIsaac, H. J. (2005). Realized vs apparent reduction in enemies of the European starling. *Biological Invasions*, **7**, 723–732.

Colautti, R. I., Ricciardi, A., Grigorovich, I. A. & MacIsaac, H. J. (2004). Is invasion success explained by the enemy release hypothesis? *Ecology Letters*, **7**, 721–733.

Collinge, S. K., Johnson, W. C., Ray, C., Matchett, R., Grensten, J., Cully, J. F., Jr, Gage, K. L., Kosoy, M. Y., Loye, J. E. & Martin, A. P. (2005). Testing the generality of a trophic-cascade model for plague. *Ecohealth*, **2**, 102–112.

Collinge, S. K. & Ray, C. (2006). *Disease Ecology: Community Structure and Pathogen Dynamics*. Oxford: Oxford University Press.

Connell, J. H. (1971). On the role of natural enemies in preventing competitive exclusion in some marine animals and in rain forest trees. In den Boer, P. J. & Gradwell, G. R. (eds), *Dynamics of Populations*. Wageningen: Centre for Agricultural Publishing and Documentation, pp. 298–312.

Cope, D. R., Iason, G. R. & Gordon, I. J. (2004). Disease reservoirs in complex systems: a comment on recent work by Laurenson et al. *Journal of Animal Ecology*, **73**, 807–810.

Cory, J. S. & Hoover, K. (2006). Plant-mediated effects in insect–pathogen interactions. *Trends in Ecology & Evolution*, **21**, 278–286.

Cory, J. S. & Myers, J. H. (2000). Direct and indirect ecological effects of biological control. *Trends in Ecology & Evolution*, **15**, 137–139.

Cossentine, J. E. & Lewis, L. C. (1988). Impact of *Nosema pyrausta, Nosema* sp. and a nuclear polyhedrosis virus on *Lydella thompsoni* within infected *Ostrinia nubilalis* hosts. *Journal of Invertebrate Pathology*, **51**, 126–132.

Cottrell, T. E. & Yeargan, K. V. (1998). Intraguild predation between an introduced lady beetle, *Harmonia axyridis* (Coleoptera: Coccinellidae), and a native lady beetle, *Coleomegilla maculata* (Coleoptera: Coccinellidae). *Journal of the Kansas Entomological Society*, **71**, 159–163.

Courchamp, F. & Sugihara, G. (1999). Modeling the biological control of an alien predator to protect island species from extinction. *Ecological Applications*, **9**, 112–123.

Craig, B. H., Pilkington, J. G., Kruuk, L. E. B. & Pemberton, J. M. (2007). Epidemiology of parasitic protozoan infections in Soay sheep (*Ovis aries* L.) on St Kilda. *Parasitology*, **134**, 9–21.

Cremer, S., Ugelvig, L. V., Drijfhout, F. P., Schlick-Steiner, B. C., Steiner, F. M., Seifert, B., Hughes, D. P., Schulz, A., Petersen, K. S., Konrad, H., Stauffer, C., Kiran, K., Espadaler, X., d'Ettorre, P., Aktac, N., Eilenberg, J., Jones, G. R., Nash, D. R., Pedersen, J. S. & Boomsma, J. J. (2008). The evolution of invasiveness in garden ants. *PLoS One*, **3**, e3838.

Cronin, J. T. (2007). Shared parasitoids in a metacommunity: indirect interactions inhibit herbivore membership in local communities. *Ecology*, **88**, 2977–2990.

Daszak, P., Cunningham, A. A. & Hyatt, A. D. (2000). Wildlife ecology: emerging infectious diseases of wildlife – threats to biodiversity and human health. *Science*, **287**, 443–449.

Daszak, P., Epstein, J. H., Kilpatrick, A. M., Aguirre, A. A., Karesh, W. B. & Cunningham, A. A. (2007). Collaborative research approaches to the role of wildlife in zoonotic disease emergence. In Childs, J. E., Mackenzie, J. S. & Richt, J. A. (eds) *Wildlife and Emerging Zoonotic Diseases: The Biology, Circumstances and Consequences of Cross-species Transmission*. Berlin: Springer, pp. 463–475.

Davies, C. M., Fairbrother, E. & Webster, J. P. (2002). Mixed strain schistosome infections of snails and the evolution of parasite virulence. *Parasitology*, **124**, 31–38.

Davies, D. M., Graves, J. D., Elias, C. O. & Williams, P. J. (1997). The impact of *Rhinanthus* spp. on sward productivity and composition: implications for the restoration of species-rich grasslands. *Biological Conservation*, **82**, 87–93.

Dawkins, R. (1982). *The Extended Phenotype*. Oxford: Oxford University Press.

Day, J. F. & Edman, J. D. (1983). Malaria renders mice susceptible to mosquito feeding when gametocytes are most infective. *Journal of Parasitology*, **69**, 163–170.

Day, T. & Gandon, S. (2007). Applying population–genetic models in theoretical evolutionary epidemiology. *Ecology Letters*, **10**, 876–888.

de Castro, F. & Bolker, B. (2005). Mechanisms of disease-induced extinction. *Ecology Letters*, **8**, 117–126.

De Moraes, C. M., Mescher, M. C. & Tumlinson, J. H. (2001). Caterpillar-induced nocturnal plant volatiles repel conspecific females. *Nature*, **410**, 577–580.

de Roode, J. C., Helinski, M. E. H., Anwar, M. A. & Read, A. F. (2005a). Dynamics of multiple infection and within-host competition in genetically diverse malaria infections. *American Naturalist*, **166**, 531–542.

de Roode, J. C., Pansini, R., Cheesman, S. J., Helinski, M. E. H., Huijben, S., Wargo, A. R., Bell, A. S., Chan, B. H. K., Walliker, D. & Read, A. F. (2005b). Virulence and competitive ability in genetically diverse malaria infections. *Proceedings of the National Academy of Sciences of the United States of America*, **102**, 7624–7628.

De Vos, M., Van Oosten, V. R., Van Poecke, R. M. P., Van Pelt, J. A., Pozo, M. J., Mueller, M. J., Buchala, A. J., Metraux, J. P., Van Loon, L. C., Dicke, M. & Pieterse, C. M. J. (2005). Signal signature and transcriptome changes of Arabidopsis during pathogen and insect attack. *Molecular Plant–Microbe Interactions*, **18**, 923–937.

Dennehy, J. J., Friedenberg, N. A., Holt, R. D. & Turner, P. E. (2006). Viral ecology and the maintenance of novel host use. *American Naturalist*, **167**, 429–439.

Diamond, J. M. (1997). *Guns, Germs and Steel: The Fates of Human Societies*. New York: W.W. Norton.

Dick, J. T. A. (1996). Post-invasion amphipod communities of Lough Neagh, Northern Ireland: influences of habitat selection and mutual predation. *Journal of Animal Ecology*, **65**, 756–767.

Dick, J. T. A. (2008). Role of behaviour in biological invasions and species distributions: lessons from interactions between the invasive *Gammarus pulex* and the native *G. duebeni* (Crustacea: Amphipoda). *Contributions to Zoology*, **77**, 91–98.

Dick, J. T. A., Armstrong, M., Clarke, H. C., Farnesworth, K. D., Hatcher, M. J., Ennis, M., Kelly A. & Dunn, A. M. (2010). Parasitism may enhance rather than reduce the predatory impact of an invader. *Biology Letters*, **6**, 636–638.

Dick, J. T. A., Montgomery, I. & Elwood, R. W. (1993). Replacement of the indigenous amphipod *Gammarus duebeni celticus* by the introduced *Gammarus pulex*: differential cannibalism and mutual predation. *Journal of Animal Ecology*, **62**, 79–88.

Dick, J. T. A. & Platvoet, D. (1996). Intraguild predation and species exclusions in amphipods: the interaction of behaviour, physiology and environment. *Freshwater Biology*, **36**, 375–383.

Dicke, M. & Baldwin, I. T. (2010). The evolutionary context for herbivore-induced plant volatiles: beyond the 'cry for help'. *Trends in Plant Science*, **15**, 167–175.

Dicke, M. & Dijkman, H. (1992). Induced defense in detached uninfested plant leaves: effects on behavior of herbivores and their predators. *Oecologia*, **91**, 554–560.

Diehl, S. & Feissel, M. (2000). Effects of enrichment on three-level food chains with omnivory. *American Naturalist*, **155**, 200–218.

Diekmann, O., Heesterbeek, J. A. P. & Metz, J. A. J. (1990). On the definition and the computation of the basic reproduction ratio R_0 in models for infectious diseases in heterogeneous populations. *Journal of Mathematical Biology*, **28**, 365–382.

Dillon, R. J., Vennard, C. T., Buckling, A. & Charnley, A. K. (2005). Diversity of locust gut bacteria protects against pathogen invasion. *Ecology Letters*, **8**, 1291–1298.

Dobson, A. (2004). Population dynamics of pathogens with multiple host species. *American Naturalist*, **164**, S64–S78.

Dobson, A. (2009). Climate variability, global change, immunity, and the dynamics of infectious diseases. *Ecology*, **90**, 920–927.

Dobson, A., Cattadori, I., Holt, R. D., Ostfeld, R. S., Keesing, F., Krichbaum, K., Rohr, J. R., Perkins, S. E. & Hudson, P. J. (2006). Sacred cows and sympathetic squirrels: the importance of biological diversity to human health. *PLoS Med*, **3**, e231.

Dobson, A. & Crawley, W. (1994). Pathogens and the structure of plant communities. *Trends in Ecology & Evolution*, **9**, 393–398.

Dobson, A. & Foufopoulos, J. (2001). Emerging infectious pathogens of wildlife. *Philosophical Transactions of the Royal Society of London Series B – Biological Sciences*, **356**, 1001–1012.

Dobson, A., Lafferty, K. D., Kuris, A. M., Hechinger, R. F. & Jetz, W. (2008). Homage to Linnaeus: how many parasites? How many hosts? *Proceedings of the National Academy of Sciences of the United States of America*, **105**, 11482–11489.

Dobson, A. P. (1985). The population dynamics of competition between parasites. *Parasitology*, **91**, 317–347.

Dobson, A. P. (1988). The population biology of parasite-induced changes in host behavior. *Quarterly Review of Biology*, **63**, 139–165.

Dobson, A. P. (1995). The ecology and epidemiology of rinderpest virus in Serengeti and Ngorongoro crater conservation area. In Sinclair, A. R. E. & Arcese, P. (eds), *Serengeti II: Research, Management and Conservation of an Ecosystem*. Chicago: University of Chicago Press, pp. 485–505.

Dobson, A. P. (2005). What links bats to emerging infectious diseases? *Science*, **310**, 628–629.

Dobson, A. P. & Carper, E. R. (1996). Infectious diseases and human population history: throughout history the establishment of disease has been a side effect of the growth of civilization. *Bioscience*, **46**, 115–126.

Dobson, A. P. & Hudson, P. J. (1986). Parasites, disease and the structure of ecological communities. *Trends in Ecology & Evolution*, **1**, 11–15.

Dobson, A. P. & Hudson, P. J. (1992). Regulation and stability of a free-living host–parasite system: *Trichostrongylus tenuis* in red grouse – 2. Population models. *Journal of Animal Ecology*, **61**, 487–498.

Dobson, A. P. & Keymer, A. E. (1985). Life history models. In Crompton, D. W. T. & Nickol, B. B. (eds), *Biology of the Acanthocephala*. Cambridge: Cambridge University Press, pp. 347–384.

Dobson, A. P. & May, R. M. (1987). The effects of parasites on fish populations: theoretical aspects. *International Journal for Parasitology*, **17**, 363–370.

Drake, J. M. (2003). The paradox of the parasites: implications for biological invasion. *Proceedings of the Royal Society of London Series B – Biological Sciences*, **270**, S133–S135.

Duffy, M. A. & Hall, S. R. (2008). Selective predation and rapid evolution can jointly dampen effects of virulent parasites on *Daphnia* populations. *American Naturalist*, **171**, 499–510.

Duffy, M. A., Hall, S. R., Tessier, A. J. & Huebner, M. (2005). Selective predators and their parasitized prey: are epidemics in zooplankton under top-down control? *Limnology and Oceanography*, **50**, 412–420.

Duffy, M. A. & Sivars-Becker, L. (2007). Rapid evolution and ecological host–parasite dynamics. *Ecology Letters*, **10**, 44–53.

Dumont, A. & Crete, M. (1996). The meningeal worm, *Parelaphostrongylus tenuis*, a marginal limiting factor for moose, *Alces alces*, in Southern Quebec. *Canadian Field-Naturalist*, **110**, 413–418.

Dunn, A. M. (2009). Parasites and biological invasions. *Advances in Parasitology*, **68**, 161–184.

Dunn, A. M. & Dick, J. T. A. (1998). Parasitism and epibiosis in native and non-native gammarids in freshwater in Ireland. *Ecography*, **21**, 593–598.

Dunn, A. M., Terry, R. S. & Smith, J. E. (2001). Transovarial transmission in the microsporidia. *Advances in Parasitology*, **48**, 57–100.

Dunn, J. C., McClymont, H. E., Christmas, M. & Dunn, A. M. (2009). Competition and parasitism in the native white clawed crayfish *Austropotamobius pallipes* and the invasive signal crayfish *Pacifastacus leniusculus* in the UK. *Biological Invasions*, **11**, 315–324.

Dunne, J. A., Williams, R. J. & Martinez, N. D. (2002a). Food-web structure and network theory: the role of connectance and size. *Proceedings of the National Academy of Sciences of the United States of America*, **99**, 12917–12922.

Dunne, J. A., Williams, R. J. & Martinez, N. D. (2002b). Network structure and biodiversity loss in food webs: robustness increases with connectance. *Ecology Letters*, **5**, 558–567.

Duron, O., Bouchon, D., Boutin, S., Bellamy, L., Zhou, L. Q., Engelstadter, J. & Hurst, G. D. (2008). The diversity of reproductive parasites among arthropods: *Wolbachia* do not walk alone. *BMC Biology*, **6**, e27.

Dwosh, H. A., Hong, H. H. L., Austgarden, D., Herman, S. & Schabas, R. (2003). Identification and containment of an outbreak of SARS in a community hospital. *Canadian Medical Association Journal*, **168**, 1415–1420.

Dwyer, G., Dushoff, J. & Yee, S. H. (2004). The combined effects of pathogens and predators on insect outbreaks. *Nature*, **430**, 341–345.

Dybdahl, M. F. & Lively, C. M. (1995). Host–parasite interactions: infection of common clones in natural populations of a fresh-water snail (*Potamopyrgus antipodarum*). *Proceedings of the Royal Society of London Series B – Biological Sciences*, **260**, 99–103.

Dye, C. & Gay, N. (2003). Modeling the SARS epidemic. *Science*, **300**, 1884–1885.

Edeline, E., Ben Ari, T., Vollestad, L. A., Winfield, I. J., Fletcher, J. M., Ben James, J. & Stenseth, N. C. (2008). Antagonistic selection from predators and pathogens alters food-web structure. *Proceedings of the National Academy of Sciences of the United States of America*, **105**, 19792–19796.

Elton, C. S. (1958). *The Ecology of Invasions by Animals and Plants.* London: Methuen.

Engelberth, J., Alborn, H. T., Schmelz, E. A. & Tumlinson, J. H. (2004). Airborne signals prime plants against insect herbivore attack. *Proceedings of the National Academy of Sciences of the United States of America,* **101,** 1781–1785.

Enscore, R. E., Biggerstaff, B. J., Brown, T. L., Fulgham, R. F., Reynolds, P. J., Engelthaler, D. M., Levy, C. E., Parmenter, R. R., Montenieri, J. A., Cheek, J. E., Grinnell, R. K., Ettestad, P. J. & Gage, K. L. (2002). Modeling relationships between climate and the frequency of human plague cases in the southwestern United States, 1960–1997. *American Journal of Tropical Medicine and Hygiene,* **66,** 186–196.

Epstein, P. R. (2000). Is global warming harmful to health? *Scientific American,* **283** (2), 50–57.

Epstein, P. R., Diaz, H. F., Elias, S., Grabherr, G., Graham, N. E., Martens, W. J. M., Mosley-Thompson, E. & Susskind, J. (1998). Biological and physical signs of climate change: focus on mosquito-borne diseases. *Bulletin of the American Meteorological Society,* **79,** 409–417.

Evans, H. C. (2008). The endophyte–enemy release hypothesis: implications for classical biological control and plant invasions. In Julien, M. H., Sforza, R., Bon, M. C., Evans, H. C., Hatcher, P. E., Hinz, H. L. & Rector, B. G. (eds), *Proceedings of the XII Symposium on Biological Control of Weeds.* Wallingford: CAB International, pp. 20–25.

Ezenwa, V. O., Godsey, M. S., King, R. J. & Guptill, S. C. (2006). Avian diversity and West Nile virus: testing associations between biodiversity and infectious disease risk. *Proceedings of the Royal Society of London Series B – Biological Sciences,* **273,** 109–117.

Ezenwa, V. O., Milheim, L. E., Coffey, M. F., Godsey, M. S., King, R. J. & Guptill, S. C. (2007). Land cover variation and West Nile virus prevalence: patterns, processes, and implications for disease control. *Vector-Borne and Zoonotic Diseases,* **7,** 173–180.

Fa, J. E., Peres, C. A. & Meeuwig, J. (2002). Bushmeat exploitation in tropical forests: an intercontinental comparison. *Conservation Biology,* **16,** 232–237.

Fauchier, J. & Thomas, F. (2001). Interaction between *Gammarinema gammari* (Nematoda), *Microphallus papillorobustus* (Trematoda) and their common host *Gammarus insensibilis* (Amphipoda). *Journal of Parasitology,* **87,** 1479–1481.

Feener, D. H. (1981). Competition between ant species: outcome controlled by parasitic flies. *Science,* **214,** 815–817.

Feener, D. H. (2000). Is the assembly of ant communities mediated by parasitoids? *Oikos,* **90,** 79–88.

Fenton, A. & Pedersen, A. B. (2005). Community epidemiology framework for classifying disease threats. *Emerging Infectious Diseases,* **11,** 1815–1821.

Fenton, A. & Rands, S. A. (2006). The impact of parasite manipulation and predator foraging behavior on predator–prey communities. *Ecology,* **87,** 2832–2841.

Ferguson, K. I. & Stiling, P. (1996). Non-additive effects of multiple natural enemies on aphid populations. *Oecologia,* **108,** 375–379.

Ferrari, N., Cattadori, I. M., Rizzoli, A. & Hudson, P. J. (2009). *Heligmosomoides polygyrus* reduces infestation of *Ixodes ricinus* in free-living yellow-necked mice, *Apodemus flavicollis.* *Parasitology,* **136,** 305–316.

Ferrer, M. & Negro, J. J. (2004). The near extinction of two large European predators: super specialists pay a price. *Conservation Biology*, **18**, 344–349.

Field, H. E., Mackenzie, J. S. & Daszak, P. (2007). Henipaviruses: emerging paramyxoviruses associated with fruit bats. In Childs, J. E., Mackenzie, J. S. & Richt, J. A. (eds) *Wildlife and Emerging Zoonotic Diseases: The Biology, Circumstances and Consequences of Cross-species Transmission*. Berlin: Springer, pp. 133–159.

Fielding, N. J., MacNeil, C., Dick, J. T. A., Elwood, R. W., Riddell, G. E. & Dunn, A. M. (2003). Effects of the acanthocephalan parasite *Echinorhynchus truttae* on the feeding ecology of *Gammarus pulex* (Crustacea: Amphipoda). *Journal of Zoology*, **261**, 321–325.

Fielding, N. J., MacNeil, C., Robinson, N., Dick, J. T. A., Elwood, R. W., Terry, R. S., Ruiz, Z. & Dunn, A. M. (2005). Ecological impacts of the microsporidian parasite *Pleistophora mulleri* on its freshwater amphipod host *Gammarus duebeni celticus*. *Parasitology*, **131**, 331–336.

Finkes, L. K., Cady, A. B., Mulroy, J. C., Clay, K. & Rudgers, J. A. (2006). Plant–fungus mutualism affects spider composition in successional fields. *Ecology Letters*, **9**, 344–353.

Firlej, A., Boivin, G., Lucas, E. & Coderre, D. (2005). First report of *Harmonia axyridis* Pallas being attacked by *Dinocampus coccinellae* Schrank in Canada. *Biological Invasions*, **7**, 553–556.

Frank, S. A. (1996). Models of parasite virulence. *Quarterly Review of Biology*, **71**, 37–78.

Fransen, J. J. & Vanlenteren, J. C. (1993). Host selection and survival of the parasitoid *Encarsia formosa* on greenhouse-whitefly, *Trialeurodes vaporariorum*, in the presence of hosts infected with the fungus *Aschersonia aleyrodis*. *Entomologia experimentalis et Applicata*, **69**, 239–249.

Fraser, C., Riley, S., Anderson, R. M. & Ferguson, N. M. (2004). Factors that make an infectious disease outbreak controllable. *Proceedings of the National Academy of Sciences of the United States of America*, **101**, 6146–6151.

Freckleton, R. P. & Lewis, O. T. (2006). Pathogens, density dependence and the coexistence of tropical trees. *Proceedings of the Royal Society of London Series B – Biological Sciences*, **273**, 2909–2916.

Fredensborg, B. L. & Poulin, R. (2005). Larval helminths in intermediate hosts: does competition early in life determine the fitness of adult parasites? *International Journal for Parasitology*, **35**, 1061–1070.

Freedman, H. I. (1990). A model of predator–prey dynamics as modified by the action of a parasite. *Mathematical Biosciences*, **99**, 143–155.

Furlong, M. J. & Pell, J. K. (1996). Interactions between the fungal entomopathogen *Zoophthora radicans* Brefeld (Entomophthorales) and two hymenopteran parasitoids attacking the diamondback moth, *Plutella xylostella* L. *Journal of Invertebrate Pathology*, **68**, 15–21.

Futerman, P. H., Layen, S. J., Kotzen, M. L., Franzen, C., Kraaijeveld, A. R. & Godfray, H. C. J. (2006). Fitness effects and transmission routes of a microsporidian parasite infecting *Drosophila* and its parasitoids. *Parasitology*, **132**, 479–492.

Galbreath, J., Smith, J. E., Terry, R. S., Becnel, J. J. & Dunn, A. M. (2004). Invasion success of *Fibrillanosema crangonycis*, n.sp., n.g.: a novel vertically transmitted microsporidian parasite from the invasive amphipod host *Crangonyx pseudogracilis*. *International Journal for Parasitology*, **34**, 235–244.

Gallup, J. L. & Sachs, J. D. (2001). The economic burden of malaria. *American Journal of Tropical Medicine and Hygiene*, **64**, 85–96.

Gandon, S. (2002). Local adaptation and the geometry of host–parasite coevolution. *Ecology Letters*, **5**, 246–256.

Gandon, S., Capowiez, Y., Dubois, Y., Michalakis, Y. & Olivieri, I. (1996). Local adaptation and gene-for-gene coevolution in a metapopulation model. *Proceedings of the Royal Society of London Series B – Biological Sciences*, **263**, 1003–1009.

Garner, T. W. J., Perkins, M. W., Govindarajulu, P., Seglie, D., Walker, S., Cunningham, A. A. & Fisher, M. C. (2006). The emerging amphibian pathogen *Batrachochytrium dendrobatidis* globally infects introduced populations of the North American bullfrog, *Rana catesbeiana*. *Biology Letters*, **2**, 455–459.

Gatherer, D. (2009). The 2009 H1N1 influenza outbreak in its historical context. *Journal of Clinical Virology*, **45**, 174–178.

Genner, M. J., Michel, E. & Todd, J. A. (2008). Resistance of an invasive gastropod to an indigenous trematode parasite in Lake Malawi. *Biological Invasions*, **10**, 41–49.

George-Nascimento, M. A. & Marin, S. L. (1992). Effects induced by two host species, the South American fur-seal *Arctocephalus australis* (Zimmerman), and the South American sea lion *Otaria Byronia* (Blainville) (Carnivora, Otariidae), on the morphology and fecundity of *Corynosoma* sp. (Acanthocephala, Polymorphidae) in Uruguay. *Revista Chilena De Historia Natural*, **65**, 183–193.

Georgiev, B. B., Sanchez, M. I., Vasileva, G. P., Nikolov, P. N. & Green, A. J. (2007). Cestode parasitism in invasive and native brine shrimps (*Artemia* spp.) as a possible factor promoting the rapid invasion of *A. franciscana* in the Mediterranean region. *Parasitology Research*, **101**, 1647–1655.

Gibbs, J. N. (1978). Intercontinental epidemiology of Dutch elm disease. *Annual Review of Phytopathology*, **16**, 287–307.

Gibson, C. C. & Watkinson, A. R. (1992). The role of the hemiparasitic annual *Rhinanthus minor* in determining grassland community structure. *Oecologia*, **89**, 62–68.

Gilbert, L., Norman, R., Laurenson, K. M., Reid, H. W. & Hudson, P. J. (2001). Disease persistence and apparent competition in a three-host community: an empirical and analytical study of large-scale, wild populations. *Journal of Animal Ecology*, **70**, 1053–1061.

Gilbert, M. T. P., Rambaut, A., Wlasiuk, G., Spira, T. J., Pitchenik, A. E. & Worobey, M. (2007). The emergence of HIV/AIDS in the Americas and beyond. *Proceedings of the National Academy of Sciences of the United States of America*, **104**, 18566–18570.

Gilligan, C. A. & Kleczkowski, A. (1997). Population dynamics of botanical epidemics involving primary and secondary infection. *Philosophical Transactions of the Royal Society of London Series B – Biological Sciences*, **352**, 591–608.

Gilligan, C. A. & van den Bosch, F. (2008). Epidemiological models for invasion and persistence of pathogens. *Annual Review of Phytopathology*, **46**, 385–418.

Godfree, R. C., Tinnin, R. O. & Forbes, R. B. (2002). The effects of dwarf mistletoe, witches' brooms, stand structure, and site characteristics on the crown architecture of lodgepole pine in Oregon. *Canadian Journal of Forest Research*, **32**, 1360–1371.

Graham, A. L. (2008). Ecological rules governing helminth–microparasite coinfection. *Proceedings of the National Academy of Sciences of the United States of America*, **105**, 566–570.

Greenman, J. V. & Hudson, P. J. (1997). Infected coexistence instability with and without density-dependent regulation. *Journal of Theoretical Biology*, **185**, 345–356.

Greenman, J. V. & Hudson, P. J. (1999). Host exclusion and coexistence in apparent and direct competition: an application of bifurcation theory. *Theoretical Population Biology*, **56**, 48–64.

Greenman, J. V. & Hudson, P. J. (2000). Parasite-mediated and direct competition in a two-host shared macroparasite system. *Theoretical Population Biology*, **57**, 13–34.

Greger, M. (2007). The human/animal interface: emergence and resurgence of zoonotic infectious diseases. *Critical Reviews in Microbiology*, **33**, 243–299.

Grenfell, B. T. (1988). Gastrointestinal nematode parasites and the stability and productivity of intensive ruminant grazing systems. *Philosophical Transactions of the Royal Society of London Series B – Biological Sciences*, **321**, 541–563.

Grenfell, B. T. (1992). Parasitism and the dynamics of ungulate grazing systems. *American Naturalist*, **139**, 907–929.

Grewell, B. J. (2008). Parasite facilitates plant species coexistence in a coastal wetland. *Ecology*, **89**, 1481–1488.

Grosholz, E. D. (1992). Interactions of intraspecific, interspecific, and apparent competition with host–pathogen population dynamics. *Ecology*, **73**, 507–514.

Grosholz, E. D. & Ruiz, G. M. (2003). Biological invasions drive size increases in marine and estuarine invertebrates. *Ecology Letters*, **6**, 700–705.

Grossman, Z., Meier-Schellersheim, M., Sousa, A. E., Victorino, R. M. M. & Paul, W. E. (2002). CD4(+) T-cell depletion in HIV infection: are we closer to understanding the cause? *Nature Medicine*, **8**, 319–323.

Gubbins, S., Gilligan, C. A. & Kleczkowski, A. (2000). Population dynamics of plant–parasite interactions: thresholds for invasion. *Theoretical Population Biology*, **57**, 219–233.

Guernier, V., Hochberg, M. E. & Guegan, J. F. O. (2004). Ecology drives the worldwide distribution of human diseases. *PLoS Biology*, **2**, 740–746.

Gumel, A. B., Ruan, S. G., Day, T., Watmough, J., Brauer, F., van den Driessche, P., Gabrielson, D., Bowman, C., Alexander, M. E., Ardal, S., Wu, J. H. & Sahai, B. M. (2004). Modelling strategies for controlling SARS outbreaks. *Proceedings of the Royal Society of London Series B – Biological Sciences*, **271**, 2223–2232.

Gurnell, J., Rushton, S. P., Lurz, P. W. W., Sainsbury, A. W., Nettleton, P., Shirley, M. D. F., Bruemmer, C. & Geddes, N. (2006). Squirrel poxvirus: landscape scale strategies for managing disease threat. *Biological Conservation*, **131**, 287–295.

Gurnell, J., Wauters, L. A., Lurz, P. W. W. & Tosi, G. (2004). Alien species and interspecific competition: effects of introduced eastern grey squirrels on red squirrel population dynamics. *Journal of Animal Ecology*, **73**, 26–35.

Hadeler, K. P. & Freedman, H. I. (1989). Predator–prey populations with parasitic infection. *Journal of Mathematical Biology*, **27**, 609–631.

Hahn, B. H., Shaw, G. M., De Cock, K. M. & Sharp, P. M. (2000). AIDS as a zoonosis – scientific and public health implications. *Science*, **287**, 607–614.

Haine, E. R., Boucansaud, K. & Rigaud, T. (2005). Conflict between parasites with different transmission strategies infecting an amphipod host. *Proceedings of the Royal Society of London Series B – Biological Sciences*, **272**, 2505–2510.

Hall, S. R., Becker, C. R., Simonis, J. L., Duffy, M. A., Tessier, A. L. & Cáceres, C. E. (2009). Friendly competition: evidence for a dilution effect among competitors in a planktonic host–parasite system. *Ecology*, **90**, 791–801.

Hall, S. R., Duffy, M. A. & Caceres, C. E. (2005). Selective predation and productivity jointly drive complex behavior in host–parasite systems. *American Naturalist*, **165**, 70–81.

Hall, S. R., Tessier, A. J., Duffy, M. A., Huebner, M. & Cáceres, C. E. (2006). Warmer does not have to mean sicker: temperature and predators can jointly drive timing of epidemics. *Ecology*, **87**, 1684–1695.

Hamback, P. A., Stenberg, J. A. & Ericson, L. (2006). Asymmetric indirect interactions mediated by a shared parasitoid: connecting species traits and local distribution patterns for two chrysomelid beetles. *Oecologia*, **148**, 475–481.

Hanley, K. A., Petren, K. & Case, T. J. (1998). An experimental investigation of the competitive displacement of a native gecko by an invading gecko: no role for parasites. *Oecologia*, **115**, 196–205.

Hanley, K. A., Vollmer, D. M. & Case, T. J. (1995). The distribution and prevalence of helminths, coccidia and blood parasites in two competing species of gecko: implications for apparent competition. *Oecologia*, **102**, 220–229.

Hanselmann, R., Rodriguez, A., Lampo, M., Fajardo-Ramos, L., Aguirre, A. A., Kilpatrick, A. M., Rodriguez, J. P. & Daszak, P. (2004). Presence of an emerging pathogen of amphibians in introduced bullfrogs *Rana catesbiana* in Venezuela. *Biological Conservation*, **120**, 115–119.

Haque, M. & Venturino, E. (2006). The role of transmissible diseases in the Holling–Tanner predator–prey model. *Theoretical Population Biology*, **70**, 273–288.

Harkonen, T., Harding, K., Rasmussen, T. D., Teilmann, J. & Dietz, R. (2007). Age- and sex-specific mortality patterns in an emerging wildlife epidemic: the phocine distemper in European harbour seals. *PLoS One*, **2**, e887.

Harmon, J. P. & Andow, D. A. (2004). Indirect effects between shared prey: predictions for biological control. *Biocontrol*, **49**, 605–626.

Hartemink, N. A., Davis, S. A., Reiter, P., Hubalek, Z. & Heesterbeek, J. A. P. (2007). Importance of bird-to-bird transmission for the establishment of West Nile virus. *Vector-Borne and Zoonotic Diseases*, **7**, 575–584.

Harvell, C. D., Mitchell, C. E., Ward, J. R., Altizer, S., Dobson, A. P., Ostfeld, R. S. & Samuel, M. D. (2002). Ecology: climate warming and disease risks for terrestrial and marine biota. *Science*, **296**, 2158–2162.

Hatcher, M. J., Dick, J. T. A. & Dunn, A. M. (2006). How parasites affect interactions between competitors and predators. *Ecology Letters*, **9**, 1253–1271.

Hatcher, M. J., Dick, J. T. A. & Dunn, A. M. (2008). A keystone effect for parasites in intraguild predation? *Biology Letters*, **4**, 534–537.

Hatcher, M. J., Taneyhill, D. E., Dunn, A. M. & Tofts, C. (1999). Population dynamics under parasitic sex ratio distortion. *Theoretical Population Biology*, **56**, 11–28.

Hatcher, M. J. & Tofts, C. (2004). Reductionism isn't functional. HP Laboratories Technical Report HPL-2004-222.

Hatcher, P. E. (1995). Three-way interactions between plant pathogenic fungi, herbivorous insects and their host plants. *Biological Reviews*, **70**, 639–694.

Hatcher, P. E., Moore, J., Taylor, J. E., Tinney, G. W. & Paul, N. D. (2004). Phytohormones and plant–herbivore–pathogen interactions: integrating the molecular with the ecological. *Ecology*, **85**, 59–69.

Hatcher, P. E. & Paul, N. D. (2000). Beetle grazing reduces natural infection of *Rumex obtusifolius* by fungal pathogens. *New Phytologist*, **146**, 325–333.

Hatcher, P. E. & Paul, N. D. (2001). Plant pathogen–herbivore interactions and their effects on weeds. In Jeger, M. J. & Spence, N. J. (eds), *Biotic Interactions in Plant–Pathogen Associations*. Wallingford: CAB International, pp. 193–225.

Hatcher, P. E., Paul, N. D., Ayres, P. G. & Whittaker, J. B. (1994a). The effect of a foliar disease (rust) on the development of *Gastrophysa viridula* (Coleoptera, Chrysomelidae). *Ecological Entomology*, **19**, 349–360.

Hatcher, P. E., Paul, N. D., Ayres, P. G. & Whittaker, J. B. (1994b). Interactions between *Rumex* spp., herbivores and a rust fungus: *Gastrophysa viridula* grazing reduces subsequent infection by *Uromyces rumicis*. *Functional Ecology*, **8**, 265–272.

Hawkins, B. A. & Cornell, H. V. (eds) (1999). *Theoretical Approaches to Biological Control*. Cambridge: Cambridge University Press.

Hay, S. I., Guerra, C. A., Tatem, A. J., Noor, A. M. & Snow, R. W. (2004). The global distribution and population at risk of malaria: past, present, and future. *Lancet Infectious Diseases*, **4**, 327–336.

Haydon, D. T., Randall, D. A., Matthews, L., Knobel, D. L., Tallents, L. A., Gravenor, M. B., Williams, S. D., Pollinger, J. P., Cleaveland, S., Woolhouse, M. E. J., Sillero-Zubiri, C., Marino, J., Macdonald, D. W. & Laurenson, M. K. (2006). Low-coverage vaccination strategies for the conservation of endangered species. *Nature*, **443**, 692–695.

Hayes, E. B., Komar, N., Nasci, R. S., Montgomery, S. P., O'Leary, D. R. & Campbell, G. L. (2005). Epidemiology and transmission dynamics of West Nile virus disease. *Emerging Infectious Diseases*, **11**, 1167–1173.

Hearne, S. J. (2009). Control: the *Striga* conundrum. *Pest Management Science*, **65**, 603–614.

Hechinger, R. F. & Lafferty, K. D. (2005). Host diversity begets parasite diversity: bird final hosts and trematodes in snail intermediate hosts. *Proceedings of the Royal Society of London Series B – Biological Sciences*, **272**, 1059–1066.

Hechinger, R. F., Lafferty, K. D., Huspeni, T. C., Brooks, A. J. & Kuris, A. M. (2007). Can parasites be indicators of free-living diversity? Relationships between species richness and the abundance of larval trematodes and of local benthos and fishes. *Oecologia*, **151**, 82–92.

Heil, M. (2008). Indirect defence via tritrophic interactions. *New Phytologist*, **178**, 41–61.

Heinz, K. M. & Nelson, J. M. (1996). Interspecific interactions among natural enemies of *Bemisia* in an inundative biological control program. *Biological Control*, **6**, 384–393.

Henneman, M. L. & Memmott, J. (2001). Infiltration of a Hawaiian community by introduced biological control agents. *Science*, **293**, 1314–1316.

Henttonen, H., Fuglei, E., Gower, C. N., Haukisalmi, V., Ims, R. A., Niemimaa, J. & Yoccoz, N. G. (2001). *Echinococcus multilocularis* on Svalbard: introduction of an intermediate host has enabled the local life-cycle. *Parasitology*, **123**, 547–552.

Hernandez, A. D. & Sukhdeo, M. V. K. (2008a). Parasite effects on isopod feeding rates can alter the host's functional role in a natural stream ecosystem. *International Journal for Parasitology*, **38**, 683–690.

Hernandez, A. D. & Sukhdeo, M. V. K. (2008b). Parasites alter the topology of a stream food web across seasons. *Oecologia*, **156**, 613–624.

Herrick, N. J., Reitz, S. R., Carpenter, J. E. & O'Brien, C. W. (2008). Predation by *Podisus maculiventris* (Hemiptera: Pentatomidae) on *Plutella xylostella* (Lepidoptera: Plutellidae) larvae parasitized by *Cotesia plutellae* (Hymenoptera: Braconidae) and its impact on cabbage. *Biological Control*, **45**, 386–395.

Hethcote, H. W., Wang, W. D., Han, L. T. & Zhien, M. (2004). A predator–prey model with infected prey. *Theoretical Population Biology*, **66**, 259–268.

Hilker, F. M. & Schmitz, K. (2008). Disease-induced stabilization of predator–prey oscillations. *Journal of Theoretical Biology*, **255**, 299–306.

Hoch, G., Schopf, A. & Maddox, J. V. (2000). Interactions between an entomopathogenic microsporidium and the endoparasitoid *Glyptapanteles liparidis* within their host, the gypsy moth larva. *Journal of Invertebrate Pathology*, **75**, 59–68.

Hochberg, M. E., Hassell, M. P. & May, R. M. (1990). The dynamics of host–parasitoid–pathogen interactions. *American Naturalist*, **135**, 74–94.

Hochberg, M. E. & Holt, R. D. (1990). The coexistence of competing parasites:1 – the role of cross-species infection. *American Naturalist*, **136**, 517–541.

Hochberg, M. E. & Lawton, J. H. (1990). Competition between kingdoms. *Trends in Ecology & Evolution*, **5**, 367–371.

Holdich, D. M. (2002). *Biology of Freshwater Crayfish*. Oxford: Blackwell Science.

Holdich, D. M. & Poeckl, M. (2007). Invasive crustaceans in European inland waters. In Gherardi, F. (ed.), *Biological Invaders in Inland Waters: Profiles, Distribution and Threats*. Netherlands: Springer, pp. 29–75.

Holdo, R. M., Sinclair, A. R. E., Dobson, A. P., Metzger, K. L., Bolker, B. M., Ritchie, M. E. & Holt, R. D. (2009). A disease-mediated trophic cascade in the Serengeti and its implications for ecosystem C. *PLoS Biology*, **7**, e1000210.

Holt, R. D. (1977). Predation, apparent competition, and structure of prey communities. *Theoretical Population Biology*, **12**, 197–229.

Holt, R. D. (1984). Spatial heterogeneity, indirect interactions, and the coexistence of prey species. *American Naturalist*, **124**, 377–406.

Holt, R. D. (1997). Community modules. In Gange, A. C. & Brown, V. K. (eds), *Multitrophic Interactions in Terrestrial Systems*. London: Blackwell Science, pp. 333–350.

Holt, R. D. & Dobson, A. P. (2006). Extending the principles of community ecology to address the epidemiology of host–pathogen systems. In Collinge, S. K. & Ray, C. (eds), *Disease Ecology: Community Structure and Pathogen Dynamics*. Oxford: Oxford University Press, pp. 6–27.

Holt, R. D., Dobson, A. P., Begon, M., Bowers, R. G. & Schauber, E. M. (2003). Parasite establishment in host communities. *Ecology Letters*, **6**, 837–842.

Holt, R. D., Grover, J. & Tilman, D. (1994). Simple rules for interspecific dominance in systems with exploitative and apparent competition. *American Naturalist*, **144**, 741–771.

Holt, R. D. & Hochberg, M. E. (1998). The coexistence of competing parasites: part II – hyperparasitism and food chain dynamics. *Journal of Theoretical Biology*, **193**, 485–495.

Holt, R. D. & Hochberg, M. E. (2001). Indirect interactions, community modules and biological control: a theoretical perspective. In Wajnberg, E., Scott, J. K. & Quimby, P. C. (eds), *Evaluating Indirect Ecological Effects of Biological Control*. Wallingford: CAB International, pp. 13–37.

Holt, R. D. & Huxel, G. R. (2007). Alternative prey and the dynamics of intraguild predation: theoretical perspectives. *Ecology*, **88**, 2706–2712.

Holt, R. D. & Kotler, B. P. (1987). Short-term apparent competition. *American Naturalist*, **130**, 412–430.

Holt, R. D. & Lawton, J. H. (1993). Apparent competition and enemy-free space in insect host–parasitoid communities. *American Naturalist*, **142**, 623–645.

Holt, R. D. & Pickering, J. (1985). Infectious disease and species coexistence: a model of Lotka–Volterra form. *American Naturalist*, **126**, 196–211.

Holt, R. D. & Polis, G. A. (1997). A theoretical framework for intraguild predation. *American Naturalist*, **149**, 745–764.

Holt, R. D. & Roy, M. (2007). Predation can increase the prevalence of infectious disease. *American Naturalist*, **169**, 690–699.

Hoogendoorn, M. & Heimpel, G. E. (2002). Indirect interactions between an introduced and a native ladybird beetle species mediated by a shared parasitoid. *Biological Control*, **25**, 224–230.

Horimoto, T. & Kawaoka, Y. (2005). Influenza: lessons from past pandemics, warnings from current incidents. *Nature Reviews Microbiology*, **3**, 591–600.

Hudault, S., Guignot, J. & Servin, A. L. (2001). *Escherichia coli* strains colonising the gastrointestinal tract protect germfree mice against *Salmonella typhimurium* infection. *GUT*, **49**, 47–55.

Hudson, P. J., Dobson, A. P. & Lafferty, K. D. (2006). Is a healthy ecosystem one that is rich in parasites? *Trends in Ecology & Evolution*, **21**, 381–385.

Hudson, P. J., Dobson, A. P. & Newborn, D. (1992a). Do parasites make prey vulnerable to predation: red grouse and parasites. *Journal of Animal Ecology*, **61**, 681–692.

Hudson, P. J., Dobson, A. P. & Newborn, D. (1998). Prevention of population cycles by parasite removal. *Science*, **282**, 2256–2258.

Hudson, P. J. & Greenman, J. (1998). Competition mediated by parasites: biological and theoretical progress. *Trends in Ecology & Evolution*, **13**, 387–390.

Hudson, P. J., Newborn, D. & Dobson, A. P. (1992b). Regulation and stability of a free-living host–parasite system: *Trichostrongylus tenuis* in red grouse – 1. Monitoring and parasite reduction experiments. *Journal of Animal Ecology*, **61**, 477–486.

Huspeni, T. C. & Lafferty, K. D. (2004). Using larval trematodes that parasitize snails to evaluate a saltmarsh restoration project. *Ecological Applications*, **14**, 795–804.

Huxham, M., Beaney, S. & Raffaelli, D. (1996). Do parasites reduce the chances of triangulation in a real food web? *Oikos*, **76**, 284–300.

Huxham, M., Raffaelli, D. & Pike, A. (1995). Parasites and food-web patterns. *Journal of Animal Ecology*, **64**, 168–176.

Inbar, M. & Gerling, D. (2008). Plant-mediated interactions between whiteflies, herbivores, and natural enemies. *Annual Review of Entomology*, **53**, 431–448.

Ishtiaq, F., Beadell, J. S., Baker, A. J., Rahmani, A. R., Jhala, Y. V. & Fleischer, R. C. (2006). Prevalence and evolutionary relationships of haematozoan parasites in native versus introduced populations of common myna *Acridotheres tristis*. *Proceedings of the Royal Society of London Series B – Biological Sciences*, **273**, 587–594.

Ives, A. R. & Murray, D. L. (1997). Can sublethal parasitism destabilize predator–prey population dynamics? A model of snowshoe hares, predators and parasites. *Journal of Animal Ecology*, **66**, 265–278.

Jaenike, J. (1995). Interactions between mycophagous *Drosophila* and their nematode parasites: from physiological to community ecology. *Oikos*, **72**, 235–244.

James, T. Y., Litvintseva, A. P., Vilgalys, R., Morgan, J. A. T., Taylor, J. W., Fisher, M. C., Berger, L., Weldon, C., du Preez, L. & Longcore, J. E. (2009). Rapid global expansion of the fungal disease chytridiomycosis into declining and healthy amphibian populations. *PLoS Pathogens*, **5**, e1000458.

Janssen, A., Sabelis, M. W., Magalhaes, S., Montserrat, M. & Van der Hammen, T. (2007). Habitat structure affects intraguild predation. *Ecology*, **88**, 2713–2719.

Janzen, D. H. (1970). Herbivores and number of tree species in tropical forests. *American Naturalist*, **104**, 501–528.

Jensen, T., Jensen, K. T. & Mouritsen, K. N. (1998). The influence of the trematode *Microphallus claviformis* on two congeneric intermediate host species (*Corophium*): infection characteristics and host survival. *Journal of Experimental Marine Biology and Ecology*, **227**, 35–48.

Johnson, N. C., Graham, J. H. & Smith, F. A. (1997). Functioning of mycorrhizal associations along the mutualism–parasitism continuum. *New Phytologist*, **135**, 575–586.

Johnson, P. T. J., Dobson, A., Lafferty, K. D., Marcogliese, D. J., Memmott, J., Orlofske, S. A., Poulin, R. & Thieltges, D. W. (2010). When parasites become prey: ecological and epidemiological significance of eating parasites. *Trends in Ecology & Evolution*, **25**, 362–371.

Joly, D. O. & Messier, F. (2004). The distribution of *Echinococcus granulosus* in moose: evidence for parasite-induced vulnerability to predation by wolves? *Oecologia*, **140**, 586–590.

Jones, C. G., Lawton, J. H. & Shachak, M. (1994). Organisms as ecosystem engineers. *Oikos*, **69**, 373–386.

Jones, K. E., Patel, N. G., Levy, M. A., Storeygard, A., Balk, D., Gittleman, J. L. & Daszak, P. (2008). Global trends in emerging infectious diseases. *Nature*, **451**, 990–993.

Joo, J., Gunny, M., Cases, M., Hudson, P., Albert, R. & Harvill, E. (2006). Bacteriophage-mediated competition in *Bordetella* bacteria. *Proceedings of the Royal Society of London Series B – Biological Sciences*, **273**, 1843–1848.

Joshi, J., Matthies, D. & Schmid, B. (2000). Root hemiparasites and plant diversity in experimental grassland communities. *Journal of Ecology*, **88**, 634–644.

Juliano, S. A. & Lounibos, L. P. (2005). Ecology of invasive mosquitoes: effects on resident species and on human health. *Ecology Letters*, **8**, 558–574.

Kaltz, O. & Shykoff, J. A. (1998). Local adaptation in host–parasite systems. *Heredity*, **81**, 361–370.

Kanno, H. & Fujita, Y. (2003). Induced systemic resistance to rice blast fungus in rice plants infested by white-backed planthopper. *Entomologia experimentalis et Applicata*, **107**, 155–158.

Kanno, H., Satoh, M., Kimura, T. & Fujita, Y. (2005). Some aspects of induced resistance to rice blast fungus, *Magnaporthe grisea*, in rice plant infested by white-backed planthopper, *Sogatella furcifera*. *Applied Entomology and Zoology*, **40**, 91–97.

Karban, R., Baldwin, I. T. (1997). *Induced Responses to Herbivory*. Chicago: University of Chicago Press.

Karban, R., Baldwin, I. T., Baxter, K. J., Laue, G. & Felton, G. W. (2000). Communication between plants: induced resistance in wild tobacco plants following clipping of neighboring sagebrush. *Oecologia*, **125**, 66–71.

Keane, R. M. & Crawley, M. J. (2002). Exotic plant invasions and the enemy release hypothesis. *Trends in Ecology & Evolution*, **17**, 164–170.

Keele, B. F., Jones, J. H., Terio, K. A., Estes, J. D., Rudicell, R. S., Wilson, M. L., Li, Y. Y., Learn, G. H., Beasley, T. M., Schumacher-Stankey, J., Wroblewski, E., Mosser, A., Raphael, J., Kamenya, S., Lonsdorf, E. V., Travis, D. A., Mlengeya, T., Kinsel, M. J., Else, J. G., Silvestri, G., Goodall, J., Sharp, P. M., Shaw, G. M., Pusey, A. E. & Hahn, B. H. (2009). Increased mortality and AIDS-like immunopathology in wild chimpanzees infected with SIVcpz. *Nature*, **460**, 515–519.

Keele, B. F., Van Heuverswyn, F., Li, Y. Y., Bailes, E., Takehisa, J., Santiago, M. L., Bibollet-Ruche, F., Chen, Y. L., Wain, L. V., Liegeois, F., Loul, S., Ngole, E. M., Bienvenue, Y., Delaporte, E., Brookfield, J. F. Y., Sharp, P. M., Shaw, G. M., Peeters, M. & Hahn, B. H. (2006). Chimpanzee reservoirs of pandemic and nonpandemic HIV-1. *Science*, **313**, 523–526.

Keesing, F., Holt, R. D. & Ostfeld, R. S. (2006). Effects of species diversity on disease risk. *Ecology Letters*, **9**, 485–498.

Kelly, D. W., Bailey, R. J., MacNeil, C., Dick, J. T. A. & McDonald, R. A. (2006). Invasion by the amphipod *Gammarus pulex* alters community composition of native freshwater macroinvertebrates. *Diversity and Distributions*, **12**, 525–534.

Kelly, D. W., Paterson, R. A., Townsend, C. R., Poulin, R. & Tompkins, D. M. (2009). Parasite spillback: a neglected concept in invasion ecology? *Ecology*, **90**, 2047–2056.

Kennedy, C. R. & Guegan, J. F. (1994). Regional versus local helminth parasite richness in British fresh-water fish: saturated or unsaturated parasite communities. *Parasitology*, **109**, 175–185.

Kessler, A. & Baldwin, I. T. (2001). Defensive function of herbivore-induced plant volatile emissions in nature. *Science*, **291**, 2141–2144.

Kfir, R., Gouws, J. & Moore, S. D. (1993). Biology of *Tetrastichus howardi* (Olliff) (Hymenoptera, Eulophidae): a facultative hyperparasitoid of stem borers. *Biocontrol Science and Technology*, **3**, 149–159.

Kiesecker, J. M. & Blaustein, A. R. (1999). Pathogen reverses competition between larval amphibians. *Ecology*, **80**, 2442–2448.

Kilpatrick, A. M., Daszak, P., Goodman, S. J., Rogg, H., Kramer, L. D., Cedeno, V. & Cunningham, A. A. (2006). Predicting pathogen introduction: West Nile virus spread to Galapagos. *Conservation Biology*, **20**, 1224–1231.

Kim, K. & Harvell, C. D. (2004). The rise and fall of a six-year coral–fungal epizootic. *American Naturalist*, **164**, S52–S63.

King, A. A. & Hastings, A. (2003). Spatial mechanisms for coexistence of species sharing a common natural enemy. *Theoretical Population Biology*, **64**, 431–438.

Kleczkowski, A., Gilligan, C. A. & Bailey, D. J. (1997). Scaling and spatial dynamics in plant–pathogen systems: from individuals to populations. *Proceedings of the Royal Society of London Series B – Biological Sciences*, **264**, 979–984.

Kloepper, J. W., Ryu, C. M. & Zhang, S. A. (2004). Induced systemic resistance and promotion of plant growth by *Bacillus* spp. *Phytopathology*, **94**, 1259–1266.

Knevel, I. C., Lans, T., Menting, F. B. J., Hertling, U. M. & Van der Putten, W. H. (2004). Release from native root herbivores and biotic resistance by soil pathogens in a new habitat both affect the alien *Ammophila arenaria* in South Africa. *Oecologia*, **141**, 502–510.

Koella, J. C., Lynch, P. A., Thomas, M. B. & Read, A. F. (2009). Towards evolution-proof malaria control with insecticides. *Evolutionary Applications*, **2**, 469–480.

Kohler, S. L. (2008). The ecology of host–parasite interactions in aquatic insects. In Lancaster, J., Briers, R. & Macadam, C. (eds), *Aquatic Insects: Challenges to Populations*. Wallingford: CAB International, pp. 55–80.

Kohler, S. L. & Wiley, M. J. (1997). Pathogen outbreaks reveal large-scale effects of competition in stream communities. *Ecology*, **78**, 2164–2176.

Kokko, H. & Lindstrom, J. (1998). Seasonal density dependence, timing of mortality, and sustainable harvesting. *Ecological Modelling*, **110**, 293–304.

Kolar, C. S. & Lodge, D. M. (2001). Progress in invasion biology: predicting invaders. *Trends in Ecology & Evolution*, **16**, 199–204.

Komar, N., Langevin, S., Hinten, S., Nemeth, N., Edwards, E., Hettler, D., Davis, B., Bowen, R. & Bunning, M. (2003). Experimental infection of North American birds with the New York 1999 strain of West Nile virus. *Emerging Infectious Diseases*, **9**, 311–322.

Kopp, K. & Jokela, J. (2007). Resistant invaders can convey benefits to native species. *Oikos*, **116**, 295–301.

Koprivnikar, J., Forbes, M. R. & Baker, R. L. (2008). Larval amphibian growth and development under varying density: are parasitized individuals poor competitors? *Oecologia*, **155**, 641–649.

Krkosek, M., Lewis, M. A., Morton, A., Frazer, L. N. & Volpe, J. P. (2006). Epizootics of wild fish induced by farm fish. *Proceedings of the National Academy of Sciences of the United States of America*, **103**, 15506–15510.

Kuiken, T., Kennedy, S., Barrett, T., Van de Bildt, M. W. G., Borgsteede, F. H., Brew, S. D., Codd, G. A., Duck, C., Deaville, R., Eybatov, T., Forsyth, M. A., Foster, G., Jepson, P. D., Kydyrmanov, A., Mitrofanov, I., Ward, C. J., Wilson, S. & Osterhaus, A. (2006). The 2000 canine distemper epidemic in Caspian seals (*Phoca caspica*): pathology and analysis of contributory factors. *Veterinary Pathology*, **43**, 321–338.

Kulmatiski, A., Beard, K. H., Stevens, J. R. & Cobbold, S. M. (2008). Plant–soil feedbacks: a meta-analytical review. *Ecology Letters*, **11**, 980–992.

Kuo, C. H., Corby-Harris, V. & Promislow, D. E. L. (2008). The unavoidable costs and unexpected benefits of parasitism: population and metapopulation models of parasite-mediated competition. *Journal of Theoretical Biology*, **250**, 244–256.

Kuris, A. M., Hechinger, R. F., Shaw, J. C., Whitney, K. L., Aguirre-Macedo, L., Boch, C. A., Dobson, A. P., Dunham, E. J., Fredensborg, B. L., Huspeni, T. C., Lorda, J., Mababa, L., Mancini, F. T., Mora, A. B., Pickering, M., Talhouk, N. L., Torchin, M. E. & Lafferty, K. D. (2008). Ecosystem energetic implications of parasite and free-living biomass in three estuaries. *Nature*, **454**, 515–518.

Kuris, A. M. & Lafferty, K. D. (1992). Modeling crustacean fisheries: effects of parasites on management strategies. *Canadian Journal of Fisheries and Aquatic Sciences*, **49**, 327–336.

Kuris, A. M. & Lafferty, K. D. (1994). Community structure: larval trematodes in snail hosts. *Annual Review of Ecology and Systematics*, **25**, 189–217.

LaDeau, S. L., Kilpatrick, A. M. & Marra, P. P. (2007). West Nile virus emergence and large-scale declines of North American bird populations. *Nature*, **447**, 710–713.

Lafferty, K. D. (1992). Foraging on prey that are modified by parasites. *American Naturalist*, **140**, 854–867.

Lafferty, K. D. (2009). The ecology of climate change and infectious diseases. *Ecology*, **90**, 888–900.

Lafferty, K. D., Allesina, S., Arim, M., Briggs, C. J., De Leo, G., Dobson, A. P., Dunne, J. A., Johnson, P. T. J., Kuris, A. M., Marcogliese, D. J., Martinez, N. D., Memmott, J., Marquet, P. A., McLaughlin, J. P., Mordecai, E. A., Pascual, M., Poulin, R. & Thieltges, D. W. (2008a). Parasites in food webs: the ultimate missing links. *Ecology Letters*, **11**, 533–546.

Lafferty, K. D., Dobson, A. P. & Kuris, A. M. (2006a). Parasites dominate food web links. *Proceedings of the National Academy of Sciences of the United States of America*, **103**, 11211–11216.

Lafferty, K. D., Hechinger, R. F., Shaw, J. C., Whitney, K. & Kuris, A. M. (2006b). Food webs and parasites in a salt marsh ecosystem. In Collinge, S. K. & Ray, C. (eds), *Disease Ecology: Community Structure and Pathogen Dynamics*. Oxford: Oxford University Press, pp. 119–134.

Lafferty, K. D. & Kuris, A. M. (2009). Parasites reduce food web robustness because they are sensitive to secondary extinction as illustrated by an invasive estuarine snail. *Philosophical Transactions of the Royal Society of London Series B – Biological Sciences*, **364**, 1659–1663.

Lafferty, K. D. & Morris, A. K. (1996). Altered behavior of parasitized killifish increases susceptibility to predation by bird final hosts. *Ecology*, **77**, 1390–1397.

Lafferty, K. D., Porter, J. W. & Ford, S. E. (2004). Are diseases increasing in the ocean? *Annual Review of Ecology Evolution and Systematics*, **35**, 31–54.

Lafferty, K. D., Shaw, J. C. & Kuris, A. M. (2008b). Reef fishes have higher parasite richness at unfished Palmyra Atoll compared to fished Kiritimati Island. *Ecohealth*, **5**, 338–345.

Lagrue, C. & Poulin, R. (2008). Intra- and interspecific competition among helminth parasites: effects on *Coitocaecum parvum* life history, strategy, size and fecundity. *International Journal for Parasitology*, **38**, 1435–1444.

Lau, S. K. P., Woo, P. C. Y., Li, K. S. M., Huang, Y., Tsoi, H. W., Wong, B. H. L., Wong, S. S. Y., Leung, S. Y., Chan, K. H. & Yuen, K. Y. (2005). Severe acute respiratory syndrome coronavirus-like virus in Chinese horseshoe bats. *Proceedings of the National Academy of Sciences of the United States of America*, **102**, 14040–14045.

Laurenson, M. K., McKendrick, I. J., Reid, H. W., Challenor, R. & Mathewson, G. K. (2007). Prevalence, spatial distribution and the effect of control measures on louping-ill virus in the Forest of Bowland, Lancashire. *Epidemiology and Infection*, **135**, 963–973.

Laurenson, M. K., Norman, R. A., Gilbert, L., Reid, H. W. & Hudson, P. J. (2003). Identifying disease reservoirs in complex systems: mountain hares as reservoirs of ticks and louping-ill virus, pathogens of red grouse. *Journal of Animal Ecology*, **72**, 177–185.

Laurenson, M. K., Norman, R. A., Gilbert, L., Reid, H. W. & Hudson, P. J. (2004). Mountain hares, louping-ill, red grouse and harvesting: complex interactions but few data. *Journal of Animal Ecology*, **73**, 811–813.

Lavigne, D. M. & Schmitz, O. J. (1990). Global warming and increasing population densities: a prescription for seal plagues. *Marine Pollution Bulletin*, **21**, 280–284.

Leath, K. T. & Byers, R. A. (1977). Interaction of fusarium root-rot with pea aphid and potato leafhopper feeding on forage legumes. *Phytopathology*, **67**, 226–229.

Lee, K. A. & Klasing, K. C. (2004). A role for immunology in invasion biology. *Trends in Ecology & Evolution*, **19**, 523–529.

Lee, K. A., Martin, L. B. & Wikelski, M. C. (2005). Responding to inflammatory challenges is less costly for a successful avian invader, the house sparrow (*Passer domesticus*), than its less-invasive congener. *Oecologia*, **145**, 244–251.

Lefevre, T., Roche, B., Poulin, R., Hurd, H., Renaud, F. & Thomas, F. (2008). Exploiting host compensatory responses: the 'must' of manipulation? *Trends in Parasitology*, **24**, 435–439.

Lehtonen, P., Helander, M., Wink, M., Sporer, F. & Saikkonen, K. (2005). Transfer of endophyte-origin defensive alkaloids from a grass to a hemiparasitic plant. *Ecology Letters*, **8**, 1256–1263.

Lello, J., Boag, B., Fenton, A., Stevenson, I. R. & Hudson, P. J. (2004). Competition and mutualism among the gut helminths of a mammalian host. *Nature*, **428**, 840–844.

Lemons, A., Clay, K. & Rudgers, J. A. (2005). Connecting plant–microbial interactions above and belowground: a fungal endophyte affects decomposition. *Oecologia*, **145**, 595–604.

Lenski, R. E. & May, R. M. (1994). The evolution of virulence in parasites and pathogens: reconciliation between two competing hypotheses. *Journal of Theoretical Biology*, **169**, 253–265.

Leroy, E. M., Kumulungui, B., Pourrut, X., Rouquet, P., Hassanin, A., Yaba, P., Delicat, A., Paweska, J. T., Gonzalez, J. P. & Swanepoel, R. (2005). Fruit bats as reservoirs of Ebola virus. *Nature*, **438**, 575–576.

Leroy, E. M., Rouquet, P., Formenty, P., Souquiere, S., Kilbourne, A., Froment, J. M., Bermejo, M., Smit, S., Karesh, W., Swanepoel, R., Zaki, S. R. & Rollin, P. E. (2004). Multiple Ebola virus transmission events and rapid decline of central African wildlife. *Science*, **303**, 387–390.

Leung, T. L. F., Keeney, D. B. & Poulin, R. (2009). Cryptic species complexes in manipulative echinostomatid trematodes: when two become six. *Parasitology*, **136**, 241–252.

Li, W. D., Shi, Z. L., Yu, M., Ren, W. Z., Smith, C., Epstein, J. H., Wang, H. Z., Crameri, G., Hu, Z. H., Zhang, H. J., Zhang, J. H., McEachern, J., Field, H., Daszak, P., Eaton, B. T., Zhang, S. Y. & Wang, L. F. (2005). Bats are natural reservoirs of SARS-like coronaviruses. *Science*, **310**, 676–679.

Lips, K. R., Diffendorfer, J., Mendelson, J. R. & Sears, M. W. (2008). Riding the wave: reconciling the roles of disease and climate change in amphibian declines. *PLoS Biology*, **6**, 441–454.

Lipsitch, M., Cohen, T., Cooper, B., Robins, J. M., Ma, S., James, L., Gopalakrishna, G., Chew, S. K., Tan, C. C., Samore, M. H., Fisman, D. & Murray, M. (2003). Transmission dynamics and control of severe acute respiratory syndrome. *Science*, **300**, 1966–1970.

Lloyd-Smith, J. O., Galvani, A. P. & Getz, W. M. (2003). Curtailing transmission of severe acute respiratory syndrome within a community and its hospital. *Proceedings of the Royal Society of London Series B – Biological Sciences*, **270**, 1979–1989.

Lloyd-Smith, J. O., George, D., Pepin, K. M., Pitzer, V. E., Pulliam, J. R. C., Dobson, A. P., Hudson, P. J. & Grenfell, B. T. (2009). Epidemic dynamics at the human–animal interface. *Science*, **326**, 1362–1367.

Lloyd-Smith, J. O., Schreiber, S. J., Kopp, P. E. & Getz, W. M. (2005). Superspreading and the effect of individual variation on disease emergence. *Nature*, **438**, 355–359.

Lockwood, J. L., Hoopes, M. F. & Marchetti, M. P. (2007). *Invasion Ecology*. Oxford: Blackwell.

LoGiudice, K., Duerr, S. T. K., Newhouse, M. J., Schmidt, K. A., Killilea, M. E. & Ostfeld, R. S. (2008). Impact of host community composition on Lyme disease risk. *Ecology*, **89**, 2841–2849.

LoGiudice, K., Ostfeld, R. S., Schmidt, K. A. & Keesing, F. (2003). The ecology of infectious disease: effects of host diversity and community composition on Lyme disease risk. *Proceedings of the National Academy of Sciences of the United States of America*, **100**, 567–571.

Loo, J. (2009). Ecological impacts of non-indigenous invasive fungi as forest pathogens. *Biological Invasions*, **11**, 81–96.

Lopez, G., Lopez-Parra, M., Fernandez, L., Martinez-Granados, C., Martinez, F., Meli, M. L., Gil-Sanchez, J. M., Viqueira, N., Diaz-Portero, M. A., Cadenas, R., Lutz, H., Vargas, A. & Simon, M. A. (2009). Management measures to control a feline leukemia virus outbreak in the endangered Iberian lynx. *Animal Conservation*, **12**, 173–182.

Losey, J. E., Ives, A. R., Harmon, J., Ballantyne, F. & Brown, C. (1997). A polymorphism maintained by opposite patterns of parasitism and predation. *Nature*, **388**, 269–272.

Lubick, N. (2010). Emergency medicine for frogs. *Nature*, **465**, 680–681.

MacArthur, R. (1955). Fluctuations of animal populations, and a measure of community stability. *Ecology*, **36**, 533–536.

MacDonald, D. W. & Laurenson, M. K. (2006). Infectious disease: inextricable linkages between human and ecosystem health. *Biological Conservation*, **131**, 143–150.

MacDonald, D. W., Riordan, P. & Mathews, F. (2006). Biological hurdles to the control of TB in cattle: a test of two hypotheses concerning wildlife to explain the failure of control. *Biological Conservation*, **131**, 268–286.

Mack, K. M. L. & Rudgers, J. A. (2008). Balancing multiple mutualists: asymmetric interactions among plants, arbuscular mycorrhizal fungi, and fungal endophytes. *Oikos*, **117**, 310–320.

Mack, R. N., Simberloff, D., Lonsdale, W. M., Evans, H., Clout, M. & Bazzaz, F. A. (2000). Biotic invasions: causes, epidemiology, global consequences, and control. *Ecological Applications*, **10**, 689–710.

MacLeod, C. J., Paterson, A. M., Tompkins, D. M. & Duncan, R. P. (2010). Parasites lost: do invaders miss the boat or drown on arrival? *Ecology Letters*, **13**, 516–527.

MacNeil, C., Dick, J. T. A., Hatcher, M. J. & Dunn, A. M. (2003a). Differential drift and parasitism in invading and native *Gammarus* spp. (Crustacea: Amphipoda). *Ecography*, **26**, 467–473.

MacNeil, C., Dick, J. T. A., Hatcher, M. J., Fielding, N. J., Hume, K. D. & Dunn, A. M. (2003b). Parasite transmission and cannibalism in an amphipod (Crustacea). *International Journal for Parasitology*, **33**, 795–798.

MacNeil, C., Dick, J. T. A., Hatcher, M. J., Terry, R. S., Smith, J. E. & Dunn, A. M. (2003c). Parasite-mediated predation between native and invasive amphipods. *Proceedings of the Royal Society of London Series B – Biological Sciences*, **270**, 1309–1314.

MacNeil, C., Fielding, N. J., Dick, J. T. A., Briffa, M., Prenter, J., Hatcher, M. J. & Dunn, A. M. (2003d). An acanthocephalan parasite mediates intraguild predation between invasive and native freshwater amphipods (Crustacea). *Freshwater Biology*, **48**, 2085–2093.

MacNeil, C., Fielding, N. J., Hume, K. D., Dick, J. T. A., Elwood, R. W., Hatcher, M. J. & Dunn, A. M. (2003e). Parasite altered micro-distribution of *Gammarus pulex* (Crustacea: Amphipoda). *International Journal for Parasitology*, **33**, 57–64.

Malmstrom, C. M., Hughes, C. C., Newton, L. A. & Stoner, C. J. (2005a). Virus infection in remnant native bunchgrasses from invaded California grasslands. *New Phytologist*, **168**, 217–230.

Malmstrom, C. M., McCullough, A. J., Johnson, H. A., Newton, L. A. & Borer, E. T. (2005b). Invasive annual grasses indirectly increase virus incidence in California native perennial bunchgrasses. *Oecologia*, **145**, 153–164.

Malmstrom, C. M., Stoner, C. J., Brandenburg, S. & Newton, L. A. (2006). Virus infection and grazing exert counteracting influences on survivorship of native bunchgrass seedlings competing with invasive exotics. *Journal of Ecology*, **94**, 264–275.

March, W. A. & Watson, D. M. (2007). Parasites boost productivity: effects of mistletoe on litterfall dynamics in a temperate Australian forest. *Oecologia*, **154**, 339–347.

Marcogliese, D. J. & Cone, D. K. (1997). Food webs: a plea for parasites. *Trends in Ecology & Evolution*, **12**, 394–394.

Marr, S. R., Mautz, W. J. & Hara, A. H. (2008). Parasite loss and introduced species: a comparison of the parasites of the Puerto Rican tree frog (*Eleutherodactylus coqui*) in its native and introduced ranges. *Biological Invasions*, **10**, 1289–1298.

Marra, P. P., Griffing, S., Caffrey, C., Kilpatrick, A. M., McLean, R., Brand, C., Saito, E., Dupuis, A. P., Kramer, L. & Novak, R. (2004). West Nile virus and wildlife. *Bioscience*, **54**, 393–402.

Marshall, K., Mamone, M. & Barclay, R. (2003). A survey of Douglas-fir dwarf mistletoe brooms used for nests by northern spotted owls on the Applegate Ranger District and Ashland Resource area in southwest Oregon. *Western Journal of Applied Forestry*, **18**, 115–117.

Martens, P., Kovats, R. S., Nijhof, S., de Vries, P., Livermore, M. T. J., Bradley, D. J., Cox, J. & McMichael, A. J. (1999). Climate change and future populations at risk of malaria. *Global Environmental Change: Human and Policy Dimensions*, **9**, S89–S107.

Martens, W. J. M., Jetten, T. H. & Focks, D. A. (1997). Sensitivity of malaria, schistosomiasis and dengue to global warming. *Climatic Change*, **35**, 145–156.

Massey, R. C., Buckling, A. & French-Constant, R. (2004). Interference competition and parasite virulence. *Proceedings of the Royal Society of London Series B – Biological Sciences*, **271**, 785–788.

Mathews, F. (2009). Zoonoses in wildlife: integrating ecology into management. *Advances in Parasitology*, **68**, 185–209.

Mathiasen, R. L., Nickrent, D. L., Shaw, D. C. & Watson, D. M. (2008). Mistletoes: pathology, systematics, ecology, and management. *Plant Disease*, **92**, 988–1006.

Matson, K. D. (2006). Are there differences in immune function between continental and insular birds? *Proceedings of the Royal Society of London Series B – Biological Sciences*, **273**, 2267–2274.

Matthies, D. (1995). Parasitic and competitive interactions between the hemiparasites *Rhinanthus serotinus* and *Odontites rubra* and their host *Medicago sativa*. *Journal of Ecology*, **83**, 245–251.

May, R. M. (1973). Stability and complexity in model ecosystems. *Monographs in Population Biology*, **6**, 1–235.

May, R. M. & Anderson, R. M. (1987). Transmission dynamics of HIV infection. *Nature*, **326**, 137–142.

May, R. M. & Anderson, R. M. (1990). Parasite–host coevolution. *Parasitology*, **100**, S89–S101.

May, R. M., Gupta, S. & McLean, A. R. (2001). Infectious disease dynamics: what characterizes a successful invader? *Philosophical Transactions of the Royal Society of London Series B – Biological Sciences*, **356**, 901–910.

McCallum, H., Barlow, N. & Hone, J. (2001). How should pathogen transmission be modelled? *Trends in Ecology & Evolution*, **16**, 295–300.

McCallum, H., Gerber, L. & Jani, A. (2005). Does infectious disease influence the efficacy of marine protected areas? A theoretical framework. *Journal of Applied Ecology*, **42**, 688–698.

McCallum, H., Tompkins, D. M., Jones, M., Lachish, S., Marvanek, S., Lazenby, B., Hocking, G., Wiersma, J. & Hawkins, C. E. (2007). Distribution and impacts of Tasmanian devil facial tumor disease. *Ecohealth*, **4**, 318–325.

McCann, K., Hastings, A. & Huxel, G. R. (1998). Weak trophic interactions and the balance of nature. *Nature*, **395**, 794–798.

McCann, K. S. (2000). The diversity–stability debate. *Nature*, **405**, 228–233.

McGeoch, M. A., Butchart, S. H. M., Spear, D., Marais, E., Kleynhans, E. J., Symes, A., Chanson, J. & Hoffmann, M. (2010). Global indicators of biological invasion: species numbers, biodiversity impact and policy responses. *Diversity and Distributions*, **16**, 95–108.

McMichael, A. J., Bouma, M. J., Mooney, H. A. & Hobbs, R. J. (2000). Global changes, invasive species, and human health. In Mooney, H. A. & Hobbs, R. J. (eds), *Invasive Species in a Changing World*. Washington, DC: Island Press, pp. 191–240.

Medoc, V. & Beisel, J. N. (2008). An acanthocephalan parasite boosts the escape performance of its intermediate host facing non-host predators. *Parasitology*, **135**, 977–984.

Medoc, V., Bollache, L. & Beisel, J. N. (2006). Host manipulation of a freshwater crustacean (*Gammarus roeseli*) by an acanthocephalan parasite (*Polymorphus minutus*) in a biological invasion context. *International Journal for Parasitology*, **36**, 1351–1358.

Memmott, J., Fowler, S. V., Paynter, Q., Sheppard, A. W. & Syrett, P. (2000). The invertebrate fauna on broom, *Cytisus scoparius*, in two native and two exotic habitats. *Acta Oecologica: International Journal of Ecology*, **21**, 213–222.

Meyling, N. V. & Hajek, A. E. (2010). Principles from community and metapopulation ecology: application to fungal entomopathogens. *Biocontrol*, **55**, 39–54.

Michaelis, M., Doerr, H. W. & Cinatl, J. (2009). Novel swine-origin influenza A virus in humans: another pandemic knocking at the door. *Medical Microbiology and Immunology*, **198**, 175–183.

Milinski, M. (1985). Risk of predation of parasitized sticklebacks (*Gasterosteus aculeatus* L.) under competition for food. *Behaviour*, **93**, 203–215.

Mills, N. J. & Gutierrez, A. P. (1996). Prospective modelling in biological control: an analysis of the dynamics of heteronomous hyperparasitism in a cotton–whitefly–parasitoid system. *Journal of Applied Ecology*, **33**, 1379–1394.

Minchella, D. J. & Scott, M. E. (1991). Parasitism: a cryptic determinant of animal community structure. *Trends in Ecology & Evolution*, **6**, 250–254.

Mitchell, C. E. & Power, A. G. (2003). Release of invasive plants from fungal and viral pathogens. *Nature*, **421**, 625–627.

Mitchell, C. E., Tilman, D. & Groth, J. V. (2002). Effects of grassland plant species diversity, abundance, and composition on foliar fungal disease. *Ecology*, **83**, 1713–1726.

Mittelbach, G. G. (2005). Parasites, communities, and ecosystems: conclusions and perspectives. In Thomas, F., Renaud, F. & Geugan, J. F. (eds) *Parasitism and Ecosystems*, Oxford: Oxford University Press, pp. 171–176.

Moleon, M., Almaraz, P. & Sanchez-Zapata, J. A. (2008). An emerging infectious disease triggering large-scale hyperpredation. *PLoS One*, **3**, e2307.

Moller, A. P. (2008). Interactions between interactions: predator–prey, parasite–host, and mutualistic interactions. *Annals of the New York Academy of Science*, **1133**, 180–186.

Molofsky, J., Bever, J. D. & Antonovics, J. (2001). Coexistence under positive frequency dependence. *Proceedings of the Royal Society of London Series B – Biological Sciences*, **268**, 273–277.

Molofsky, J., Bever, J. D., Antonovics, J. & Newman, T. J. (2002). Negative frequency dependence and the importance of spatial scale. *Ecology*, **83**, 21–27.

Money, N. P. (2007). *Triumph of the Fungi: A Rotten History*. Oxford: Oxford University Press.

Moore, J. (2002). *Parasites and the Behavior of Animals*. Oxford: Oxford University Press.

Moore, S. M., Borer, E. T. & Hosseini, P. R. (2010). Predators indirectly control vector-borne disease: linking predator–prey and host–pathogen models. *Journal of the Royal Society Interface*, **7**, 161–176.

Morens, D. M., Folkers, G. K. & Fauci, A. S. (2008). Emerging infections: a perpetual challenge. *Lancet Infectious Diseases*, **8**, 710–719.

Moret, Y., Bollache, L., Wattier, R. & Rigaud, T. (2007). Is the host or the parasite the most locally adapted in an amphipod–acanthocephalan relationship? A case study in a biological invasion context. *International Journal for Parasitology*, **37**, 637–644.

Morris, R. J., Lewis, O. T. & Godfray, H. C. J. (2004). Experimental evidence for apparent competition in a tropical forest food web. *Nature*, **428**, 310–313.

Morris, R. J., Muller, C. B. & Godfray, H. C. J. (2001). Field experiments testing for apparent competition between primary parasitoids mediated by secondary parasitoids. *Journal of Animal Ecology*, **70**, 301–309.

Morris, W. F., Hufbauer, R. A., Agrawal, A. A., Bever, J. D., Borowicz, V. A., Gilbert, G. S., Maron, J. L., Mitchell, C. E., Parker, I. M., Power, A. G., Torchin, M. E. & Vazquez, D. P. (2007). Direct and interactive effects of enemies and mutualists on plant performance: a meta-analysis. *Ecology*, **88**, 1021–1029.

Morrison, L. W. (1999). Indirect effects of phorid fly parasitoids on the mechanisms of interspecific competition among ants. *Oecologia*, **121**, 113–122.

Mouillot, D., Krasnov, B. R. & Poulin, R. (2008a). High intervality explained by phylogenetic constraints in host–parasite webs. *Ecology*, **89**, 2043–2051.

Mouillot, D., Krasnov, B. R., Shenbrot, G. I. & Poulin, R. (2008b). Connectance and parasite diet breadth in flea–mammal webs. *Ecography*, **31**, 16–20.

Mouritsen, K. N. (2001). Hitch-hiking parasite: a dark horse may be the real rider. *International Journal for Parasitology*, **31**, 1417–1420.

Mouritsen, K. N. & Haun, S. C. B. (2008). Community regulation by herbivore parasitism and density: trait-mediated indirect interactions in the intertidal. *Journal of Experimental Marine Biology and Ecology*, **367**, 236–246.

Mouritsen, K. N. & Jensen, T. (2006). The effect of *Sacculina carcini* infections on the fouling, burying behaviour and condition of the shore crab, *Carcinus maenas*. *Marine Biology Research*, **2**, 270–275.

Mouritsen, K. N., Mouritsen, L. T. & Jensen, K. T. (1998). Change of topography and sediment characteristics on an intertidal mud-flat following mass-mortality of the amphipod *Corophium volutator*. *Journal of the Marine Biological Association of the United Kingdom*, **78**, 1167–1180.

Mouritsen, K. N. & Polml, R. (2006). A parasite indirectly impacts both abundance of primary producers and biomass of secondary producers in an intertidal benthic community. *Journal of the Marine Biological Association of the United Kingdom*, **86**, 221–226.

Mouritsen, K. N. & Poulin, R. (2003). Parasite-induced trophic facilitation exploited by a non-host predator: a manipulator's nightmare. *International Journal for Parasitology*, **33**, 1043–1050.

Mouritsen, K. N. & Poulin, R. (2005). Parasitism can influence the intertidal zonation of non-host organisms. *Marine Biology*, **148**, 1–11.

Mueller, R. C. & Gehring, C. A. (2006). Interactions between an above-ground plant parasite and below-ground ectomycorrhizal fungal communities on pinyon pine. *Journal of Ecology*, **94**, 276–284.

Muller, C. B. & Brodeur, J. (2002). Intraguild predation in biological control and conservation biology. *Biological Control*, **25**, 216–223.

Murray, D. L., Cary, J. R. & Keith, L. B. (1997). Interactive effects of sublethal nematodes and nutritional status on snowshoe hare vulnerability to predation. *Journal of Animal Ecology*, **66**, 250–264.

Musselman, L. J. & Press, M. C. (1995). Introduction to parasitic plants. In Press, M. C. & Graves, J. D. (eds), *Parasitic Plants*. London: Chapman & Hall, pp. 1–13.

Myers, J. H. & Bazely, D. R. (2003). *Ecology and Control of Introduced Plants*. Cambridge: Cambridge University Press.

Mylius, S. D., Klumpers, K., de Roos, A. M. & Persson, L. (2001). Impact of intraguild predation and stage structure on simple communities along a productivity gradient. *American Naturalist*, **158**, 259–276.

Naugle, D. E., Aldridge, C. L., Walker, B. L., Cornish, T. E., Moynahan, B. J., Holloran, M. J., Brown, K., Johnson, G. D., Schmidtmann, E. T., Mayer, R. T., Kato, C. Y., Matchett, M. R., Christiansen, T. J., Cook, W. E., Creekmore, T., Falise, R. D., Rinkes, E. T. & Boyce, M. S. (2004). West Nile virus: pending crisis for greater sage-grouse. *Ecology Letters*, **7**, 704–713.

Nel, L. H. & Rupprecht, C. E. (2007). Emergence of Lyssaviruses in the old world: the case of Africa. In Childs, J. E., Mackenzie, J. S. & Richt, J. A. (eds), *Wildlife and Emerging Zoonotic Diseases: The Biology, Circumstances and Consequences of Cross-species Transmission*. Berlin: Springer, pp. 161–193.

Neutel, A. M., Heesterbeek, J. A. P. & de Ruiter, P. C. (2002). Stability in real food webs: weak links in long loops. *Science*, **296**, 1120–1123.

New, T. R. (2005). *Invertebrate Conservation and Agricultural Ecosystems*. Cambridge: Cambridge University Press.

Nickrent, D. L., Malecot, V., Vidal-Russell, R. & Der, J. P. (2010). A revised classification of Santalales. *Taxon*, **59**, 538–558.

Nogales, M., Martin, A., Tershy, B. R., Donlan, C. J., Witch, D., Puerta, N., Wood, B. & Alonso, J. (2004). A review of feral cat eradication on islands. *Conservation Biology*, **18**, 310–319.

Norman, R., Bowers, R. G., Begon, M. & Hudson, P. J. (1999). Persistence of tick-borne virus in the presence of multiple host species: tick reservoirs and parasite mediated competition. *Journal of Theoretical Biology*, **200**, 111–118.

Norton, D. A. & Reid, N. (1997). Lessons in ecosystem management from management of threatened and pest loranthaceous mistletoes in New Zealand and Australia. *Conservation Biology*, **11**, 759–769.

Novembre, J., Galvani, A. P. & Slatkin, M. (2005). The geographic spread of the CCR5 Delta32 HIV-resistance allele. *PLoS Biology*, **3**, e339.

Odum, E. P. (1953). *Fundamentals of Ecology*. Philadelphia: Saunders.

Oidtmann, B., El-Matbouli, M., Fischer, H., Hoffmann, R., Klarding, K., Schmid, I. & Schmidt, R. (1997). *Light Microscopy of Astacus astacus L. Under Normal and Selected Pathological Conditions, with Special Emphasis to Porcelain Disease and Crayfish Plague*. Lafayette: International Association of Astacology.

Oksanen, L., Fretwell, S. D., Arruda, J. & Niemela, P. (1981). Exploitation ecosystems in gradients of primary productivity. *American Naturalist*, **118**, 240–261.

Olff, H., Hoorens, B., de Goede, R. G. M., Van der Putten, W. H. & Gleichman, J. M. (2000). Small-scale shifting mosaics of two dominant grassland species: the possible role of soil-borne pathogens. *Oecologia*, **125**, 45–54.

Oliveira, N. M. & Hilker, F. M. (2010). Modelling disease introduction as biological control of invasive predators to preserve endangered prey. *Bulletin of Mathematical Biology*, **72**, 444–468.

Olsen, B., Munster, V. J., Wallensten, A., Waldenstrom, J., Osterhaus, A. & Fouchier, R. A. M. (2006). Global patterns of influenza A virus in wild birds. *Science*, **312**, 384–388.

Omacini, M., Chaneton, E. J., Ghersa, C. M. & Muller, C. B. (2001). Symbiotic fungal endophytes control insect host–parasite interaction webs. *Nature*, **409**, 78–81.

Omacini, M., Chaneton, E. J., Ghersa, C. M. & Otero, P. (2004). Do foliar endophytes affect grass litter decomposition? A microcosm approach using *Lolium multiflorum*. *Oikos*, **104**, 581–590.

Omacini, M., Eggers, T., Bonkowski, M., Gange, A. C. & Jones, T. H. (2006). Leaf endophytes affect mycorrhizal status and growth of co-infected and neighbouring plants. *Functional Ecology*, **20**, 226–232.

Onstad, D. W. (1992). Evaluation of epidemiological thresholds and asymptotes with variable plant densities. *Phytopathology*, **82**, 1028–1032.

Onstad, D. W. & Kornkven, E. A. (1992). Persistence and endemicity of pathogens in plant populations over time and space. *Phytopathology*, **82**, 561–566.

Orr, M. R., De Camargo, R. X. & Benson, W. W. (2003). Interactions between ant species increase arrival rates of an ant parasitoid. *Animal Behaviour*, **65**, 1187–1193.

Ostfeld, R. S. (2009a). Biodiversity loss and the rise of zoonotic pathogens. *Clinical Microbiology and Infection*, **15**, 40–43.

Ostfeld, R. S. (2009b). Climate change and the distribution and intensity of infectious diseases. *Ecology*, **90**, 903–905.

Ostfeld, R. S., Canham, C. D., Oggenfuss, K., Winchcombe, R. J. & Keesing, F. (2006). Climate, deer, rodents, and acorns as determinants of variation in Lyme-disease risk. *PloS Biology*, **4**, 1058–1068.

Ostfeld, R. S., Glass, G. E. & Keesing, F. (2005). Spatial epidemiology: an emerging (or re-emerging) discipline. *Trends in Ecology & Evolution*, **20**, 328–336.

Ostfeld, R. S. & Holt, R. D. (2004). Are predators good for your health? Evaluating evidence for top-down regulation of zoonotic disease reservoirs. *Frontiers in Ecology and the Environment*, **2**, 13–20.

Ostfeld, R. S. & Keesing, F. (2000). Biodiversity and disease risk: the case of Lyme disease. *Conservation Biology*, **14**, 722–728.

Ostfeld, R. S. & LoGiudice, K. (2003). Community disassembly, biodiversity loss, and the erosion of an ecosystem service. *Ecology*, **84**, 1421–1427.

Otterstatter, M. C. & Thomson, J. D. (2008). Does pathogen spillover from commercially reared bumble bees threaten wild pollinators? *PLoS One*, **3**, e2771.

Ouellet, M., Mikaelian, I., Pauli, B. D., Rodrigue, J. & Green, D. M. (2005). Historical evidence of widespread chytrid infection in North American amphibian populations. *Conservation Biology*, **19**, 1431–1440.

Paaijmans, K. P., Read, A. F. & Thomas, M. B. (2009). Understanding the link between malaria risk and climate. *Proceedings of the National Academy of Sciences of the United States of America*, **106**, 13844–13849.

Packer, A. & Clay, K. (2000). Soil pathogens and spatial patterns of seedling mortality in a temperate tree. *Nature*, **404**, 278–281.

Packer, C., Altizer, S., Appel, M., Brown, E., Martenson, J., O'Brien, S. J., Roelke-Parker, M., Hofmann-Lehmann, R. & Lutz, H. (1999). Viruses of the Serengeti: patterns of infection and mortality in African lions. *Journal of Animal Ecology*, **68**, 1161–1178.

Packer, C., Holt, R. D., Hudson, P. J., Lafferty, K. D. & Dobson, A. P. (2003). Keeping the herds healthy and alert: implications of predator control for infectious disease. *Ecology Letters*, **6**, 797–802.

Paddock, C. D. & Yabsley, M. J. (2007). Ecological havoc, the rise of white-tailed deer, and the emergence of *Amblyomma americanum* – associated zoonoses in the United States. *Current Topics in Microbiology and Immunology*, **315**, 289–324.

Park, T. (1948). Experimental studies of interspecies competition: 1 – competition between populations of the flour beetles, *Tribolium confusum* Duval and *Tribolium castaneum* Herbst. *Ecological Monographs*, **18**, 265–307.

Parker, C. (2009). Observations on the current status of *Orobanche* and *Striga* problems worldwide. *Pest Management Science*, **65**, 453–459.

Parker, G. A., Chubb, J. C., Ball, M. A. & Roberts, G. N. (2003). Evolution of complex life cycles in helminth parasites. *Nature*, **425**, 480–484.

Parrish, C. R., Holmes, E. C., Morens, D. M., Park, E. C., Burke, D. S., Calisher, C. H., Laughlin, C. A., Saif, L. J. & Daszak, P. (2008). Cross-species virus transmission and the emergence of new epidemic diseases. *Microbiology and Molecular Biology Reviews*, **72**, 457–470.

Pascual, M., Ahumada, J. A., Chaves, L. F., Rodo, X. & Bouma, M. (2006). Malaria resurgence in the East African highlands: temperature trends revisited. *Proceedings of the National Academy of Sciences of the United States of America*, **103**, 5829–5834.

Pascual, M. & Bouma, M. J. (2009). Do rising temperatures matter? *Ecology*, **90**, 906–912.

Pasternak, Z., Diamant, A. & Abelson, A. (2007). Co-invasion of a Red Sea fish and its ectoparasitic monogenean, *Polylabris* cf. *mamaevi* into the Mediterranean: observations on oncomiracidium behavior and infection levels in both seas. *Parasitology Research*, **100**, 721–727.

Pattison, J. (1998). The emergence of bovine spongiform encephalopathy and related diseases. *Emerging Infectious Diseases*, **4**, 390–394.

Paul, N. D., Hatcher, P. E. & Taylor, J. E. (2000). Coping with multiple enemies: an integration of molecular and ecological perspectives. *Trends in Plant Science*, **5**, 220–225.

Pedersen, A. B. & Fenton, A. (2007). Emphasizing the ecology in parasite community ecology. *Trends in Ecology & Evolution*, **22**, 133–139.

Pedersen, A. B., Jones, K. E., Nunn, C. L. & Altizer, S. (2007). Infectious diseases and extinction risk in wild mammals. *Conservation Biology*, **21**, 1269–1279.

Pell, J. K., Hannam, J. J. & Steinkraus, D. C. (2010). Conservation biological control using fungal entomopathogens. *Biocontrol*, **55**, 187–198.

Pennings, S. C. & Callaway, R. M. (1996). Impact of a parasitic plant on the structure and dynamics of salt marsh vegetation. *Ecology*, **77**, 1410–1419.

Pennings, S. C. & Callaway, R. M. (2002). Parasitic plants: parallels and contrasts with herbivores. *Oecologia*, **131**, 479–489.

Petermann, J. S., Fergus, A. J. F., Turnbull, L. A. & Schmid, B. (2008). Janzen–Connell effects are widespread and strong enough to maintain diversity in grasslands. *Ecology*, **89**, 2399–2406.

Peterson, R. O., Thomas, N. J., Thurber, J. M., Vucetich, J. A. & Waite, T. A. (1998). Population limitation and the wolves of Isle Royale. *Journal of Mammalogy*, **79**, 828–841.

Philpott, S. M. (2005). Trait-mediated effects of parasitic phorid flies (Diptera: Phoridae) on ant (Hymenoptera: Formicidae) competition and resource access in coffee agro-ecosystems. *Environmental Entomology*, **34**, 1089–1094.

Phoenix, G. K. & Press, M. C. (2005). Linking physiological traits to impacts on community structure and function: the role of root hemiparasitic Orobanchaceae (ex-Scrophulariaceae). *Journal of Ecology*, **93**, 67–78.

Pieterse, C. M. J., Van Wees, S. C. M., Ton, J., Van Pelt, J. A. & Van Loon, L. C. (2002). Signalling in rhizobacteria-induced systemic resistance in *Arabidopsis thaliana*. *Plant Biology*, **4**, 535–544.

Pimentel, D., Lach, L., Zuniga, R. & Morrison, D. (2000). Environmental and economic costs of nonindigenous species in the United States. *Bioscience*, **50**, 53–65.

Pimm, S. L. (1982). *Food Webs*. London: Chapman & Hall.

Pimm, S. L. & Lawton, J. H. (1978). Feeding on more than one trophic level. *Nature*, **275**, 542–544.

Pinzon, J. E., Wilson, J. M., Tucker, C. J., Arthur, R., Jahrling, P. B. & Formenty, P. (2004). Trigger events: enviroclimatic coupling of Ebola hemorrhagic fever outbreaks. *American Journal of Tropical Medicine and Hygiene*, **71**, 664–674.

Plantier, J. C., Leoz, M., Dickerson, J. E., De Oliveira, F., Cordonnier, F., Lemee, V., Damond, F., Robertson, D. L. & Simon, F. (2009). A new human immunodeficiency virus derived from gorillas. *Nature Medicine*, **15**, 871–872.

Ploetz, R. C. (1994). Panama disease: return of the first banana menace. *International Journal of Pest Management*, **40**, 326–336.

Polis, G. A. & Holt, R. D. (1992). Intraguild predation: the dynamics of complex trophic interactions. *Trends in Ecology & Evolution*, **7**, 151–154.

Polis, G. A., Myers, C. A. & Holt, R. D. (1989). The ecology and evolution of intraguild predation: potential competitors that eat each other. *Annual Review of Ecology and Systematics*, **20**, 297–330.

Poulin, R. (1995). 'Adaptive' changes in the behaviour of parasitized animals: a critical review. *International Journal for Parasitology*, **25**, 1371–1383.

Poulin, R. (1997). Species richness of parasite assemblages: evolution and patterns. *Annual Review of Ecology and Systematics*, **28**, 341–358.

Poulin, R. (1999). The functional importance of parasites in animal communities: many roles at many levels? *International Journal for Parasitology*, **29**, 903–914.

Poulin, R. & Mouritsen, K. N. (2006). Climate change, parasitism and the structure of intertidal ecosystems. *Journal of Helminthology*, **80**, 183–191.

Poulin, R., Nichol, K. & Latham, A. D. A. (2003). Host sharing and host manipulation by larval helminths in shore crabs: cooperation or conflict? *International Journal for Parasitology*, **33**, 425–433.

Pounds, J. A., Bustamante, M. R., Coloma, L. A., Consuegra, J. A., Fogden, M. P. L., Foster, P. N., La Marca, E., Masters, K. L., Merino-Viteri, A., Puschendorf, R., Ron, S. R., Sanchez-Azofeifa, G. A., Still, C. J. & Young, B. E. (2006). Widespread amphibian extinctions from epidemic disease driven by global warming. *Nature*, **439**, 161–167.

Power, A. G. & Mitchell, C. E. (2004). Pathogen spillover in disease epidemics. *American Naturalist*, **164**, S79–S89.

Prenter, J., MacNeil, C., Dick, J. T. A. & Dunn, A. M. (2004). Roles of parasites in animal invasions. *Trends in Ecology & Evolution*, **19**, 385–390.

Press, M. C., Nour, J. J., Bebawi, F. F. & Stewart, G. R. (1989). Antitranspirant-induced heat-stress in the parasitic plant *Striga hermonthica* – a novel method of control. *Journal of Experimental Botany*, **40**, 585–591.

Press, M. C. & Phoenix, G. K. (2005). Impacts of parasitic plants on natural communities. *New Phytologist*, **166**, 737–751.

Price, P. W., Bouton, C. E., Gross, P., McPheron, B. A., Thompson, J. N. & Weis, A. E. (1980). Interactions among three trophic levels – influence of plants on interactions between insect herbivores and natural enemies. *Annual Review of Ecology and Systematics*, **11**, 41–65.

Price, P. W., Westoby, M., Rice, B., Atsatt, P. R., Fritz, R. S., Thompson, J. N. & Mobley, K. (1986). Parasite mediation in ecological interactions. *Annual Review of Ecology and Systematics*, **17**, 487–505.

Prins, H. H. T. & Vanderjeugd, H. P. (1993). Herbivore population crashes and woodland structure in East Africa. *Journal of Ecology*, **81**, 305–314.

Prins, H. H. T. & Weyerhaeuser, F. J. (1987). Epidemics in populations of wild ruminants: anthrax and impala, rinderpest and buffalo in Lake Manyara national park, Tanzania. *Oikos*, **49**, 28–38.

Pywell, R. F., Bullock, J. M., Walker, K. J., Coulson, S. J., Gregory, S. J. & Stevenson, M. J. (2004). Facilitating grassland diversification using the hemiparasitic plant *Rhinanthus minor*. *Journal of Applied Ecology*, **41**, 880–887.

Quested, H. M. (2008). Parasitic plants: impacts on nutrient cycling. *Plant and Soil*, **311**, 269–272.

Rand, T. A. & Louda, S. A. (2006). Spillover of agriculturally subsidized predators as a potential threat to native insect herbivores in fragmented landscapes. *Conservation Biology*, **20**, 1720–1729.

Randolph, S. E. (2009). Perspectives of climate change impacts on infectious diseases. *Ecology*, **90**, 927–931.

Rasmann, S., Kollner, T. G., Degenhardt, J., Hiltpold, I., Toepfer, S., Kuhlmann, U., Gershenzon, J. & Turlings, T. C. J. (2005). Recruitment of entomopathogenic nematodes by insect-damaged maize roots. *Nature*, **434**, 732–737.

Rasmann, S. & Turlings, T. C. J. (2007). Simultaneous feeding by aboveground and belowground herbivores attenuates plant-mediated attraction of their respective natural enemies. *Ecology Letters*, **10**, 926–936.

Ray, C. & Collinge, S. K. (2006). Potential effects of a keystone species on the dynamics of sylvatic plague. In Collinge, S. K. & Ray, C. (eds), *Disease Ecology: Community Structure and Pathogen Dynamics*. Oxford: Oxford University Press, pp. 202–216.

Read, A. F. & Taylor, L. H. (2001). The ecology of genetically diverse infections. *Science*, **292**, 1099–1102.

Reinhart, K. O. & Callaway, R. M. (2006). Soil biota and invasive plants. *New Phytologist*, **170**, 445–457.

Reinhart, K. O., Packer, A., Van der Putten, W. H. & Clay, K. (2003). Plant–soil biota interactions and spatial distribution of black cherry in its native and invasive ranges. *Ecology Letters*, **6**, 1046–1050.

Reiter, P. (2010). West Nile virus in Europe: understanding the present to gauge the future. *Eurosurveillance*, **15**, 19508.

Rejmanek, M. & Richardson, D. M. (1996). What attributes make some plant species more invasive? *Ecology*, **77**, 1655–1661.

Reynolds, H. L., Packer, A., Bever, J. D. & Clay, K. (2003). Grassroots ecology: plant–microbe–soil interactions as drivers of plant community structure and dynamics. *Ecology*, **84**, 2281–2291.

Reynolds, J. C. (1985). Details of the geographic replacement of the red squirrel (*Sciurus vulgaris*) by the grey squirrel (*Sciurus carolinensis*) in eastern England. *Journal of Animal Ecology*, **54**, 149–162.

Rigaud, T. & Moret, Y. (2003). Differential phenoloxidase activity between native and invasive gammarids infected by local acanthocephalans: differential immuno-suppression? *Parasitology*, **127**, 571–577.

Riley, S., Fraser, C., Donnelly, C. A., Ghani, A. C., Abu-Raddad, L. J., Hedley, A. J., Leung, G. M., Ho, L. M., Lam, T. H., Thach, T. Q., Chau, P., Chan, K. P., Leung, P. Y., Tsang, T., Ho, W., Lee, K. H., Lau, E. M. C., Ferguson, N. M. & Anderson, R. M. (2003). Transmission dynamics of the etiological agent of SARS in Hong Kong: impact of public health interventions. *Science*, **300**, 1961–1966.

Roberts, M. G. (1995). A pocket guide to host–parasite models. *Parasitology Today*, **11**, 172–177.

Roelke-Parker, M. E., Munson, L., Packer, C., Kock, R., Cleaveland, S., Carpenter, M., Obrien, S. J., Pospischil, A., Hofmann-Lehmann, R., Lutz, H., Mwamengele, G. L. M., Mgasa, M. N., Machange, G. A., Summers, B. A. & Appel, M. J. G. (1996). A canine distemper virus epidemic in Serengeti lions (*Panthera leo*). *Nature*, **379**, 441–445.

Rogers, D. J. & Randolph, S. E. (2000). The global spread of malaria in a future, warmer world. *Science*, **289**, 2283–2284.

Rohde, K. (1998). Is there a fixed number of niches for endoparasites of fish? *International Journal for Parasitology*, **28**, 1861–1865.

Rohlfs, M. (2008). Host–parasitoid interaction as affected by interkingdom competition. *Oecologia*, **155**, 161–168.

Rohr, J. R., Raffel, T. R., Romansic, J. M., McCallum, H. & Hudson, P. J. (2008). Evaluating the links between climate, disease spread, and amphibian declines. *Proceedings of the National Academy of Sciences of the United States of America*, **105**, 17436–17441.

Rosenblum, E. B., Voyles, J., Poortne, T. J. & Stajich, J. E. (2010). The deadly chytrid fungus: a story of an emerging pathogen. *PLoS Pathogens*, **6**, e1000550.

Rosenheim, J. A. (1998). Higher-order predators and the regulation of insect herbivore populations. *Annual Review of Entomology*, **43**, 421–447.

Rosenheim, J. A. (2007). Intraguild predation: new theoretical and empirical perspectives. *Ecology*, **88**, 2679–2680.

Rosenheim, J. A., Kaya, H. K., Ehler, L. E., Marois, J. J. & Jaffee, B. A. (1995). Intraguild predation among biological-control agents: theory and evidence. *Biological Control*, **5**, 303–335.

Rosenzweig, M. L. (1973). Exploitation in three trophic levels. *American Naturalist*, **107**, 275–294.

Roy, H. E. & Cottrell, T. E. (2008). Forgotten natural enemies: interactions between coccinellids and insect-parasitic fungi. *European Journal of Entomology*, **105**, 391–398.

Roy, H. E. & Pell, J. K. (2000). Interactions between entomopathogenic fungi and other natural enemies: implications for biological control. *Biocontrol Science and Technology*, **10**, 737–752.

Roy, H. E., Pell, J. K. & Alderson, P. G. (1999). Effects of fungal infection on the alarm response of pea aphids. *Journal of Invertebrate Pathology*, **74**, 69–75.

Roy, H. E., Pell, J. K. & Alderson, P. G. (2001). Targeted dispersal of the aphid pathogenic fungus *Erynia neoaphidis* by the aphid predator *Coccinella septempunctata*. *Biocontrol Science and Technology*, **11**, 99–110.

Roy, H. E., Pell, J. K., Clark, S. J. & Alderson, P. G. (1998). Implications of predator foraging on aphid pathogen dynamics. *Journal of Invertebrate Pathology*, **71**, 236–247.

Roy, M. & Holt, R. D. (2008). Effects of predation on host–pathogen dynamics in SIR models. *Theoretical Population Biology*, **73**, 319–331.

Rudgers, J. A., Holah, J., Orr, S. P. & Clay, K. (2007). Forest succession suppressed by an introduced plant–fungal symbiosis. *Ecology*, **88**, 18–25.

Rudgers, J. A., Koslow, J. M. & Clay, K. (2004). Endophytic fungi alter relationships between diversity and ecosystem properties. *Ecology Letters*, **7**, 42–51.

Rudgers, J. A., Mattingly, W. B. & Koslow, J. M. (2005). Mutualistic fungus promotes plant invasion into diverse communities. *Oecologia*, **144**, 463–471.

Rudgers, J. A. & Orr, S. (2009). Non-native grass alters growth of native tree species via leaf and soil microbes. *Journal of Ecology*, **97**, 247–255.

Rudolf, V. H. W. (2007). The interaction of cannibalism and omnivory: consequences for community dynamics. *Ecology*, **88**, 2697–2705.

Rudolf, V. H. W. (2008). Consequences of size structure in the prey for predator–prey dynamics: the composite functional response. *Journal of Animal Ecology*, **77**, 520–528.

Rudolf, V. H. W. & Antonovics, J. (2005). Species coexistence and pathogens with frequency-dependent transmission. *American Naturalist*, **166**, 112–118.

Ruggieri, E. & Schreiber, S. J. (2005). The dynamics of the Schoener–Polis–Holt model of intra-guild predation. *Mathematical Biosciences and Engineering*, **2**, 279–288.

Runyon, J. B., Mescher, M. C. & De Moraes, C. M. (2008). Parasitism by *Cuscuta pentagona* attenuates host plant defenses against insect herbivores. *Plant Physiology*, **146**, 987–995.

Rushton, S. P., Lurz, P. W. W., Gurnell, J. & Fuller, R. (2000). Modelling the spatial dynamics of parapoxvirus disease in red and grey squirrels: a possible cause of the decline in the red squirrel in the UK? *Journal of Applied Ecology*, **37**, 997–1012.

Rushton, S. P., Lurz, P. W. W., Gurnell, J., Nettleton, P., Bruemmer, C., Shirley, M. D. F. & Sainsbury, A. W. (2006). Disease threats posed by alien species: the role of a poxvirus in the decline of the native red squirrel in Britain. *Epidemiology and Infection*, **134**, 521–533.

Sachs, J. & Malaney, P. (2002). The economic and social burden of malaria. *Nature*, **415**, 680–685.

Sage, R. B., Woodburn, M. I. A., Davis, C. & Aebischer, N. J. (2002). The effect of an experimental infection of the nematode *Heterakis gallinarum* on hand-reared grey partridges *Perdix perdix*. *Parasitology*, **124**, 529–535.

Saikkonen, K., Faeth, S. H., Helander, M. & Sullivan, T. J. (1998). Fungal endophytes: a continuum of interactions with host plants. *Annual Review of Ecology and Systematics*, **29**, 319–343.

Sainsbury, A. W., Nettleton, P., Gilray, J. & Gurnell, J. (2000). Grey squirrels have high seroprevalence to a parapoxvirus associated with deaths in red squirrels. *Animal Conservation*, **3**, 229–233.

Sato, T., Arizono, M., Sone, R. & Harada, Y. (2008). Parasite-mediated allochthonous input: do hairworms enhance subsidized predation of stream salmonids on crickets? *Canadian Journal of Zoology*, **86**, 231–235.

Schall, J. J. (1992). Parasite-mediated competition in anolis lizards. *Oecologia*, **92**, 58–64.

Schloegel, L. M., Hero, J. M., Berger, L., Speare, R., McDonald, K. & Daszak, P. (2006). The decline of the sharp-snouted day frog (*Taudactylus acutirostris*): the first documented case of extinction by infection in a free-ranging wildlife species? *Ecohealth*, **3**, 35–40.

Schmitz, O. J. & Nudds, T. D. (1994). Parasite-mediated competition in deer and moose: how strong is the effect of meningeal worm on moose? *Ecological Applications*, **4**, 91–103.

Settle, W. H. & Wilson, L. T. (1990). Invasion by the variegated leafhopper and biotic interactions: parasitism, competition, and apparent competition. *Ecology*, **71**, 1461–1470.

Sharp, P. M. & Hahn, B. H. (2008). Aids: prehistory of HIV-1. *Nature*, **455**, 605–606.

Shoemaker, D. D., Ross, K. G., Keller, L., Vargo, E. L. & Werren, J. H. (2000). *Wolbachia* infections in native and introduced populations of fire ants (*Solenopsis* spp.). *Insect Molecular Biology*, **9**, 661–673.

Siegel, J. P., Maddox, J. V. & Ruesink, W. G. (1986). Impact of *Nosema pyrausta* on a braconid, *Macrocentrus grandii*, in central Illinois. *Journal of Invertebrate Pathology*, **47**, 271–276.

Sisterson, M. S. & Averill, A. L. (2003). Interactions between parasitized and unparasitized conspecifics: parasitoids modulate competitive dynamics. *Oecologia*, **135**, 362–371.

Slippers, B., Stenlid, J. & Wingfield, M. J. (2005). Emerging pathogens: fungal host jumps following anthropogenic introduction. *Trends in Ecology & Evolution*, **20**, 420–421.

Slothouber Galbreath, J. G. M., Smith, J. E., Becnel, J. J., Butlin, R. K. & Dunn, A. M. (2010). Reduction in post-invasion genetic diversity in *Crangonyx pseudogracilis* (Amphipoda: Crustacea): a genetic bottleneck or the work of hitchhiking vertically transmitted microparasites? *Biological Invasions*, **12**, 191–209.

Smith, D. (2000). The population dynamics and community ecology of root hemiparasitic plants. *American Naturalist*, **155**, 13–23.

Smith, G. J. D., Vijaykrishna, D., Bahl, J., Lycett, S. J., Worobey, M., Pybus, O. G., Ma, S. K., Cheung, C. L., Raghwani, J., Bhatt, S., Peiris, J. S. M., Guan, Y. & Rambaut, A. (2009a). Origins and evolutionary genomics of the 2009 swine-origin H1N1 influenza A epidemic. *Nature*, **459**, 1122–1125.

Smith, K. F., Acevedo-Whitehouse, K. & Pedersen, A. B. (2009b). The role of infectious diseases in biological conservation. *Animal Conservation*, **12**, 1–12.

Smith, K. F., Behrens, M., Schloegel, L. M., Marano, N., Burgiel, S. & Daszak, P. (2009c). Reducing the risks of the wildlife trade. *Science*, **324**, 594–595.

Smith, K. F., Sax, D. F. & Lafferty, K. D. (2006). Evidence for the role of infectious disease in species extinction and endangerment. *Conservation Biology*, **20**, 1349–1357.

Smith, R. M. M., Drobniewski, F., Gibson, A., Montague, J. D. E., Logan, M. N., Hunt, D., Hewinson, G., Salmon, R. L. & O'Neill, B. (2004). *Mycobacterium bovis* infection, United Kingdom. *Emerging Infectious Diseases*, **10**, 539–541.

Snyder, W. E., Ballard, S. N., Yang, S., Clevenger, G. M., Miller, T. D., Ahn, J. J., Hatten, T. D. & Berryman, A. A. (2004). Complementary biocontrol of aphids by the ladybird beetle *Harmonia axyridis* and the parasitoid *Aphelinus asychis* on greenhouse roses. *Biological Control*, **30**, 229–235.

Sodora, D. L., Allan, J. S., Apetrei, C., Brenchley, J. M., Douek, D. C., Else, J. G., Estes, J. D., Hahn, B. H., Hirsch, V. M., Kaur, A., Kirchhoff, F., Muller-Trutwin, M., Pandrea, I., Schmitz, J. E. & Silvestri, G. (2009). Toward an AIDS vaccine: lessons from natural simian immunodeficiency virus infections of African nonhuman primate hosts. *Nature Medicine*, **15**, 861–865.

Sokolow, S. (2009). Effects of a changing climate on the dynamics of coral infectious disease: a review of the evidence. *Diseases of Aquatic Organisms*, **87**, 5–18.

Sorensen, R. E. & Minchella, D. J. (2001). Snail–trematode life history interactions: past trends and future directions. *Parasitology*, **123**, S3–S18.

Stahlhut, J. K., Liebert, A. E., Starks, P. T., Dapporto, L. & Jaenike, J. (2006). *Wolbachia* in the invasive European paper wasp *Polistes dominulus*. *Insectes Sociaux*, **53**, 269–273.

Steen, H., Taitt, M. & Krebs, C. J. (2002). Risk of parasite-induced predation: an experimental field study on Townsend's voles (*Microtus townsendii*). *Canadian Journal of Zoology*, **80**, 1286–1292.

Stout, M. J., Fidantsef, A. L., Duffey, S. S. & Bostock, R. M. (1999). Signal interactions in pathogen and insect attack: systemic plant-mediated interactions between pathogens and herbivores of the tomato, *Lycopersicon esculentum*. *Physiological and Molecular Plant Pathology*, **54**, 115–130.

Stout, M. J., Thaler, J. S. & Thomma, B. (2006). Plant-mediated interactions between pathogenic microorganisms and herbivorous arthropods. *Annual Review of Entomology*, **51**, 663–689.

Straub, C. S., Finke, D. L. & Snyder, W. E. (2008). Are the conservation of natural enemy biodiversity and biological control compatible goals? *Biological Control*, **45**, 225–237.

Stuart, S. N., Chanson, J. S., Cox, N. A., Young, B. E., Rodrigues, A. S. L., Fischman, D. L. & Waller, R. W. (2004). Status and trends of amphibian declines and extinctions worldwide. *Science*, **306**, 1783–1786.

Sumption, K. J. & Flowerdew, J. R. (1985). The ecological effects of the decline in rabbits (*Oryctolagus cuniculus* L.) due to myxomatosis. *Mammal Review*, **15**, 151–186.

Tain, L., Perrot-Minnot, M. J. & Cezilly, F. (2007). Differential influence of *Pomphorhynchus laevis* (Acanthocephala) on brain serotonergic activity in two congeneric host species. *Biology Letters*, **3**, 68–71.

Tanabe, K. & Namba, T. (2005). Omnivory creates chaos in simple food web models. *Ecology*, **86**, 3411–3414.

Tatem, A. J., Hay, S. I. & Rogers, D. J. (2006). Global traffic and disease vector dispersal. *Proceedings of the National Academy of Sciences of the United States of America*, **103**, 6242–6247.

Tattoni, C., Preatoni, D. G., Lurz, P. W. W., Rushton, S. P., Tosi, G., Bertolino, S., Martinoli, A. & Wauters, L. A. (2006). Modelling the expansion of a grey squirrel population: implications for squirrel control. *Biological Invasions*, **8**, 1605–1619.

Taylor, J. E., Hatcher, P. E. & Paul, N. D. (2004). Crosstalk between plant responses to pathogens and herbivores: a view from the outside in. *Journal of Experimental Botany*, **55**, 159–168.

Taylor, L. H., Latham, S. M. & Woolhouse, M. E. J. (2001). Risk factors for human disease emergence. *Philosophical Transactions of the Royal Society of London Series B – Biological Sciences*, **356**, 983–989.

Telfer, S., Birtles, R., Bennett, M., Lambin, X., Paterson, S. & Begon, M. (2008). Parasite interactions in natural populations: insights from longitudinal data. *Parasitology*, **135**, 767–781.

Telfer, S., Bown, K. J., Sekules, R., Begon, I., Hayden, T. & Birtles, R. (2005). Disruption of a host–parasite system following the introduction of an exotic host species. *Parasitology*, **130**, 661–668.

Terry, R. S., Smith, J. E., Sharpe, R. G., Rigaud, T., Littlewood, D. T. J., Ironside, J. E., Rollinson, D., Bouchon, D., MacNeil, C., Dick, J. T. A. & Dunn, A. M. (2004). Widespread vertical transmission and associated host sex-ratio distortion within the eukaryotic phylum Microspora. *Proceedings of the Royal Society of London Series B – Biological Sciences*, **271**, 1783–1789.

Thieltges, D. W., de Montaudouin, X., Fredensborg, B., Jensen, K. T., Koprivnikar, J. & Poulin, R. (2008a). Production of marine trematode cercariae: a potentially overlooked path of energy flow in benthic systems. *Marine Ecology Progress Series*, **372**, 147–155.

Thieltges, D. W., Jensen, K. T. & Poulin, R. (2008b). The role of biotic factors in the transmission of free-living endohelminth stages. *Parasitology*, **135**, 407–426.

Thomas, F., Adamo, S. & Moore, J. (2005a). Parasitic manipulation: where are we and where should we go? *Behavioural Processes*, **68**, 185–199.

Thomas, F., Fauchier, J. & Lafferty, K. D. (2002a). Conflict of interest between a nematode and a trematode in an amphipod host: test of the 'sabotage' hypothesis. *Behavioral Ecology and Sociobiology*, **51**, 296–301.

Thomas, F., Mete, K., Helluy, S., Santalla, F., Verneau, O., DeMeeus, T., Cezilly, F. & Renaud, F. (1997). Hitch-hiker parasites or how to benefit from the strategy of another parasite. *Evolution*, **51**, 1316–1318.

Thomas, F., Poulin, R., de Meeus, T., Guegan, J. F. & Renaud, F. (1999). Parasites and ecosystem engineering: what roles could they play? *Oikos*, **84**, 167–171.

Thomas, F., Poulin, R. & Renaud, F. (1998). Nonmanipulative parasites in manipulated hosts: 'hitch-hikers' or simply 'lucky passengers'? *Journal of Parasitology*, **84**, 1059–1061.

Thomas, F., Renaud, F. & Guegan, J. F. (2005b). *Parasitism and Ecosystems*. Oxford: Oxford University Press.

Thomas, F., Schmidt-Rhaesa, A., Martin, G., Manu, C., Durand, P. & Renaud, F. (2002b). Do hairworms (Nematomorpha) manipulate the water seeking behaviour of their terrestrial hosts? *Journal of Evolutionary Biology*, **15**, 356–361.

Thomas, K., Tompkins, D. M., Sainsbury, A. W., Wood, A. R., Dalziel, R., Nettleton, P. F. & McInnes, C. J. (2003). A novel poxvirus lethal to red squirrels (*Sciurus vulgaris*). *Journal of General Virology*, **84**, 3337–3341.

Thompson, R. M., Hemberg, M., Starzomski, B. M. & Shurin, J. B. (2007). Trophic levels and trophic tangles: the prevalence of omnivory in real food webs. *Ecology*, **88**, 612–617.

Thompson, R. M., Mouritsen, K. N. & Poulin, R. (2005). Importance of parasites and their life cycle characteristics in determining the structure of a large marine food web. *Journal of Animal Ecology*, **74**, 77–85.

Thrall, P. H. & Burdon, J. J. (2003). Evolution of virulence in a plant host–pathogen metapopulation. *Science*, **299**, 1735–1737.

Thrall, P. H. & Jarosz, A. M. (1994). Host–pathogen dynamics in experimental populations of *Silene alba* and *Ustilago violacea*: II – Experimental tests of theoretical models. *Journal of Ecology*, **82**, 561–570.

Tilman, D. (1982). *Resource Competition and Community Structure*. Princeton: Princeton University Press.

Tilman, D., May, R. M., Lehman, C. L. & Nowak, M. A. (1994). Habitat destruction and the extinction debt. *Nature*, **371**, 65–66.

Tompkins, D. M., Dickson, G. & Hudson, P. J. (1999). Parasite-mediated competition between pheasant and grey partridge: a preliminary investigation. *Oecologia*, **119**, 378–382.

Tompkins, D. M., Draycott, R. A. H. & Hudson, P. J. (2000a). Field evidence for apparent competition mediated via the shared parasites of two gamebird species. *Ecology Letters*, **3**, 10–14.

Tompkins, D. M., Dunn, A. M., Smith, M. J. & Telfer, S. (2010). Wildlife diseases: from individuals to ecosystems. *Journal of Animal Ecology*, **80**, 19–38.

Tompkins, D. M., Greenman, J. V. & Hudson, P. J. (2001). Differential impact of a shared nematode parasite on two gamebird hosts: implications for apparent competition. *Parasitology*, **122**, 187–193.

Tompkins, D. M., Greenman, J. V., Robertson, P. A. & Hudson, P. J. (2000b). The role of shared parasites in the exclusion of wildlife hosts: *Heterakis gallinarum* in the ring-necked pheasant and the grey partridge. *Journal of Animal Ecology*, **69**, 829–840.

Tompkins, D. M., Sainsbury, A. W., Nettleton, P., Buxton, D. & Gurnell, J. (2002). Parapoxvirus causes a deleterious disease in red squirrels associated with UK population declines. *Proceedings of the Royal Society of London Series B – Biological Sciences*, **269**, 529–533.

Tompkins, D. M., White, A. R. & Boots, M. (2003). Ecological replacement of native red squirrels by invasive greys driven by disease. *Ecology Letters*, **6**, 189–196.

Torchin, M. E., Byers, J. E. & Huspeni, T. C. (2005). Differential parasitism of native and introduced snails: replacement of a parasite fauna. *Biological Invasions*, **7**, 885–894.

Torchin, M. E., Lafferty, K. D., Dobson, A. P., McKenzie, V. J. & Kuris, A. M. (2003). Introduced species and their missing parasites. *Nature*, **421**, 628–630.

Torchin, M. E., Lafferty, K. D. & Kuris, A. M. (2002). Parasites and marine invasions. *Parasitology*, **124**, S137–S151.

Tounou, A. K., Kooyman, C., Douro-Kpindou, O. K. & Poehling, H. M. (2008a). Combined field efficacy of *Paranosema locustae* and *Metarhizium anisopliae* var. *acridum* for the control of Sahelian grasshoppers. *Biocontrol*, **53**, 813–828.

Tounou, A. K., Kooyman, C., Douro-Kplndou, O. K. & Poehling, H. M. (2008b). Interaction between *Paranosema locustae* and *Metarhizium anisopliae* var. *acridum*, two pathogens of the desert locust, *Schistocerca gregaria* under laboratory conditions. *Journal of Invertebrate Pathology*, **97**, 203–210.

Tsutsui, N. D., Kauppinen, S. N., Oyafuso, A. F. & Grosberg, R. K. (2003). The distribution and evolutionary history of *Wolbachia* infection in native and introduced populations of the invasive Argentine ant (*Linepithema humile*). *Molecular Ecology*, **12**, 3057–3068.

Tumpey, T. M., Basler, C. F., Aguilar, P. V., Zeng, H., Solorzano, A., Swayne, D. E., Cox, N. J., Katz, J. M., Taubenberger, J. K., Palese, P. & Garcia-Sastre, A. (2005). Characterization of the reconstructed 1918 Spanish influenza pandemic virus. *Science*, **310**, 77–80.

Turlings, T. C. J. & Tumlinson, J. H. (1992). Systemic release of chemical signals by herbivore-injured corn. *Proceedings of the National Academy of Sciences of the United States of America*, **89**, 8399–8402.

Twu, S. J., Chen, T. J., Chen, C. J., Olsen, S. J., Lee, L. T., Fisk, T., Hsu, K. H., Chang, S. C., Chen, K. T., Chiang, I. H., Wu, Y. C., Wu, J. S. & Dowell, S. F. (2003). Control measures for severe acute respiratory syndrome (SARS) in Taiwan. *Emerging Infectious Diseases*, **9**, 718–720.

van Dam, N. M., Raaijmakers, C. E. & Van der Putten, W. H. (2005). Root herbivory reduces growth and survival of the shoot feeding specialist *Pieris rapae* on *Brassica nigra*. *Entomologia experimentalis et Applicata*, **115**, 161–170.

Van der Putten, W. H., Van Dijk, C. & Peters, B. A. M. (1993). Plant-specific soil-borne diseases contribute to succession in foredune vegetation. *Nature*, **362**, 53–56.

Van der Putten, W. H. & Peters, B. A. M. (1997). How soil-borne pathogens may affect plant competition. *Ecology*, **78**, 1785–1795.

Van der Putten, W. H., Vet, L. E. M., Harvey, J. A. & Wackers, F. L. (2001). Linking above- and belowground multitrophic interactions of plants, herbivores, pathogens, and their antagonists. *Trends in Ecology & Evolution*, **16**, 547–554.

van Loon, L. C., Bakker, P. & Pieterse, C. M. J. (1998). Systemic resistance induced by rhizosphere bacteria. *Annual Review of Phytopathology*, **36**, 453–483.

van Nouhuys, S. & Hanski, I. (2000). Apparent competition between parasitoids mediated by a shared hyperparasitoid. *Ecology Letters*, **3**, 82–84.

van Ommeren, R. J. & Whitham, T. G. (2002). Changes in interactions between juniper and mistletoe mediated by shared avian frugivores: parasitism to potential mutualism. *Oecologia*, **130**, 281–288.

Van Reeth, K. (2007). Avian and swine influenza viruses: our current understanding of the zoonotic risk. *Veterinary Research*, **38**, 243–260.

Van Veen, F. J. F., Morris, R. J. & Godfray, H. C. J. (2006). Apparent competition, quantitative food webs, and the structure of phytophagous insect communities. *Annual Review of Entomology*, **51**, 187–208.

Van Veen, F. J. F., Mueller, C. B., Pell, J. K. & Godfray, H. C. J. (2008). Food web structure of three guilds of natural enemies: predators, parasitoids and pathogens of aphids. *Journal of Animal Ecology*, **77**, 191–200.

Vance-Chalcraft, H. D., Rosenheim, J. A., Vonesh, J. R., Osenberg, C. W. & Sih, A. (2007). The influence of intraguild predation on prey suppression and prey release: a meta-analysis. *Ecology*, **88**, 2689–2696.

Vial, F., Cleaveland, S., Rasmussen, G. & Haydon, D. T. (2006). Development of vaccination strategies for the management of rabies in African wild dogs. *Biological Conservation*, **131**, 180–192.

Vicari, M., Hatcher, P. E. & Ayres, P. G. (2002). Combined effect of foliar and mycorrhizal endophytes on an insect herbivore. *Ecology*, **83**, 2452–2464.

Vidal, N., Peeters, M., Mulanga-Kabeya, C., Nzilambi, N., Robertson, D., Ilunga, W., Sema, H., Tshimanga, K., Bongo, B. & Delaporte, E. (2000). Unprecedented degree of human immunodeficiency virus type 1 (HIV-1) group M genetic diversity in the Democratic Republic of Congo suggests that the HIV-1 pandemic originated in Central Africa. *Journal of Virology*, **74**, 10498–10507.

Vinale, F., Sivasithamparam, K., Ghisalberti, E. L., Marra, R., Woo, S. L. & Lorito, M. (2008). Trichoderma–plant–pathogen interactions. *Soil Biology & Biochemistry*, **40**, 1–10.

Vredenburg, V. T., Knapp, R. A., Tunstall, T. S. & Briggs, C. J. (2010). Dynamics of an emerging disease drive large-scale amphibian population extinctions. *Proceedings of the National Academy of Sciences of the United States of America*, **107**, 9689–9694.

Vucetich, J. A. & Peterson, R. O. (2004). The influence of top-down, bottom-up and abiotic factors on the moose (*Alces alces*) population of Isle Royale. *Proceedings of the Royal Society of London Series B – Biological Sciences*, **271**, 183–189.

Wang, L. F. & Eaton, B. T. (2007). Bats, civets and the emergence of SARS. *Current Topics in Microbiology and Immunology*, **315**, 325–344.

Ward, J. R., Kim, K. & Harvell, C. D. (2007). Temperature affects coral disease resistance and pathogen growth. *Marine Ecology Progress Series*, **329**, 115–121.

Ward, J. R. & Lafferty, K. D. (2004). The elusive baseline of marine disease: are diseases in ocean ecosystems increasing? *PLoS Biology*, **2**, 542–547.

Warkentin, I. G., Bickford, D., Sodhi, N. S. & Bradshaw, C. J. A. (2009). Eating frogs to extinction. *Conservation Biology*, **23**, 1056–1059.

Washburn, J. O., Mercer, D. R. & Anderson, J. R. (1991). Regulatory role of parasites: impact on host population shifts with resource availability. *Science*, **253**, 185–188.

Watson, D. M. (2001). Mistletoe: a keystone resource in forests and woodlands worldwide. *Annual Review of Ecology and Systematics*, **32**, 219–249.

Wattier, R. A., Haine, E. R., Beguet, J., Martin, G., Bollache, L., Musko, I. B., Platvoet, D. & Rigaud, T. (2007). No genetic bottleneck or associated microparasite loss in invasive populations of a freshwater amphipod. *Oikos*, **116**, 1941–1953.

Webster, R. G., Bean, W. J., Gorman, O. T., Chambers, T. M. & Kawaoka, Y. (1992). Evolution and ecology of influenza-A viruses. *Microbiological Reviews*, **56**, 152–179.

Werner, E. E. & Peacor, S. D. (2003). A review of trait-mediated indirect interactions in ecological communities. *Ecology*, **84**, 1083–1100.

Werren, J. H., Baldo, L. & Clark, M. E. (2008). *Wolbachia*: master manipulators of invertebrate biology. *Nature Reviews Microbiology*, **6**, 741–751.

Werren, J. H. & Beukeboom, L. W. (1993). Population genetics of a parasitic chromosome: theoretical analysis of PSR in subdivided populations. *American Naturalist*, **142**, 224–241.

Wertheim, J. O. (2009). When pigs fly: the avian origin of a 'swine flu'. *Environmental Microbiology*, **11**, 2191–2192.

Westbury, D. B. & Dunnett, N. P. (2008). The promotion of grassland forb abundance: a chemical or biological solution? *Basic and Applied Ecology*, **9**, 653–662.

White, S. M., Sait, S. M. & Rohani, P. (2007). Population dynamic consequences of parasitised-larval competition in stage-structured host–parasitoid systems. *Oikos*, **116**, 1171–1185.

Whitehouse, A. T., Peay, S. & Kindemba, V. (2009). *Ark Sites for White-clawed Crayfish: Guidance for the Aggregates Industry*. Peterborough: Buglife – The Invertebrate Conservation Trust.

Whitlaw, H. A. & Lankester, M. W. (1994). The cooccurrence of moose, white-tailed deer, and *Parelaphostrongylus tenuis* in Ontario. *Canadian Journal of Zoology*, **72**, 819–825.

Wilcove, D. S., Rothstein, D., Dubow, J., Phillips, A. & Losos, E. (1998). Quantifying threats to imperiled species in the United States. *Bioscience*, **48**, 607–615.

Williams, P. D. & Day, T. (2001). Interactions between sources of mortality and the evolution of parasite virulence. *Proceedings of the Royal Society of London Series B – Biological Sciences*, **268**, 2331–2337.

Wilmers, C. C., Post, E., Peterson, R. O. & Vucetich, J. A. (2006). Predator disease out-break modulates top-down, bottom-up and climatic effects on herbivore population dynamics. *Ecology Letters*, **9**, 383–389.

Wodarz, D. & Sasaki, A. (2004). Apparent competition and recovery from infection. *Journal of Theoretical Biology*, **227**, 403–412.

Wolfe, L. M. (2002). Why alien invaders succeed: support for the escape-from-enemy hypothesis. *American Naturalist*, **160**, 705–711.

Wolfe, N. D., Daszak, P., Kilpatrick, A. M. & Burke, D. S. (2005). Bushmeat hunting deforestation, and prediction of zoonoses emergence. *Emerging Infectious Diseases*, **11**, 1822–1827.

Wolfe, N. D., Dunavan, C. P. & Diamond, J. (2007). Origins of major human infectious diseases. *Nature*, **447**, 279–283.

Wood, C. L., Byers, J. E., Cottingham, K. L., Altman, I., Donahue, M. J. & Blakeslee, A. M. H. (2007). Parasites alter community structure. *Proceedings of the National Academy of Sciences of the United States of America*, **104**, 9335–9339.

Woolhouse, M. & Gaunt, E. (2007). Ecological origins of novel human pathogens. *Critical Reviews in Microbiology*, **33**, 231–242.

Woolhouse, M. E. J. (2002). Population biology of emerging and re-emerging pathogens. *Trends in Microbiology*, **10**, S3–S7.

Woolhouse, M. E. J., Dye, C., Etard, J. F., Smith, T., Charlesworth, J. D., Garnett, G. P., Hagan, P., Hii, J. L., Ndhlovu, P. D., Quinnell, R. J., Watts, C. H., Chandiwana, S. K. & Anderson, R. M. (1997). Heterogeneities in the transmission of infectious agents: implications for the design of control programs. *Proceedings of the National Academy of Sciences of the United States of America*, **94**, 338–342.

Woolhouse, M. E. J. & Gowtage-Sequeria, S. (2005). Host range and emerging and reemerging pathogens. *Emerging Infectious Diseases*, **11**, 1842–1847.

Woolhouse, M. E. J., Haydon, D. T. & Antia, R. (2005). Emerging pathogens: the epidemiology and evolution of species jumps. *Trends in Ecology & Evolution*, **20**, 238–244.

Woolhouse, M. E. J., Webster, J. P., Domingo, E., Charlesworth, B. & Levin, B. R. (2002). Biological and biomedical implications of the co-evolution of pathogens and their hosts. *Nature Genetics*, **32**, 569–577.

Worobey, M., Gemmel, M., Teuwen, D. E., Haselkorn, T., Kunstman, K., Bunce, M., Muyembe, J. J., Kabongo, J. M. M., Kalengayi, R. M., Van Marck, E., Gilbert, M. T. P. & Wolinsky, S. M. (2008). Direct evidence of extensive diversity of HIV-1 in Kinshasa by 1960. *Nature*, **455**, 661–664.

Wright, H. A., Wootton, R. J. & Barber, I. (2006). The effect of *Schistocephalus solidus* infection on meal size of three-spined stickleback. *Journal of Fish Biology*, **68**, 801–809.

Yan, G. Y. (1996). Parasite-mediated competition: a model of directly transmitted macroparasites. *American Naturalist*, **148**, 1089–1112.

Yan, G. Y., Stevens, L., Goodnight, C. J. & Schall, J. J. (1998). Effects of a tapeworm parasite on the competition of *Tribolium* beetles. *Ecology*, **79**, 1093–1103.

Yates, T. L., Mills, J. N., Parmenter, C. A., Ksiazek, T. G., Parmenter, R. R., Vande Castle, J. R., Calisher, C. H., Nichol, S. T., Abbott, K. D., Young, J. C., Morrison, M. L., Beaty, B. J., Dunnum, J. L., Baker, R. J., Salazar-Bravo, J. & Peters, C. J. (2002). The ecology and evolutionary history of an emergent disease: hantavirus pulmonary syndrome. *Bioscience*, **52**, 989–998.

Yoshida, T., Ellner, S. P., Jones, L. E., Bohannan, B. J. M., Lenski, R. E. & Hairston, N. G. (2007). Cryptic population dynamics: rapid evolution masks trophic interactions. *PLoS Biology*, **5**, 1868–1879.

Zhang, P. J., Zheng, S. J., van Loon, J. J. A., Boland, W., David, A., Mumm, R. & Dicke, M. (2009). Whiteflies interfere with indirect plant defense against spider mites in Lima bean. *Proceedings of the National Academy of Sciences of the United States of America*, **106**, 21202–21207.

Index

440 · **Index**